VOLUME THREE HUNDRED AND FIFTY TWO

INTERNATIONAL REVIEW OF CELL AND MOLECULAR BIOLOGY

Cell Death Regulation in Health and Disease - Part B

INTERNATIONAL REVIEW OF CELL AND MOLECULAR BIOLOGY

VOLUME THREE HUNDRED AND FIFTY TWO

INTERNATIONAL REVIEW OF CELL AND MOLECULAR BIOLOGY

Cell Death Regulation in Health and Disease - Part B

Edited by

JOHAN K.E. SPETZ
Harvard T.H. Chan School of Public Health, Boston, MA, USA

LORENZO GALLUZZI
Weill Cornell Medical College, New York, NY, United States

ELSEVIER

ACADEMIC PRESS
An imprint of Elsevier

Academic Press is an imprint of Elsevier
50 Hampshire Street, 5th Floor, Cambridge, MA 02139, United States
525 B Street, Suite 1650, San Diego, CA 92101, United States
The Boulevard, Langford Lane, Kidlington, Oxford OX5 1GB, United Kingdom
125 London Wall, London, EC2Y 5AS, United Kingdom

First edition 2020

ISBN: 978-0-12-819929-9
ISSN: 1937-6448

For information on all Academic Press publications
visit our website at https://www.elsevier.com/books-and-journals

Publisher: Zoe Kruze
Acquisitions Editor: Ashlie M. Jackman
Editorial Project Manager: Leticia M. Lima
Production Project Manager: Denny Mansingh
Cover Designer: Matthew Limbert

Typeset by SPi Global, India

Contents

Contributors

Ivana Barraza
Department of Biological Sciences, Nova Southeastern University, Fort Lauderdale, FL, United States

Maria A. Bauer
Institute of Molecular Biosciences, NAWI Graz, University of Graz, Graz, Austria

Gregory L. Bellot
Department of Hand & Reconstructive Microsurgery, University Orthopedic, Hand & Reconstructive Microsurgery Cluster, National University Health System, Singapore

Catherine Brenner
Université Paris-Saclay, Institut Gustave Roussy, CNRS, Metabolic and Systemic Aspects of Oncogenesis, Villejuif, France

Didac Carmona-Gutierrez
Institute of Molecular Biosciences, NAWI Graz, University of Graz, Graz, Austria

Scott D. Cramer
Department of Pharmacology, University of Colorado Anschutz Medical Campus, Aurora, CO, United States

W. Douglas Fairlie
Olivia Newton-John Cancer Research Institute, Heidelberg; School of Cancer Medicine; La Trobe Institute for Molecular Science, La Trobe University, Melbourne, VIC, Australia

Marla C. Fortoul
Department of Biological Sciences, Nova Southeastern University, Fort Lauderdale, FL, United States

Sebastian J. Hofer
Institute of Molecular Biosciences, NAWI Graz, University of Graz, Graz, Austria

Katharina Kainz
Institute of Molecular Biosciences, NAWI Graz, University of Graz, Graz, Austria

Tracy A. Larson
Department of Biology, University of Virginia, Charlottesville, VA, United States

Erinna F. Lee
Olivia Newton-John Cancer Research Institute, Heidelberg; School of Cancer Medicine; La Trobe Institute for Molecular Science, La Trobe University, Melbourne, VIC, Australia

Antoinette Lemoine
Departement of Oncogenetics, APHP, GH Paris-Sud, Hôpital Paul Brousse, Inserm UMR-S 1193, Université Paris-Saclay, Villejuif, France

Frank Madeo
Institute of Molecular Biosciences, NAWI Graz, University of Graz; BioTechMed Graz, Graz, Austria

Dannah R. Miller
Department of Pharmacology, University of Colorado Anschutz Medical Campus, Aurora, CO, United States

Shazib Pervaiz
Department of Physiology and Medical Science Cluster Cancer Program, Yong Loo Lin School of Medicine; NUS Graduate School for Integrative Sciences and Engineering, National University of Singapore; National University Cancer Institute, National University Health System, Singapore

Rebecca J. Quinn
Department of Biological Sciences, Nova Southeastern University, Fort Lauderdale, FL, United States

Robert P. Smith
Department of Biological Sciences, Nova Southeastern University, Fort Lauderdale, FL, United States

Jelena Tadic
Institute of Molecular Biosciences, NAWI Graz, University of Graz; Division of Immunology and Pathophysiology, Otto Loewi Research Center, Medical University of Graz, Graz, Austria

Andrew Thorburn
Department of Pharmacology, University of Colorado Anschutz Medical Campus, Aurora, CO, United States

Sharon Tran
Olivia Newton-John Cancer Research Institute, Heidelberg; School of Cancer Medicine, La Trobe University, Melbourne, VIC, Australia

Andreas Zimmermann
Institute of Molecular Biosciences, NAWI Graz, University of Graz, Graz, Austria

Preface: Life through death—Key role of cellular suicide for colonial and organismal homeostasis

Johan Spetz[a,b,*], **Lorenzo Galluzzi**[c,d,e,f,g,*]
[a]John B. Little Center for Radiation Sciences, Harvard School of Public Health, Boston, MA, United States
[b]Laboratory of Systems Pharmacology, Harvard Program in Therapeutic Science, Department of Systems Biology, Harvard Medical School, Boston, MA, United States
[c]Department of Radiation Oncology, Weill Cornell Medical College, New York, NY, United States
[d]Sandra and Edward Meyer Cancer Center, New York, NY, United States
[e]Caryl and Israel Englander Institute for Precision Medicine, New York, NY, United States
[f]Department of Dermatology, Yale School of Medicine, New Haven, CT, United States
[g]Université de Paris, Paris, France
[*]Corresponding authors: e-mail address: spetz@hsph.harvard.edu; deadoc80@gmail.com

The term "regulated cell death" (RCD) is commonly employed to indicate a form of cell death that is controlled by a dedicated molecular machinery, and hence can be modulated (at least to some degree) with specific pharmacological or genetic interventions (Galluzzi et al., 2015, 2018a). RCD can occur in the context of two diametrically opposed scenarios. On the one hand, RCD can represent the ultimate response of cells to microenvironmental perturbations that are excessively intense or prolonged for cells to adapt and recover homeostasis (Spetz et al., 2019). On the other hand, RCD can contribute to purely physiological programs of development and adult tissue homeostasis (Bock and Tait, 2020). The expression "programmed cell death" (PCD) is often used to refer to these latter variants of RCD that occur at a predetermined point in time, irrespective of potentially cytotoxic alterations of the intracellular or extracellular microenvironment (Fuchs and Steller, 2011, 2015).

For multicellular organisms, the advantages provided by genetically encoded programs for cell death are obvious. On the one hand, PCD can be harnessed as a morphogenetic mechanism that shapes first (and maintains then) tissue and organ architecture (Tuzlak et al., 2016). On the other hand, stress-driven RCD constitutes a robust mechanism for the elimination first (and replacement then) of postmitotic cells that are damaged beyond recovery, and hence functionless and/or prone to constitute a threat for the entire organism (e.g., prone to malignant transformation) (Cao and Tait, 2018). Less obvious are the advantages provided by RCD to unicellular organisms, such as bacteria and yeast. However, both prokaryotes and unicellular eukaryotes have rudimentary RCD machineries that allow individual cells to suicide (Carmona-Gutierrez et al., 2018). This happens, as an example, when nutrients for the entire colony are limiting, and the demise of old cells increases the chances of younger, fitter cells to survive dwindling nutrient levels and ultimately reconstitute the colony (Fabrizio et al., 2004; Herker et al., 2004).

In higher eukaryotes, the molecular machinery for RCD has considerably diversified (Galluzzi et al., 2017a; Stockwell et al., 2017; Vanden Berghe et al., 2014; Weinlich et al., 2017) to enable variants of RCD that (1) occur according to diverse kinetics (Buque et al., 2019; Grootjans et al., 2016) and (2) are accompanied by the emission of specific signals that engage or suppress immune reactions (Galluzzi et al., 2017b; Rodriguez-Ruiz et al., 2019; Vanpouille-Box et al., 2018). Thus, dying mammalian cells can deliver signals in the form of damage-associated molecular patterns (DAMPs) or cytokines that are responsible for immune modulation at sites of active RCD (Galluzzi et al., 2018b; Green et al., 2009; Krysko et al., 2012). Many of the enzymes that have long been considered as pure RCD regulators, such as apoptotic caspases, have now been shown to mediate multipronged immunomodulatory functions (Galluzzi et al., 2016; Rodriguez-Ruiz et al., 2019; Rongvaux et al., 2014; White et al., 2014), and to support tissue regeneration, at least in some settings (Cheng et al., 2019; Huang et al., 2011; Ichim and Tait, 2016).

Given the key role of RCD in the maintenance of organismal homeostasis and the control of immune responses against dying cells, it is not surprising that deregulated RCD contributes to the pathology of a variety of human disorders. On the one hand, uncontrolled RCD associated with inflammatory reactions is involved in the pathogenesis of ischemic, cardiac, renal and neurodegenerative conditions, among others (Del Re et al., 2019; Fang et al., 2019; Hetz and Saxena, 2017; Kers et al., 2016; Quinlan et al., 2017). On the other hand, insufficient RCD (as such or in the context of immunosuppressive signaling) is etiologically linked to autoimmunity, proliferative syndromes, and cancer (Cao and Tait, 2018; Czabotar et al., 2014; Khan and Ghazanfar, 2018; Nagata and Tanaka, 2017; Singh et al., 2019; Spetz et al., 2019). In summary, RCD in its various forms stands out as a key mechanism for the preservation of colonial and organismal homeostasis, thus constituting an attractive target for the development of pharmacological interventions against a plethora of human disorders for more than 2 decades (Degterev et al., 2019; Huang and Ichikawa, 2008; Naderer and Fulcher, 2018; Valentin et al., 2018). As of today, however, very few drugs that target the molecular machinery for RCD have successfully entered clinical development, including the BCL2 inhibitor venetoclax (currently approved for the therapy of some hematological tumors) (Jain et al., 2019), and the pan-caspase inhibitor emricasan (which was recently granted fast track status by the US Food and Drug Administration for the therapy of cirrhosis) (Mehta et al., 2018; Shiffman et al., 2019). It is tempting to speculate, yet remains to be formally demonstrated, that translating RCD modulation from the bench to the bedside has been difficult because RCD has been often investigated (especially when the molecular machinery for RCD has begun to emerge) irrespective of its immunological context.

In *Cell Death Regulation in Health and Disease* (a special issue of the *International Review of Cell and Molecular Biology* series) a group of world-renowned experts in the field gathered to discuss mechanistic features and physiopathological implications of RCD in a variety of model organisms. Each of the chapters of *Cell Death Regulation in Health and Disease* addresses a specific aspect of cell death biology, ranging from RCD control in prokaryotes and lower eukaryotes to the contribution of cell death deregulation to various human disorders. The book (which comes in three separate volumes, i.e., parts A, B, and C) is equally addressed to experienced investigators who may want to acquire in-depth insights into a specific aspect of cell death biology, as well as to students or neophytes who are moving their first steps in this broad field of research.

Acknowledgments

L.G. is supported by a Breakthrough Level 2 grant from the US Department of Defense (DoD), Breast Cancer Research Program (BRCP) (#BC180476P1), by the 2019 Laura Ziskin Prize in Translational Research (#ZP-6177, PI: Formenti) from the Stand up to Cancer (SU2C), by a Mantle Cell Lymphoma Research Initiative (MCL-RI, PI: Chen-Kiang) grant from the Leukemia and Lymphoma Society (LLS), by a startup grant from the Department of Radiation Oncology at Weill Cornell Medicine (New York, USA), by a Rapid Response Grant from the Functional Genomics Initiative (New York, USA), by industrial collaborations with Lytix (Oslo, Norway) and Phosplatin (New York, USA), and by donations from Phosplatin (New York, USA), the Luke Heller TECPR2 Foundation (Boston, USA) and Sotio a.s. (Prague, Czech Republic).

Conflict of interest statement

LG received consulting fees from OmniSEQ, Astra Zeneca, Inzen and the Luke Heller TECPR2 Foundation, and he is member of the Scientific Advisory Committee of Boehringer Ingelheim, The Longevity Labs and OmniSEQ.

References

Bock, F.J., Tait, S.W.G., 2020. Mitochondria as multifaceted regulators of cell death. Nat. Rev. Mol. Cell Biol. 21, 85–100.

Buque, A., et al., 2019. Apoptotic caspases cut down the immunogenicity of radiation. Onco. Targets. Ther. 8 (11), e1655364.

Cao, K., Tait, S.W.G., 2018. Apoptosis and cancer: force awakens, phantom menace, or both? Int. Rev. Cell Mol. Biol. 337, 135–152.

Carmona-Gutierrez, D., et al., 2018. Guidelines and recommendations on yeast cell death nomenclature. Microb. Cell 5 (1), 4–31.

Cheng, J., et al., 2019. The caspase-3/PKCdelta/Akt/VEGF-A signaling pathway mediates tumor repopulation during radiotherapy. Clin. Cancer Res. 25 (12), 3732–3743.

Czabotar, P.E., et al., 2014. Control of apoptosis by the BCL-2 protein family: implications for physiology and therapy. Nat. Rev. Mol. Cell Biol. 15 (1), 49–63.

Degterev, A., et al., 2019. Targeting RIPK1 for the treatment of human diseases. Proc. Natl. Acad. Sci. U. S. A. 116 (20), 9714–9722.

Del Re, D.P., et al., 2019. Fundamental mechanisms of regulated cell death and implications for heart disease. Physiol. Rev. 99 (4), 1765–1817.

Fabrizio, P., et al., 2004. Superoxide is a mediator of an altruistic aging program in Saccharomyces cerevisiae. J. Cell Biol. 166 (7), 1055–1067.

Fang, X., et al., 2019. Ferroptosis as a target for protection against cardiomyopathy. Proc. Natl. Acad. Sci. U. S. A. 116 (7), 2672–2680.

Fuchs, Y., Steller, H., 2011. Programmed cell death in animal development and disease. Cell 147 (4), 742–758.

Fuchs, Y., Steller, H., 2015. Live to die another way: modes of programmed cell death and the signals emanating from dying cells. Nat. Rev. Mol. Cell Biol. 16 (6), 329–344.

Galluzzi, L., et al., 2015. Essential versus accessory aspects of cell death: recommendations of the NCCD 2015. Cell Death Differ. 22 (1), 58–73.

Galluzzi, L., et al., 2016. Caspases connect cell-death signaling to organismal homeostasis. Immunity 44 (2), 221–231.

Galluzzi, L., et al., 2017a. Necroptosis: mechanisms and relevance to disease. Annu. Rev. Pathol. 12, 103–130.

Galluzzi, L., et al., 2017b. Immunogenic cell death in cancer and infectious disease. Nat. Rev. Immunol. 17 (2), 97–111.

Galluzzi, L., et al., 2018a. Molecular mechanisms of cell death: recommendations of the nomenclature committee on cell death 2018. Cell Death Differ. 25 (3), 486–541.

Galluzzi, L., et al., 2018b. Linking cellular stress responses to systemic homeostasis. Nat. Rev. Mol. Cell Biol. 19 (11), 731–745.

Green, D.R., et al., 2009. Immunogenic and tolerogenic cell death. Nat. Rev. Immunol. 9 (5), 353–363.

Grootjans, S., et al., 2016. A real-time fluorometric method for the simultaneous detection of cell death type and rate. Nat. Protoc. 11 (8), 1444–1454.

Herker, E., et al., 2004. Chronological aging leads to apoptosis in yeast. J. Cell Biol. 164 (4), 501–507.

Hetz, C., Saxena, S., 2017. ER stress and the unfolded protein response in neurodegeneration. Nat. Rev. Neurol. 13 (8), 477–491.

Huang, D., Ichikawa, K., 2008. Drug discovery targeting apoptosis. Nat. Rev. Drug Discov. 7 (12), 1041.

Huang, Q., et al., 2011. Caspase 3-mediated stimulation of tumor cell repopulation during cancer radiotherapy. Nat. Med. 17 (7), 860–866.

Ichim, G., Tait, S.W., 2016. A fate worse than death: apoptosis as an oncogenic process. Nat. Rev. Cancer 16 (8), 539–548.

Jain, N., et al., 2019. Ibrutinib and venetoclax for first-line treatment of CLL. N. Engl. J. Med. 380 (22), 2095–2103.
Kers, J., et al., 2016. An overview of pathways of regulated necrosis in acute kidney injury. Semin. Nephrol. 36 (3), 139–152.
Khan, U., Ghazanfar, H., 2018. T lymphocytes and autoimmunity. Int. Rev. Cell Mol. Biol. 341, 125–168.
Krysko, D.V., et al., 2012. Immunogenic cell death and DAMPs in cancer therapy. Nat. Rev. Cancer 12 (12), 860–875.
Mehta, G., et al., 2018. A placebo-controlled, multicenter, double-blind, phase 2 randomized trial of the pan-caspase inhibitor emricasan in patients with acutely decompensated cirrhosis. J. Clin. Exp. Hepatol. 8 (3), 224–234.
Naderer, T., Fulcher, M.C., 2018. Targeting apoptosis pathways in infections. J. Leukoc. Biol. 103 (2), 275–285.
Nagata, S., Tanaka, M., 2017. Programmed cell death and the immune system. Nat. Rev. Immunol. 17 (5), 333–340.
Quinlan, S., et al., 2017. MicroRNAs in neurodegenerative diseases. Int. Rev. Cell Mol. Biol. 334, 309–343.
Rodriguez-Ruiz, M.E., et al., 2019. Apoptotic caspases inhibit abscopal responses to radiation and identify a new prognostic biomarker for breast cancer patients. Onco. Targets. Ther. 8 (11), e1655964.
Rongvaux, A., et al., 2014. Apoptotic caspases prevent the induction of type I interferons by mitochondrial DNA. Cell 159 (7), 1563–1577.
Shiffman, M., et al., 2019. Randomised clinical trial: emricasan versus placebo significantly decreases ALT and caspase 3/7 activation in subjects with non-alcoholic fatty liver disease. Aliment. Pharmacol. Ther. 49 (1), 64–73.
Singh, R., et al., 2019. Regulation of apoptosis in health and disease: the balancing act of BCL-2 family proteins. Nat. Rev. Mol. Cell Biol. 20 (3), 175–193.
Spetz, J., et al., 2019. T cells and regulated cell death: kill or be killed. Int. Rev. Cell Mol. Biol. 342, 27–71.
Stockwell, B.R., et al., 2017. Ferroptosis: a regulated cell death nexus linking metabolism, redox biology, and disease. Cell 171 (2), 273–285.
Tuzlak, S., et al., 2016. Interrogating the relevance of mitochondrial apoptosis for vertebrate development and postnatal tissue homeostasis. Genes Dev. 30 (19), 2133–2151.
Valentin, R., et al., 2018. The rise of apoptosis: targeting apoptosis in hematologic malignancies. Blood 132 (12), 1248–1264.
Vanden Berghe, T., et al., 2014. Regulated necrosis: the expanding network of non-apoptotic cell death pathways. Nat. Rev. Mol. Cell Biol. 15 (2), 135–147.
Vanpouille-Box, C., et al., 2018. Cytosolic DNA sensing in organismal tumor control. Cancer Cell 34 (3), 361–378.
Weinlich, R., et al., 2017. Necroptosis in development, inflammation and disease. Nat. Rev. Mol. Cell Biol. 18 (2), 127–136.
White, M.J., et al., 2014. Apoptotic caspases suppress mtDNA-induced STING-mediated type I IFN production. Cell 159 (7), 1549–1562.

CHAPTER ONE

The mechanisms and cell signaling pathways of programmed cell death in the bacterial world

Robert P. Smith*, Ivana Barraza, Rebecca J. Quinn, Marla C. Fortoul
Department of Biological Sciences, Nova Southeastern University, Fort Lauderdale, FL, United States
*Corresponding author: e-mail address: rsmith@nova.edu

Contents

International Review of Cell and Molecular Biology, Volume 352
ISSN 1937-6448
https://doi.org/10.1016/bs.ircmb.2019.12.002

Abstract

While programmed cell death was once thought to be exclusive to eukaryotic cells, there are now abundant examples of well regulated cell death mechanisms in bacteria. The mechanisms by which bacteria undergo programmed cell death are diverse, and range from the use of toxin-antitoxin systems, to prophage-driven cell lysis. Moreover, some bacteria have learned how to coopt programmed cell death systems in competing bacteria. Interestingly, many of the potential reasons as to why bacteria undergo programmed cell death may parallel those observed in eukaryotic cells, and may be altruistic in nature. These include protection against infection, recycling of nutrients, to ensure correct morphological development, and in response to stressors. In the following chapter, we discuss the molecular and signaling mechanisms by which bacteria undergo programmed cell death. We conclude by discussing the current open questions in this expanding field.

1. Introduction

Programmed cell death (PCD) is critical for the survival of eukaryotic, multicellular organisms. For example, it plays critical roles in morphological development where PCD is partially responsible for digit formation in vertebrates (Lindsten et al., 2000). PCD has numerous protective functions. During lymphocyte development, progenitor cells that react strongly to self-antigens, or that cannot respond to antigens, are removed using PCD. Pathogen infected immune cells can undergo PCD to protect uninfected cells (Nagata and Tanaka, 2017). Cells that are faulty in cell cycle regulation will undergo PCD (Pucci et al., 2000). Damage by noxious agents will also drive PCD (Norbury and Hickson, 2001). Finally, PCD can help to redistribute important nutrients. For example, during the development of plants, PCD can lead to nutrient recycling, which ensures continued plant development and survival (Guiboileau et al., 2010). From the perspective of the multicellular organism, PCD in these scenarios has a benefit—the sacrifice of a few cells will benefit the many, remaining cells. Thus, it has been hypothesized that PCD has been well established inside multicellular organisms to ensure their survival.

While PCD was initially thought to be exclusively found in eukaryotic cells, the past 20 years of research have led to the discovery and description of PCD in bacteria. Interestingly, the hypothesized functional roles of PCD in bacteria share many common similarities with eukaryotic cells. PCD in bacteria plays roles in morphological development, protection against pathogens, and the redistribution of nutrients. But why would seemingly independently growing and acting bacteria undergo PCD? To address this question, one has to look no further than the increasing number of examples where bacteria can behave as "multicellular" organisms. There is ample evidence demonstrating that they can form complex structures (e.g., biofilms), act as a group to locate food or fend off predators using cooperative defense strategies, and generally work together to ensure the survival of the population. Like their eukaryotic counterparts, PCD in bacteria may appear to have altruistic aspects where the sacrifice of a few bacteria serves to benefit the remaining bacteria, thus ensuring the survival of the entire population.

Indeed, the mechanisms by which PCD is initiated, regulated, and executed are as diverse as the hypothesized functional roles of PCD. Given the importance of bacteria in a wide range of scientific realms (health, biotechnology, ecology, evolution, etc.) discovery of these mechanisms has been, and is currently, the focus of a large body of research. The rational manipulation of the presence and absence of bacteria is incredibly desirable across scientific disciplines. This includes maintaining the survival of beneficial bacteria, such as those found in the gut microbiome, and the elimination of pathogens in the clinical setting, including antibiotic resistant bacteria.

In the following chapter, we focus on a variety of mechanisms that result in PCD in bacteria. Importantly, we define PCD as a well regulated and specifically initiated process whose goal is to result in cell death. While the mechanisms that drive PCD are still being discovered, we focus on our current understanding of the molecular mechanisms and signaling pathways that result in PCD. We focus on bacterial PCD that can be used to prevent and promote infections, allow for the release of DNA, and during morphological development. We also touch upon the mechanisms by which bacteria can manipulate PCD in their competing counterparts. Finally, we conclude by discussing unresolved questions in the study of bacterial PCD. It is important to note that PCD in bacteria remains an open evolutionary question. In some instances, examples of PCD in bacteria may or may not capture the ideal conceptualization of PCD.

While the evolutionary significance of PCD in multicellular eukaryotes is well understood (Ameisen, 2002), the significance of PCD in bacteria remains unclear. In many of the examples highlighted below, it would appear that PCD evolved in bacteria as an altruistic trait. That is, the sacrifice of a few members of a bacterial population via PCD directly benefits any remaining bacteria. While this is plausible, it is challenging to determine whether or not a selective pressure specifically gave rise to this behavior, or if we are assigning this behavior to non-altruistic traits. In fact, PCD may have arisen in bacteria due to selective pressures that are completely unrelated to any altruistic benefit. We note that this notion can be extended to several other behaviors in bacteria (Redfield, 2001). It is important to keep this in mind as we outline the current hypotheses to explain the evolutionary significance of the mechanisms of PCD described in this chapter.

2. Plasmids induce PCD to ensure their maintenance

In Richard Dawkin's book "The Selfish Gene," Dawkin's argues that all genes have selfish tendencies that influence their replication into the next generation (Dawkins, 1976). Successful replication, although often costly to the organism, results in the selection of genes that are considered "immortal" within the population. After Dawkins developed this selfish gene hypothesis, researchers have attempted to locate genes that replicate "selfishly" within their hosts. Perhaps one of the first instances of a "selfish gene" was identified on the R1 conjugative plasmid (Gerdes, 1988; Gerdes et al., 1986). This plasmid contains a region called the *parB*+ locus, which contains two genes: *hok* (for host killing) and *sok* (for suppressor of killing). Together, these genes ensure that daughter cells derived from a mother cell containing the R1 conjugative plasmid also contain the R1 conjugative plasmid. Stated otherwise, any daughter cell that loses the R1 plasmid during cell division is killed. This is called post-segregational killing, which has been observed to occur during the partitioning of conjugative plasmids, non-conjugative plasmids, mobile genetic elements (e.g., transposons) and chromosomes (reviewed in Díaz-Orejas et al., 2017). In the case of plasmids, loss can occur due to stochastic (random) plasmid partitioning during cell division (Nordström et al., 1980; Summers, 1991). That is, there is the probability that a daughter cell does not receive a plasmid upon cell division. However, plasmid loss may also be due to additional mechanisms including conflicts between the transcription of plasmid genes and replication of the plasmid (Wein et al., 2019). While the R1 plasmid does contain the ability to resist some antibiotics, in the absence of those antibiotics, it places a burden on cells that maintain it. Having to replicate and maintain a plasmid that currently has no fitness benefit (e.g., in the absence of antibiotics) makes plasmid-containing bacteria slightly less fit than their plasmid-less counterparts. By forcing daughter cells to maintain the R1 conjugative plasmid through the *hok*/*sok* TA system, the plasmid is enforcing its transmission to daughter cells. Thus, the R1 plasmid appears to behave like a selfish genetic element.

Since the initial discovery of the *hok/sok* system, several other systems that have the same general behavior have been identified. These systems are named toxin-antitoxin systems (which we will abbreviate as TA systems) and have been found to occur in both plasmids and bacterial chromosomes. In fact, they are sufficiently common that they have now been categorized based on their mechanism of action. In general, the toxin of these systems is very stable, and can interfere with a critical cellular function. If the interference with this function occurs for a sufficiently long time, then the bacterium can die. To prevent constant interference from the toxin, a less stable antitoxin is produced. The antitoxin is designed to interfere with the toxin, and thus acts as a protective mechanism against toxin-driven death. As there remains some debate as to whether all types of TA systems are involved in PCD, they will not be discussed in depth in this chapter. The reader is referred to a comprehensive set of reviews that summarize TA system types and their mechanisms of action (Hall et al., 2017; Page and Peti, 2016; Yamaguchi and Inouye, 2016).

The *hok/sok* system is a Type I TA system. In general, Type I TA systems consist of antitoxins that are unstable, small regulatory RNAs (sRNA) that are normally 50–200 nucleotides in length (Fozo et al., 2010; Wen et al., 2014). In contrast, Type I toxins are small, stable hydrophobic proteins designed to damage the cell by compromising membrane integrity (Fozo et al., 2010). To prevent the toxin from interfering with the cell membrane, the antitoxin sRNA will base pair to the toxin's mRNA, creating a region of double-stranded mRNA (Gerdes et al., 1990; Thisted and Gerdes, 1992). Double-stranded mRNA can prevent the binding and assembly of ribosomes, thus preventing translation of the toxin protein. Double-stranded mRNA might also be degraded using RNaseIII, which is the case in the *hok/sok* system (Gerdes et al., 1992). In either scenario, the mRNA for the toxin protein is rendered untranslatable by the antitoxin mRNA. In the case of the *hok/sok* system that regulates post-segregational killing, if the R1 plasmid is probabilistically lost following cell division, the genes encoding *hok* and *sok* are also lost. Thus, additional *hok* and *sok* transcripts cannot be produced. Due to its unstable nature, and the fact that additional *sok* mRNA cannot be produced in the cell, the concentration of *sok* mRNA declines in the cell as it is quickly degraded. This allows the *hok* mRNA, which is more stable and in substantial excess relative to the *sok* mRNA, to undergo translation. When the Hok protein is produced, it creates damaging pores within the inner bacterial membrane. This causes the cell to lose membrane potential, which abolishes, among other things, its ability to generate ATP (Pecota et al., 2003). As a result, the cell will eventually become non-viable. Overall, the *hok/sok* type 1 TA system induces PCD if a daughter cell fails to maintain the R1 plasmid.

3. PCD to protect against infection by bacteriophage

Bacteria have developed mechanisms to prevent the replication of bacteriophages. One mechanism is through *ab*ortive phage *i*nfection (ABI) systems. Here, the presence of an entering or replicating bacteriophage activates the ABI system. The ABI system then causes PCD in the bacteria, which prevents any further replication of the bacteriophage (Fig. 1A). ABI systems are different from bacteriophage-induced lysis during

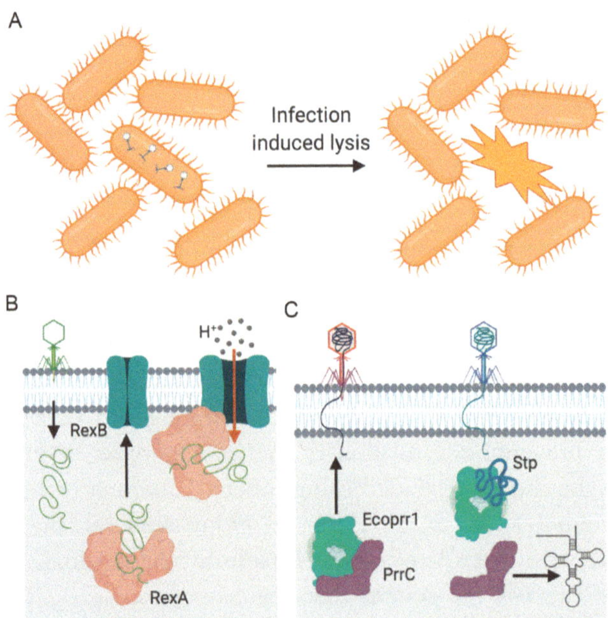

Fig. 1 PCD to prevent replication of bacteriophage in bacteria. (A) In some bacteria, if the presence of a bacteriophage is detected, the infected bacterium will undergo PCD. This prevents the bacteriophage from spreading in the population. (B) The RexA/RexB ABI system. When bacteriophage DNA (green) is detected by RexA, it interacts with the membrane bound RexB channel protein. Opening of the RexB channel protein results in depolarization of the cell. This results in the inability to produce ATP and eventually cell death. (C) The PrrC ABI system. Ecoprr1 normally recognizes bacteriophage DNA and cuts it (left). However, some bacteriophage contain the Stp peptide, which renders the Ecoprr1 protein inactive (right). In this case, PrrC becomes activated and cleaves tRNAs carrying lysine. This stops global translation and results in cell death.

the lytic phase of viral replication. In the former, ABI systems result in PCD regulated by the bacterium to prevent the replication of the bacteriophage. In the latter, the bacteriophage causes the bacterium to lyse in order to facilitate their spread. As we will detail below, many ABI systems use prophages (dormant/non-functional phages that have entered into the chromosomes of bacteria through the lysogenic phase) to regulate and ultimately cause PCD regulated by the bacterium. It has been suggested that ABIs might have altruistic properties. For example, a bacterium undergoing PCD to prevent the replication of a bacteriophage would appear to benefit any remaining bacteria as the bacteriophage would not be able to spread in the population. Evidence of this can be derived from studies that demonstrate that the deletion of ABI components results in increased bacteriophage production (e.g., Hazan et al., 2004). However, whether or not ABIs evolved as true altruistic traits remains unknown.

3.1 The RexA/RexB and Lit systems

The best characterized ABI system is the Rex system (Fig. 1B). This system can be found in *E. coli* that carry λ phage in its lysogenic stage (Labrie et al., 2010). λ Phage will activate in a subset of *E. coli* cells infected with a competing phage, which ultimately results in cell lysis. RexA serves as a sensor for competing phage replication, whereas RexB is a membrane bound channel protein that upon activation, initiates PCD (Parma et al., 1992). Genes encoding both proteins are regulated by the cI repressor protein (Matz et al., 1982). Once a competing phage begins its replication process in the cytoplasm, RexA proteins form a complex with the competing bacteriophage's DNA. This complex moves to the cell membrane where it interacts with and activates RexB. Once activated, RexB is thought to open its channel pore and allow the free movement of protons across the cell membrane. This causes the bacterium to lose the proton motive force required for ATP synthesis resulting in a rapid reduction in total ATP in the bacterium. This reduction in ATP shuts off replication, transcription, and translation, thus preventing any competing phage from being produced (Parma et al., 1992). Ultimately, the lack of ATP, and thus cellular metabolism, results in the death of the bacterium.

While the RexA/RexB ABI system inhibits translation indirectly by depleting global ATP levels, another ABI system found in *E. coli*, driven by the Lit protein, directly inhibits translation (Snyder, 1995). Lit is encoded by a prophage called e14, which is embedded in the chromosomally encoded isocitrate dehydrogenase gene. Interestingly, this ABI system appears to specifically target bacteriophage T4's replication. During the near final stages of assembly, T4 phage begins to manufacturer a major head protein in the cytoplasm of an infected bacterium. As this head protein, called Gol, emerges from the ribosome, a specific amino acid sequence on the Gol protein forms a complex with the EF-Tu elongation factor protein (Bingham et al., 2000). The formation of this complex appears to trigger the activation of the Lit protein. Upon activation, Lit cleaves a consensus sequence in the EF-Tu protein, rendering it inactive. Inactivation of EF-Tu completely represses global translation (Yu and Snyder, 1994), thus preventing further T4 phage assembly. Activated Lit also represses transcription of the Gol protein and any downstream elements, likely by binding to a specific sequence in the *gol* gene (Bergsland et al., 1990). Taken together, the inhibition of global translation eventually results in death of the T4 phage infected bacterium. Thus, T4 phage cannot replicate.

3.2 PrrC represents a backup system to prevent bacteriophage infection

The PrrC ABI system in *E. coli* is unique in that it represents a "back up" system to prevent T4 phage infection if the first phage protection system fails (Fig. 1C). The chromosomally encoded *prrC* gene is part of a four gene system that appears to have originated from a now inactive phage (Jabbar and Snyder, 1984; Labrie et al., 2010). One of the other genes in this genetic element, *prrD*, encodes a type Ic restriction endonuclease called Ecoprr1 (Linder et al., 1990). This endonuclease system serves as the first line of defense against T4 phage. In the absence of T4 phage, Ecoprr1 is complexed with PrrC. In this complex, PrrC is inactive, but Ecoprr1 is functional (Tyndall et al., 1994). When T4 phage injects its nucleic acid, the first line of defense, Ecoprr1, cuts the phage DNA in half by recognizing phage specific DNA sequences. This stops T4 phage replication. However, some T4 phages encode a short, non-essential peptide called Stp, which has been shown to interact with Ecoprr1 (Penner et al., 1995; Snyder, 1995). It has been hypothesized that interactions between Stp and Ecoprr1 render Ecoprr1 inactive, and this could allow the T4 infection to proceed. However, inactivation of Ecoprr1 also affects its interaction with PrrC, likely resulting in the separation of the two. Now uninhibited by Ecoprr1, PrrC becomes activated and, using its own ribonuclease activity, cleaves the anticodon loop of the bacterium produced $tRNA^{LYS}$ (Amitsur et al., 1987). As lysine is required for proper translation of proteins, cleavage of this pool of tRNAs arrests global translation. This ultimately prevents T4 phage from replicating, and eventually results in death of the bacterium.

3.3 TA systems to prevent bacteriophage infection

There is some evidence that TA systems can also serve as a mechanism to prevent the replication of bacteriophages via PCD. One of the most well studied TA systems is the Type II TA system *MazEF*, which can be found in diverse bacteria including *E. coli, Staphylococcus aureus, Bacillus subtilis, B. anthracis, and Mycobacterium tuberculosis* (Engelberg-Kulka et al., 2006; Fu et al., 2007; Hayes, 2003; Kumar et al., 2013; Mittenhuber, 1999; Pandey and Gerdes, 2005; Ramage et al., 2009). Like other Type II TA systems, both the toxin and antitoxin are proteins. The toxin protein is proteolytically stable, whereas the antitoxin protein is more unstable. To prevent the toxin from initiating cell death, the antitoxin protein physically binds to the toxin protein. However, given its labile nature, the cell must continuously synthesize antitoxin protein to prevent the more stable toxin protein from reaching its target. Stressful conditions that either decrease transcription/translation of antitoxin protein, or conditions that increase the degradation rate of the antitoxin protein, allow the toxin to initiate cell death. Most of our understanding of the *mazEF* system has come from studies performed in *E. coli* (Aizenman et al., 1996; Engelberg-Kulka et al., 2006). *E. coli mazF* codes for the toxin MazF, a sequence specific endoribonuclease (Zhang et al., 2003). The *mazE* gene encodes for the labile antitoxin, MazE, which complexes with MazF to prevent its activity. The degradation of MazE occurs via the ATP-dependent ClpAP serine protease (Aizenman et al., 1996). Stressful conditions, including the inhibition of transcription or translation by antibiotics (Hazan et al., 2004) and severe amino acid starvation, leads to the reduction of MazE in the cell, allowing MazF to act freely (Aizenman et al., 1996). When uninhibited by MazE, MazF cleaves mRNAs at ACA sites closely positioned to the AUG start codon, creating a leaderless mRNA (Vesper et al., 2011). These mRNA are rendered untranslatable. Given the frequency of ACA sites in the mRNA pool, translation of multiple mRNAs that are important for survival is stopped.

Evidence that the MazEF system can protect against phage infection comes from experiments performed with phage P1. Phage P1 has two forms: virulent particles that result in cell lysis and lysogenic prophages that replicate intracellularly, but do not normally cause lysis. However, given the right set of signals, including heat and intracellular stressors, the lysogenic virus will enter its lytic, virulent stage. Infection of *E. coli* lacking the *mazEF* TA system results in significant production of virulent P1 and near total lysis of the entire population. In contrast, in *E. coli* that contain a functional copy of *mazEF*, phage production is attenuated, and lysis is not observed. One of the current hypotheses to explain this observation is that the concentration of MazE in P1 phage infected cells is reduced upon infection. In the absence of MazE, MazF becomes activated and results in a metabolically inactive cell. This ultimately prevents PI phage from replicating in the bacterium, thus preventing its spread in the population (Engelberg-Kulka et al., 2005; Hazan and Engelberg-Kulka, 2004).

Type III TA systems can also prevent the replication of bacteriophage. In general, Type III TA systems consist of a toxin protein and an antitoxin sRNA. The antitoxin sRNA binds to the toxin protein and forms structures called pseudoknots. This prevents the toxin protein from reaching its target. The first identified Type III TA system that may be involved in preventing the replication of bacteriophages is the ToxIN system, which was discovered in the phytopathogen *Pectobacterium atrosepticum* (previously called *Erwinia carotovora*) (Allocati et al., 2015). In this system, ToxN serves as a sequence specific endoribonuclease that cleaves mRNA, thus preventing translation. ToxI, has a dual function as it is both a substrate and inhibitor of ToxN. Specific sequences in ToxI are cleaved by ToxN into 36 individual nucleotide units, which then adopt a pseudoknot fold. ToxN then self assembles with these pseudoknot products to form a self-closing triangular complex comprised of three ToxN proteins interlaced by three ToxI RNAs. The interaction between ToxI and ToxN in this complex results in toxin inhibition (Short et al., 2018).

Evidence that the ToxIN system prevents bacteriophage infection is derived from *P. atrosepticum* that do not contain functional copies of *toxN*, and engineered strains of bacteria that contain exogenous copies of *toxIN*. Specifically, exogenous expression of *toxIN* can protect against phages φA2 and φM1. In contrast, non-functional copies of *toxN* render *P. atrosepticums* susceptible to these phages. Currently, it is hypothesized that infection with a bacteriophage releases ToxN from its interaction with ToxI by preventing the transcription of additional *toxI*, by preventing the proper folding of ToxI, or by interfering with the interactions between ToxI and ToxN. Regardless of which mechanism, or combination of mechanisms, activates ToxN, the ultimate result is a reduction in translation, which prevents bacteriophage replication.

It is important to note that, while the aforementioned TA systems have clear roles in preventing the replication of bacteriophage, there is ambiguity in determining whether or not the effect of these TA systems on bacteria fits the ideal conceptualization of PCD. On one hand, both *MazF* and *ToxN* do not result in lysis of the cell, and there is evidence that the effects of these toxins are reversible. Moreover, in some experiments where true cell death is observed, strong promoters are used to drive copies of these systems, thus leading to questions as to whether or not PCD occurs under normal physiological conditions. On the other hand, it is clear that cells where the toxins are activated do not go through cell division, and appear metabolically inactive (thus it appears that these toxins are bacteriostatic, at least initially). If these bacteria are indeed metabolically inactive, but truly not "dead," these systems would continue to prevent the replication of phage (Song and Wood, 2018). Nevertheless, open questions remain as to if systems such as *mazEF* and *toxIN* truly result in PCD to prevent the replication of phage. For example, how long does a cell have to be metabolically inactive before being declared dead? Are all cells that enter into a state of metabolic decline due to toxins able to exit this state? Is the defining characteristic of PCD to protect against bacteriophage infection with complete and irreversible cell lysis? These questions warrant answers toward furthering our understanding of how bacteria can use PCD to limit the replication of bacteriophage.

4. PCD as a mechanism to promote the infection process

Given the ubiquity of bacteria that can cause infection, the mechanisms by which bacterial pathogens infect a host have been a central focus of study in clinical microbiology. There are diverse mechanisms that bacteria can use to enter and colonize a host, and eventually evade or combat the host's immune system. These strategies range from producing toxins that lyse host cells, to harvesting critical nutrients from the host, to manipulating their appearance to trick the immune system. Interestingly, there is growing evidence that bacterial PCD is important in the infection process. In general, a small subpopulation of bacteria will undergo cell lysis. This results in the release of molecules or proteins that help invade and colonize the host (Fig. 2A). Lysis of a subpopulation of bacteria for the purposes of releasing virulence factors would therefore appear to provide a benefit for any remaining, non-lysed bacteria. As such, it would appear that PCD to release virulence factors may be an altruistic trait. While several studies show that abolishing the ability of bacteria to undergo cell lysis for the purposes of

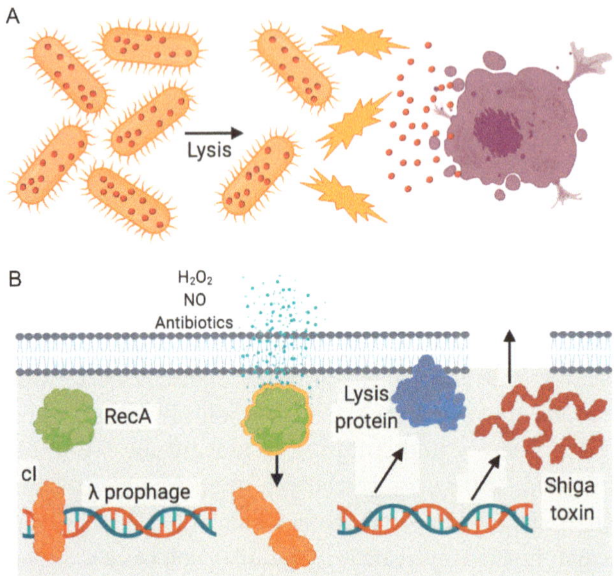

Fig. 2 PCD to release virulence factors. (A) For some bacteria, the release of non-exportable virulence factors is critical for the infection process. In this case, a subpopulation of cells will undergo lysis and release the virulence factor(s). These released virulence factors can harm host cells. In some cases, nutrients are released during lysis of host cells, which may benefit the surviving bacterial population. (B) Shiga toxin release from *E. coli*. Upon sensing a stressor (H_2O_2, NO or antibiotics), the RecA protein from the SOS response pathway cleaves the cI repressor protein. A lysis protein and the non-exportable Shiga toxin from the λ prophage are then expressed. The lysis protein allows the release of the Shiga toxin.

releasing virulence factors attenuates pathogenicity (e.g., Berry and Paton, 2000), it remains unclear as to if the systems described in this section below evolved as true altruistic traits.

4.1 Bacteria driven PCD to release virulence factors

Streptococcus pneumoniae is an important human pathogen that can cause a variety of infections ranging from pneumonia to meningitis (Hirst et al., 2004). One of the key virulence factors produced by *S. pneumoniae* is a pore forming toxin called pneumolysin. Once released by *S. pneumoniae*, pneumolysin affects host cells in a concentration dependent manner. At low concentrations it can trigger apoptosis in host cells, induce inflammation, and activate the host complement system. At high concentrations, it causes lysis of host cell membranes by interacting with membrane embedded cholesterol (Nöllmann et al., 2004). The release of pneumolysin is critical for pathogenesis of *S. pneumoniae* as knockout strains that lack the ability to produce this virulence factor are far less virulent (Berry and Paton, 2000). Interestingly, pneumolysin cannot be secreted by live bacteria as it lacks secretion signal sequences that are required for export. Instead, this virulence factor can only be released if the bacterial cell undergoes lysis, which is characterized by the breakdown of the bacterial cell wall. Damage to the cell wall and subsequent cell lysis is mediated by an autolysin called LytA. This autolysin cleaves important structures in the peptidoglycan cell wall of *S. pneumoniae*. Specifically, LytA cuts the lactyl-amid bond that holds stem peptides and glycan strands together in peptidoglycan (Howard and Gooder, 1974).

Activity of LytA, and thus cell lysis, is tightly controlled. During exponential cell growth, LytA is primarily localized in the cytoplasm, with very little associated with the cell wall or extracellular space. However, once the population enters stationary phase growth, LytA is permitted to interact with and cause degradation of the cell wall. It is currently thought that during exponential growth, ongoing cell wall synthesis prevents the association of LytA with the cell wall. However, once the stationary phase is entered, the cell wall production machinery becomes inactivated. This exposes nascent and uncapped peptidoglycan of which LytA can readily bind to and begin the cell wall degradation process. Release of a preliminary amount of LytA from the cell triggers a lysis cascade. Here, the release of LytA from one cell activates cell lysis in neighboring cells (Mellroth et al., 2012). These newly lysed neighboring cells also release any LytA from their cytoplasm, thus causing additional neighboring cells to undergo lysis. The result is the lysis of a large number of *S. pneumoniae* cells that release pneumolysin into the surrounding environment. Pneumolysin then furthers the infection process by interacting with host cells, and either causing apoptosis or complete cell lysis. To our knowledge, the mechanism that dictates which cells survive and which cells undergo LytA-driven lysis remains unknown for the purposes of pneumolysin release. As host cells are cleared out of the way and release nutrients due to pneumolysin, the remaining *S. pneumoniae* drives deeper into the host's tissues and the infection continues. It is important to note that this is not the only proposed theory as to why LytA causes cell lysis in *S. pneumoniae*. It has also been hypothesized that LytA mediated lysis may facilitate the exchange of DNA (Eldholm et al., 2009), and/or may be used to interfere with proper immune system functioning (Martner et al., 2009).

The leading cause of clinically acquired diarrhea in Western society, *Clostridium difficile* (Polage et al., 2012), also uses cell lysis as a mechanism of toxin delivery. *C. difficile* contains two clostridial toxins, TcdA and TcdB. Like pneumolysin, TcdA and TcdB do not have secretion sequences, and thus *C. difficile* bacteria must undergo cell lysis to release these toxins. Once released, these toxins gain entry into host cells and interact with Rho GTPases involved in the regulation of the actin cytoskeleton. During this interaction, both TcdA and TcdB irreversibly inactivate their targets through glycosylation. As a result, the host cells become deformed, membrane blebbing is observed, and eventually the host cells undergo apoptosis. In the long-term, this results in tissue degradation and inflammation (Govind and Dupuy, 2012).

Lysis of *C. difficile* appears to be regulated by two proteins, TcdE and Cwp19. TcdE is a holin-like protein whose role in mediating cell lysis remains somewhat unclear. In a recent study, it was demonstrated that a functional copy of TcdE is required to release the TcdA and TcdB toxins but that complete cell lysis was not required. This suggests that small holes may open in the cell membrane to allow the passage of these toxins (Govind and Dupuy, 2012). In contrast, other studies have found that TcdE does in fact result in complete cell lysis allowing the release of these toxins (Govind et al., 2015). Under this scenario, and like other holin-like proteins, TcdE is thought to form pores in the cell membrane (Saier and Reddy, 2015; Tan et al., 2001). It is possible that differences between growth conditions, amounts of TcdE isoforms, and *C. difficile* strains may account for the lack of congruence on the role that TcdE plays in cell lysis. Conversely, the role of Cwp19 appears to be clearer. Cwp19 is a lytic transglycosylase that degrades peptidoglycan in the cell wall during stationary phase growth (Wydau-Dematteis et al., 2018). The regulation, and exact mechanism, of Cwp19 have yet to be elucidated. To our knowledge, it currently remains unclear as to what determines which cells undergo lysis. However, on a broad scale, it has been hypothesized that stochastic interactions that regulate holin-mediated cell lysis can impose time delays in cell death (Ryan and Rutenberg, 2007). It is possible that such stochastic interactions can cause delays in *C. difficile* cell lysis. This, in theory, could allow for different subpopulations to develop under scenarios where TcdE is implicated in causing cell lysis. We note that asynchrony in activating Cwp19 has been proposed as a mechanism to develop subpopulations of lysing and non-lysing cells (Wydau-Dematteis et al., 2018).

4.2 Phage driven PCD to release virulence factors

Interestingly, induction of lysis via a prophage can also help to further the infection process. Lysis through prophages has been considered to be a form of bacterial PCD in previous work (Tanouchi et al., 2013). For example, embedded in the genome of some enterohemorrhagic *E. coli* is λ phage (Fig. 2B). Genes encoding λ prophage include those conferring the ability to lyse the bacterial cell membrane as well as the ability to produce a potent virulence factor called Shiga toxin. While there is quite a bit of diversity in the types of Shiga toxins produced by *E. coli* and other pathogens, in general, these toxins harm mammalian host cells by interfering with translation, which can result in inflammation and apoptosis (Jandhyala et al., 2011). Release of Shiga toxin occurs after *E. coli* undergoes prophage-driven cell lysis. Under normal growth conditions, the genes of prophage λ are repressed by the cI repressor protein. However, if *E. coli* encounters a cellular stressor, activity of the cI repressor is repressed, which results in the expression of λ prophage genes.

To survive stressful conditions, *E. coli* will partially rely on a network of proteins that together comprise the SOS stress response pathway. There are two master regulators of the SOS response pathway; a repressor protein called LexA, and an activator protein called RecA. Under non-stressful conditions, the LexA repressor protein prevents the expression of a suite of genes that are activated in the SOS response pathway. Once RecA detects a stressor, it becomes activated and cleaves the LexA repressor protein (Little, 1993). Cleavage of LexA prevents it from repressing the expression of genes in the SOS response pathway. Thus, RecA ultimately activates expression of a series of genes that help *E. coli* combat a stressor. In the case of Shiga toxin release, a stressor will activate the RecA protein. In its activated form RecA cleaves the cI repressor (Roberts et al., 1978), rendering it unable to repress the expression of λ prophage. This results in the production of large amounts of Shiga toxin, and the genes capable of causing cell lysis. Eventually, cell lysis due to λ prophage occurs, thus releasing large quantities of Shiga toxin. While the mechanism that determines which cells undergo lysis, and which cells do not, has yet to be determined in the specific case of Shiga toxin release, it is apparent that there is significant heterogeneity in the expression of SOS response controlled genes in isogenic *E. coli* populations (Kamenšek et al., 2010; Mrak et al., 2007). Heterogeneity is thought to be due to stochastic fluctuations in LexA protein concentrations and LexA binding affinity to SOS responsive promoters. These factors may contribute to heterogeneity in cell lysis for the purposes of releasing Shigella toxin.

While several exogenously applied chemicals, including fluoroquinolone antibiotics, can activate the SOS stress response in *E. coli*, it is clear that the release of Shiga toxin occurs in the absence of these antibiotics. Interestingly, human neutrophils produce H_2O_2 and NO, both of which lead to SOS stress response by triggering DNA damage. Accordingly, it has been proposed that the host's immune system cells trigger the release of Shiga toxin by activating the expression of prophage λ (Wagner et al., 2001). The number of bacteria lysed by prophage λ increases as a function of H_2O_2. Therefore, small concentrations of H_2O_2, perhaps those along the lines of what is observed in human neutrophils, trigger lysis of only a small amount of *E. coli*. Moreover, because Shiga toxin production is driven by strong phage promoters, even a small percentage of bacteria undergoing lysis will release a large amount of toxin (Los et al., 2013).

4.3 TA systems that drive the release of virulence factors

S. aureus uses a TA system to regulate PCD during the infection process. Similar to other Type I TA systems, this system contains a toxic protein called PepA1 (encoded by the gene called *spr1*), and an RNA antitoxin $SprA1_{AS}$. Under normal growth conditions, the $SprA1_{AS}$ RNA binds to a region of the mRNA encoding PepA1. This serves to prevent production of the PepA1 toxin. However, under conditions of oxidative stress and/or acidic stress, the concentration of the $SprA1_{AS}$ mRNA is reduced, which allows the production of the PepA1 toxin protein (Sayed et al., 2012). A more recently discovered TA system, SprA2 (encoding a toxin peptide, PepA2) and $SprA2_{AS}$ (encoding an antitoxin mRNA) works in a similar fashion. However, in this system, a reduction in the production of the antitoxin mRNA occurs during osmotic or nutrient stress (Germain-Amiot et al., 2018). Interestingly, the production of either PepA1 and/or PepA2 can lead to cell lysis of *S. aureus*. In the case of PepA1, this protein inserts itself into the bacterial cell membrane, and undergoes a conformational change. This presumably results in the formation of pores in the bacterial cell membrane, which leads to cell lysis.

There are currently several hypotheses to explain why cell lysis would benefit *S. aureus*. First, extracellular PepA1 and PepA2 can damage the cell membrane of host cells, which may help *S. aureus* to acquire additional nutrients, such as iron, from erythrocytes (Germain-Amiot et al., 2018; Sayed et al., 2012). Interestingly, extracellular PepA1 cannot damage *S. aureus* cells as the cell wall may prevent access to the cell membrane. Second, the conditions of oxidative stress and acidity are often encountered in phagolysosomes after bacteria are internalized by host's immune systems defenses. Lysis of a small subpopulation of *S. aureus* captured in phagolysosomes may release sufficient PepA protein to damage the phagolysosome and surrounding cell membrane, thus allowing escape of the surviving bacteria (Sayed et al., 2012). As this TA system has only recently been described, it currently remains unclear as to how subpopulations of lysing and non-lysing cells are established. Moreover, whether either or both of these hypotheses are correct, mounting evidence suggests that regulated cell lysis by *S. aureus* helps to promote the infection process by damaging host cells.

4.4 Host-cell driven bacterial cell lysis to release virulence factors

Salmonella typhimurium, a pathogen that most often causes gastroenteritis in humans, also uses PCD to promote infection. However, unlike *S. pneumoniae* and *C. difficile* that use proteins encoded by the bacterium to drive cell lysis, *S. typhimurium* relies on host cells to cause cell lysis. One of the main barriers preventing the colonization of the human gastrointestinal tract by *S. typhimurium* is the resident microbiota. However, the resident microbiota can be removed if *S. typhimurium* can trigger an inflammatory response from the host. Removal of resident microbiota competitors facilitates colonization of *S. typhimurium*, which serves to drive the infection process forward (Stecher et al., 2007).

Two of the major virulence factors associated with triggering an inflammatory response from *S. typhimurium* are a type III secretion system (TTSS) and flagella (encoded by the gene *fliC*). Interestingly, these two virulence factors are not expressed equally in all members of an *S. typhimurium* population. Instead, two subpopulations can be observed: one that produces large quantities of these virulence factors, and one that does not (Cummings et al., 2006). It has been proposed that stochastic fluctuations in gene expression serve to establish these two subpopulations. Specifically, it has been proposed that stochastic fluctuations in the regulators of *fliC* expression can establish two populations of *S. typhimurium*; those that express *fliC*, and thus produce flagella, and those that do not (Stewart and Cookson, 2014; Stewart et al., 2011). Expression of *fliC* relies on two mutually repressing regulators, YdiV and FliZ, as well as an additional complex, $FlhD_4C_2$. In this system, $FlhD_4C_2$ controls expression of operons that ultimately lead to the expression of *fliC*. YdiV represses the activity of $FlhD_4C_2$, whereas FliZ promotes the activity of $FlhD_4C_2$. It has been proposed that heterogeneity in amount of $FlhD_4C_2$ complex in each bacterium due to stochastic fluctuations in its many regulators, including transcription and translation of its components, determine *fliC* expression. Bacteria with a low abundance of $FlhD_4C_2$ require more FliZ to drive expression of *fliC*. Conversely, bacteria with a high abundance of $FlhD_4C_2$ require more YdiV to repress expression of *fliC* (Stewart and Cookson, 2014). These differences allow *S. typhimurium* to establish populations of cells that do, and do not, produce flagella.

Similarly, expression of the heterogeneously distributed type III secretion system (TTSS) in *S. typhimurium* is regulated by the SPI-1 regulatory network (Schlumberger et al., 2005). Members of the regulatory network activate as a transcriptional cascade, and there are several intertwined feed forward and positive feedback loops embedded in the network (Saini et al., 2010; Temme et al., 2008). Further complicating the regulation of SPI-1 are additional post-transcriptional regulation mechanisms (Golubeva et al., 2012, 2016; López-Garrido and Casadesús, 2010). While it is apparent that two subpopulations of *S. typhimurium* exist (those that express the SPI-1 regulon, and those that do not), the exact mechanism that leads to differential expression remains under investigation.

Experimental evidence shows that the expression of TTSS and flagella take an important toll on the bacteria that produce them. While these bacteria are able to invade gut epithelial cells and cause an inflammatory response, the vast majority of these bacteria are killed by the host's immune system. The inflammatory response caused by these bacteria serves to clear out resident microbiota competitors, which may facilitate colonization of the non-TTSS and flagella producing bacterial subpopulation (Ackermann et al., 2008; Stecher et al., 2007). Thus, in this system, *S. typhimurium* specifically triggers its own death via the host's immune system, and thus serves to fit within our broad definition of PCD.

5. PCD as a mechanism to alleviate environmental stressors

In their natural environment, bacteria can face a variety of environmental stressors. Chief among these is the presence of noxious chemicals, including antibiotics (Perry et al., 2016). Bacteria have been encountering antibiotics for thousands of years (D'Costa et al., 2011; Perry et al., 2016). On an evolutionary timescale, it is only recently that we have developed the use of antibiotics in the clinical and agricultural settings. Given the rise of antibiotic resistant bacteria, understanding the mechanisms by which bacteria die during the application of antibiotics is important for developing new treatment protocols and drugs. While the specific targets of most antibiotics have been well understood for a considerable amount of time, more recently common mechanisms of antibiotic driven cell death have been described (e.g., Kohanski et al., 2008; Yang et al., 2019). Interestingly, bacteria may use PCD mechanisms to either release their own antibiotics or in response to the presence of antibiotics. Understanding how PCD occurs in these two scenarios may allow us to develop novel approaches to treat bacterial infections. Interestingly, a recent manuscript using engineered bacteria has provided evidence that bacteria can use PCD to collectively degrade an antibiotic stressor (Tanouchi et al., 2012, 2013).

5.1 Antimicrobials produced by bacteria can be released through PCD

Bacteria can produce their own antibiotics in response to a stressor. One broad class of antibiotics produced by bacteria are antimicrobial peptides. These peptides are produced by one population of bacteria to eliminate other microbial competitors. In certain instances, the antimicrobial peptide cannot be secreted from the bacterium that produces it. Instead, export of such peptides is mediated by cell lysis. An example of this is the production and release of the antimicrobial peptide colicin, which is produced by *E. coli*. Upon its release, colicin can kill bacterial competitors, which may benefit the remaining *E. coli* population (Gardner et al., 2004; Mavridou et al., 2018). Colicin does not have the ability to be secreted, and can only be released by *E. coli* that undergo cell lysis.

The operon encoding *E. coli*'s colicin E2 system is housed on a plasmid. Its expression is regulated by the SOS response pathway, which is activated after DNA damage or during times of nutrient stress. The operon consists of three genes: *cea* (a colicin activity gene), *cei* (the gene conferring self-immunity to colicin), and *cel* (the gene that encodes a protein responsible for cell lysis). There are two mRNAs that are produced from the operon: a short mRNA consisting of the *cea* and *cei* genes, and a longer mRNA consisting of all three genes. When *E. coli* is growing at an exponential rate, LexA of the SOS response pathway, prevents transcription of the colicin operon. Upon DNA damage or nutrient stress, LexA is cleaved by RecA rendering LexA inactive. Accordingly, genes in the colicin operon can now be expressed, including the Cea protein, which produces colicin, and the Cel protein, which drives cell lysis (Cascales et al., 2007). However, not all *E. coli* cells will activate expression of the colicin E2 operon at the same time, which leads to heterogeneity in the production colicin. Under low levels of stress, heterogeneity in the expression of colicin genes is highest. That is, there exist clear subpopulations that do, and do not, express the colicin genes. In contrast, high levels of stress result in nearly all of the population expressing colicin genes (Mader et al., 2015).

If all bacteria in the population were to lyse and release colicin at the same moment, then there would not be any remaining *E. coli* bacteria to take advantage of the reduction in bacterial competitors. Accordingly, *E. coli* has developed a mechanism to prevent simultaneous lysis of the entire population. To regulate production of the Cel lysis protein, a protein called CsrA binds to the long mRNA (which contains the *cel* gene) produced by the colicin E2 operon. Specifically, CsrA binds to a region near the ribosomal binding site, thus preventing ribosome assembly and subsequent translation of the *cel* gene (Liu et al., 1995). CsrA itself is negatively regulated by two sRNAs, CsrB and CsrC, which prevent CsrA from binding to the long mRNA (Gutiérrez et al., 2005). When growth occurs under non-stressful environmental conditions, a high abundance of CsrA is present. Thus, translation of any long mRNA that might be produced by triggering the SOS response is near immediately prevented. However, during a sustained period of stress or after significant DNA damage, the SOS response pathway is activated near continuously. This leads to the production of an abundance of both the short and long mRNAs. The short mRNA is not regulated by CsrA. Therefore, the Cei and Cea proteins are produced, and colicin accumulates inside the cell. Initially, translation of the long mRNA is prevented by CsrA. However, as the concentration of long mRNA increases, binding of CsrA to each of these mRNAs serve to essentially titrate out the CsrA regulator. Once the CsrA regulator dips below a threshold, it can no longer prevent translation of the *cel* gene. The Cel protein is then produced, and the bacterium undergoes cell lysis. Ultimately, colicin is released from lysed cells where it can damage competitor microbes.

The titration mechanism that regulates the release of colicin provides several advantages to the cell. First, it provides an extended window where colicin can be produced from the short mRNA via the *cea* gene. This ensures that a large concentration of colicin is released from the cell upon lysis. Second, it ensures that the cell does not undergo lysis in response to transient or stochastic activation of the SOS response pathway. Third, heterogeneity across the timing and intensity in the activation of the SOS response pathway generates heterogeneity in the time that cells undergo lysis. But what is the role of the sRNAs (CsrB and CsrC) regulating the activity of CsrA? Interestingly, mathematical modeling suggests that these sRNAs act as an additional mechanism to buffer against fluctuations in the activation of the SOS response pathway. They can also serve to regulate the timing between when the SOS response pathway is triggered, and when cell lysis occurs (Götz et al., 2018; Lechner et al., 2016).

When taken together, *E. coli* has developed a well regulated and noise buffered system that results in the lysis of a subpopulation of cells under periods of stress. This releases an antimicrobial peptide, colicin, that serves to kill competitor bacteria, which may provide a benefit (increased access to nutrients, increased access to substrate binding sites) to the remaining subpopulation of *E. coli*. That is, production and release of colicin may be regarded as an altruistic behavior (Riley and Chavan, 2007). Interestingly, recent experimental evidence has suggested that bacteria can coordinate the production of colicin after the detection of competitors. This would perceivably allow *E. coli* to coordinate "attacks" on colonies of competitors, or defend established colonies against invasion by competitors (Mavridou et al., 2018).

5.2 The production of ghost cells in response to antibiotics

Some bacteria create "ghost cells" when encountering environmental stressors, including antibiotics (Cheng et al., 2014). Ghost cells do not contain cytoplasmic materials, including DNA, and are thus metabolically inactive. However, they maintain numerous immunogenic properties as their cell wall and cell membrane structures remain largely intact. One mechanism that *E. coli* can use to produce ghost cells is through the aptly named *ghoST* TA system, which a member of the Type V TA systems (Wang et al., 2012). In this system, *ghoS* encodes an RNase that serves as the antitoxin. In contrast, *ghoT* encodes a hydrophobic peptide that serves as the toxin. When *E. coli* are grown under non-stressful conditions, GhoS degrades the mRNA produced by the *ghoT* locus. Thus, translation of *ghoT* mRNA is prevented, and the toxin protein is not produced. However, when *E. coli* are grown under stressful conditions, a second TA system, *mqsRA*, regulates the functionality of the *ghoST* system. Here, the toxin protein of the *mqsRA* system, MqsR, preferentially degrades *ghoS* mRNA, thus preventing the translation of new GhoS proteins. As a result, the *ghoT* mRNA is not degraded, and the GhoT toxin is produced. The GhoT protein localizes to the cell membrane where it forms a multimer. Upon embedding itself in the membrane, GhoT multimers produce pores (Kim et al., 2018). This ultimately results in a loss of proton motive force across the membrane, and a significant reduction in intracellular levels of ATP (Cheng et al., 2014). This results in the formation of ghost cells.

The evolutionary significance of ghost cells, from the perspective of the *ghoST* system, remains somewhat unresolved. There is evidence suggesting that the *ghoST* system is involved in the formation of antibiotic resistance and metabolically dormant persister cells (Wang et al., 2012). Importantly, persister cells can metabolically reactivate, and thus do not represent true "ghost cells" that are devoid of all intracellular contents. Interestingly, a recent study found that dead cells can serve a protective function against antibiotics. Here, the dead cells serve as a "sponge" and protect living cells by removing antibiotic from the environment (Wu and Tan, 2019). However, it is currently unclear as to if ghost cells have a similar "protective" function by soaking up stressor from the environment.

6. PCD during developmental processes in bacteria

Similar to their eukaryotic counterparts, bacteria have well regulated mechanisms by which subpopulations of cells undergo morphological differentiation to fulfill a role in a developmental process. These include the development of biofilms (Kostakioti et al., 2013) and endospores (Driks, 2002), and the generation of highly elaborate reproductive structures (Ward and Zusman, 2000). In many, but not all cases, bacteria will undergo development during stressful environmental conditions most often as a means to extend the survival of the bacterial population. Interestingly, there are instances where PCD of a subpopulation of cells is important in the developmental process. A common hypothesis used to explain PCD during development is that lysed cells release nutrients that are consumed by the remaining living cells (Allocati et al., 2015). The living cells use these nutrients to continue, or delay, the development process. Therefore, it may appear that PCD in the following examples is an altruistic trait where a subset of cells sacrifice themselves to ensure the long-term survival of the bacterial population. However, to our knowledge, this hypothesis has not yet been explicitly tested in the systems detailed below.

6.1 Endospore development in *B. subtilis*

Under conditions of severe nutritional limitation, *B. subtilis* forms endospores that remain inactive for extended periods of time in an attempt to resist extreme environmental stress and prolong survival of the population. Although the benefits of sporulation to survival are evident, there is a large energy and time commitment involved in the sporulation process. Moreover, after a certain stage, the process becomes nearly irreversible. If nutrients do become available again, spores do not return to active growth as efficiently as vegetative cells (Ellermeier et al., 2006). Thus, the spores can be outcompeted by other microbial competitors that did not undergo a sporulation process. Therefore, the sporulation strategy is considered to be a "last resort" mechanism for survival. Populations of *B. subtilis* have developed strategies to delay the sporulation process as long as possible.

One delay tactic is through "bacterial cannibalism" (Fig. 3A). In this delay strategy, two subpopulations of *B. subtilis* are formed. One subpopulation produces a toxic peptide and regulatory elements that confer resistance against the peptide (this subpopulation is henceforth referred to as "producers"). The second subpopulation does not produce the toxic peptide and is susceptible to it (this subpopulation is henceforth referred to as "non-producers"). The toxic peptide results in lysis of the non-producers. This releases nutrients from the non-producers, which may be consumed by the producers. This likely serves to delay sporulation behavior (González-Pastor et al., 2003; López et al., 2009).

Given the uniqueness of this microbial behavior, the pathways that lead to cannibalism in *B. subtilis* have been well studied. In general, cannibalism processes are triggered by conditions of starvation or low nutrient availability through activation of the master regulator Spo0A (Engelberg-Kulka and Hazan, 2003). Spo0A is post-translationally activated through phosphorylation (Spo0A-P). In its activated form, Spo0A-P acts a transcription factor to regulate gene expression. Phosphorylation of Spo0A-P is dependent upon a four component phosphorylation cascade system (Burbulys et al., 1991). Activation of downstream elements controlled by Spo0A-P occurs in a concentration dependent manner. Low concentrations of Spo0A-P result in the activation of genes involved in cannibalism whereas high concentrations of Spo0A-P result in the activation of genes involved in sporulation. Accordingly, only sustained production of Spo0A-P is sufficient to activate sporulation.

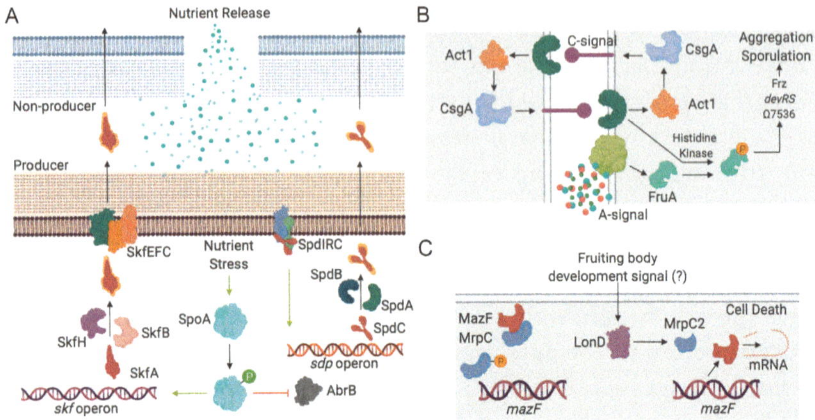

Fig. 3 Bacteria that use programmed cell death for development. (A) Bacterial cannibalism in *B. subtilis*. Nutrient stress creates two populations: producers and non-producers. In producer cells, SpoA becomes phosphorylated due to nutrient stress. SpoA-P directly drives expression of the *skf* operon, and inhibits the AbrB repressor, allowing the *sdp* operon to be expressed. The SkfH and SkfB proteins process the SkfA lysis protein, which is subsequently exported from the producer bacteria by SkfEFC. The SpdC lysis protein is modified by SpdA and SpdB, and is exported from the producer bacteria via the general secretory pathway. Both the SkfA and SpdA proteins result in lysis of non-producer cells. This releases nutrients, which may benefit the producer cells. SkfEF and SpdI confer immunity from SkfA and SpdC, respectively. SpdI, SpdIR and SpdC interact to form a positive feedback loop, increasing expression of genes from the *sdp* operon. (B) Signaling involved in fruiting body development and PCD in *M. xanthus*. A-signal activates expression of FruA and the production of the surface attached C-signal though CsgA. C-signal interacts with an adjacent aggregating cell, which promotes activation of Act1 and CsgA. This leads to the production of more C-signal. This forms a positive feedback loop in C-signal production between the two adjacent cells. C-signal also phosphorylates the FruA protein via a putative histidine kinase, which activates downstream elements involved in aggregation and sporulation. (C) Cell death in *M. xanthus*. In the absence of a yet to be determined signal, the activity and expression of MazF is prevented by MrpC. In the presence of a signal to drive fruiting body development, the LonD protease is activated. LonD cleaves MrpC to create MrpC2. Expression and activity of MazF are uninhibited. MazF cleaves mRNAs, which ultimately results in cell lysis and the release of nutrients. The released nutrients drive the sporulation process. *Panel B figure logic adapted from Jelsbak, L., Søgaard-Andersen, L., 2000. Pattern formation: fruiting body morphogenesis in Myxococcus xanthus. Curr. Opin. Microbiol. 3, 637–642.*

To establish both the producer and non-producer populations, it has been proposed that the flux of phosphate through the phosphorylation cascade responsible for activating Spo0A-P is a noise generator. This leads to a heterogeneous distribution of cells in the population that differ in their concentration of SpoA-P (Chastanet et al., 2010). Cells that reach a low level threshold of Spo0A-P first activate expression of two operons involved in the cannibalism process: the sporulation delaying protein operon (*sdp*) and the sporulation killing factor operon (*skf*)(González-Pastor et al., 2003). The *skf* operon is composed of eight genes (*skfABCDEFGH*) whereas the *sdp* operon is composed of five genes (*sdpABCRI*). Together, expression of these operons results in the producer cannibalistic phenotype.

Expression of the *skf* operon is directly activated by Spo0A-P. The promoter region of *skf* has been shown to have a high affinity-binding site for Spo0A-P. The *skfA* gene produces a 56 amino acid antimicrobial-like peptide, which serves as a killing factor precursor (Claverys and Håvarstein, 2007; Popp and Mascher, 2019). This precursor is modified by the SkfB protein, and eventually processed into an active state by the SkfH protein (Popp and Mascher, 2019). The ATP binding cassette transporter SkfEF has a dual role. First, and along with SkfC, it is thought to be involved in the export of mature SkfA across the cell membrane (Höfler et al., 2016). Second, SkfEF is thought to provide immunity against mature SkfA (Popp and Mascher, 2019). While little is known about the exact mechanism by which SkfEF protects against SkfA, it is thought that additional immunity proteins are present within the *skf* operon. Specifically, *skfD* has been observed to show significant similarities to type II CAAX prenyl endopeptidases (Claverys and Håvarstein, 2007), thus suggesting a strong resemblance to bacteriocin immunity proteins that undergo proteolytic cleavage to inactivate toxins. However, this form of immunity and its mechanism has yet to be fully explored.

In contrast to the *skf* operon, Spo0A-P does not directly regulate transcription of the *sdp* operon. Instead, activation of Spo0A allows indirect activation of *sdp* genes by repressing the expression of a repressor protein, AbrB (Claverys and Håvarstein, 2007). Under conditions of vegetative growth, AbrB represses the *sdp* operon. However, when low levels of Spo0A-P are found in the cell, expression of *abrB* genes is attenuated, resulting in the decay of the AbrB repressor protein. This allows half of the *sdp* operon (*sdpABC*) to undergo transcription (Popp and Mascher, 2019). Upon translation, the toxin protein, SdpC, is post-transcriptionally modified into its active form by SdpA and SdpB (Claverys and Håvarstein, 2007). SdpC is subsequently exported from the cell via the general secretory pathway (Linde et al., 2003), where it can cause lysis of non-producers. SpdC regulates the expression of the corresponding immunity protein, SpdI, which is contained in the second half of the operon (*sdpRI*). SpdI serves a dual role in that not only does it confer immunity to SpdC, but it also acts as a membrane bound signal transduction protein that regulates its own expression. SpdR is an autorepressor that blocks expression of itself and SpdI. When the SpdC toxin protein interacts with SpdI, a conformation change in SpdI ensues, which allows it to sequester the SpdR repressor protein. Once sequestered, the SpdR protein can no longer block expression of SpdI and SpdR at the transcriptional level, and thus expression of *spdRI* is increased. While the exact mechanism by which SpdI confers immunity has yet to be determined, it is currently hypothesized that complexes between SpdC and SpdI neutralize the effects of SpdC toxicity (Ellermeier et al., 2006).

Both SkfA and SpdC are thought to differ in their mechanism of cell lysis. SkfA has a structure similar to other bacteriocins. However, the specific mechanism by which SkfA kills non-producers has yet to be described. SpdC is thought to induce lysis through two mechanisms. The primary mode of action of SpdC is to cause pore forming channels in the cell membrane, which consequently causes the loss of proton motive force. As a secondary effect, SpdC appears to activate autolysin driven autolysis of *B. subtilis*. Here, autolysins, which specifically cleave peptidoglycan, are either regulated or hyperactivated as a consequence of the loss of the proton motive force. Autolysin driven degradation of peptidoglycan would result in large pores forming in the cell wall, which in turn would further allow permeabilization of the cell (Lamsa et al., 2012). When taken together, the combined modes of action of SkfA and SpdC result in rapid and efficient lysis of non-producers. It is hypothesized that this releases nutrients contained in the non-producers, which can then be consumed by the remaining producers. The influx of nutrients delays the sporulation process in producers in hopes that additional nutrients become available in the environment.

6.2 PCD drives fruiting body development in *M. xanthus*

In *Myxococcus xanthus*, lysis of a subset of the population releases nutrients that aid in the development of a structure called a fruiting body (Wireman and Dworkin, 1977). During times of nutritional stress, *M. xanthus* will swarm together to form an aggregate. Formation of these aggregates eventually leads to the formation of a fruiting body, which is designed to form and release spores that are able to resist harsh environmental conditions. During the aggregation process, *M. xanthus* cells will divide into three subpopulations. Approximately 30% of the cells differentiate into structures called peripheral rods. These structures surround the fruiting body and remain in a persister-like, or dormant, state owing to the lack of nutrients (Connor and Zusman, 1991b). It is thought that the role of these cells is to repopulate the area should additional nutrients be found in the environment (Connor and Zusman, 1991a). Only 10% of cells differentiate into spores in the fruiting body (Connor and Zusman, 1991b). These spores are highly resistant to environmental stressors, and can serve to start a new population in a new environment. The remaining cells (~60%) undergo PCD, which is thought to release nutrients, thus providing energy to the sporulating cells. In general, the cell lysis process can occur slightly before the aggregation process, but can also continue as the sporulation process occurs (Muñoz-Dorado et al., 2016). While some of the interactions that initiate and drive the development process are known (reviewed in Kroos, 2017), it is currently unclear as to how aggregating cells are partitioned into subpopulations (Arias Del Angel et al., 2018).

The formation of the initial aggregate appears to be regulated through quorum sensing. Quorum sensing involves the production and subsequent secretion of small molecules or peptides (called autoinducers) that are diffusible and can be shared by all members of a microbial population. Once the concentration of autoinducer reaches a sufficiently high concentration, the microbial population responds by altering gene expression. In general, each member of the microbial population produces autoinducers, and thus its concentration serves as a measure of population density (Miller and Bassler, 2001). In *M. xanthus*, nutritional stress leads to the rapid accumulation of (p) ppGpp (Fig. 3B). Its accumulation results in the expression of the quorum sensing A-signal, which consists of a mixture of peptides, amino acids, and proteases (Kaplan and Plamann, 1996). The produced A-signal is subsequently exported from the cell (Harris et al., 1998). Once the A-signal reaches a sufficiently high density, *M. xanthus* begins to aggregate. The aggregation process is promoted by a surface attached signal called the C-signal, which is produced by the protein product of the *csgA* gene (Shimkets et al., 1983). Activation of downstream effectors in response to the C-signal is contact dependent. As *M. xanthus* cells begin to aggregate, the C-signal from one adjacent cell will contact a second adjacent cell. This promotes the production of C-signal in the second cell via the *act1* protein product (Gorski et al., 2000). Once the C-signal on the second cell is presented to the first cell, additional C-signal is produced in the first cell. This series of interactions serve as a positive feedback loop that rapidly increases the amount of C-signal being produced. This serves to promote aggregation of more cells. At the same time, the C-signal activates a putative histidine protein kinase, which phosphorylates and subsequently activates a protein called FruA (Ellehauge et al., 1998). FruA is initially produced in response to the original A-signaling event. Phosphorylated FruA activates its downstream targets (Frz, transcription of *dsvRV* genes and ω7536), which promotes further aggregation, and eventual sporulation (Licking et al., 2000; Søgaard-Andersen and Kaiser, 1996). The C-signal appears to act in a concentration dependent manner; intermediate concentrations of C-signal induce aggregation, whereas the sporulation process is triggered by a high concentration of C-signal (Kim and Kaiser, 1991). An excellent review of the additional aspects of pattern formation in *M. xanthus* has been provided by Jelsbak and Søgaard-Andersen (2000)

Currently, it remains unclear as to how lysis is triggered and regulated in the non-sporulating population. To date, two mechanisms have been proposed. First, it has been proposed that members of the *M. xanthus* population releases five autocides, numbered AMI through AMV, which result in lysis of non-sporulating cells (Varon et al., 1984). AMV appears to work in a concentration dependent manner. That is, for the same initial concentration of autocide, the lethality of AMV increases as a function of initial cell density. In contrast, lethality of AMI appears to be independent of initial cell density (Gelvan et al., 1987). Second, it has been proposed that lysis is mediated by the MazF-MrpC TA system (Fig. 3C). Unlike other bacterial counterparts, *mazE* is absent in *M. xanthus*. Instead, it appears that the developmental transcription factor MrpC plays a role in regulating cell lysis. MazF acts as an mRNA interferase, which cleaves a ubiquitous consensus sequence in mRNA, thus preventing global translation. On the other hand, MrpC acts as an antitoxin of MazF, as well as a transcriptional regulator of *mazF* expression. During vegetative growth, MrpC plays two roles. First, it complexes with available MazF protein, thus attenuating its mRNA interferase activity. Second, any uncomplexed MrpC is phosphorylated by a regulatory phosphorylation cascade, which prevents it from activating expression of *mazF*. During fruiting body development, MrpC is proteolytically cleaved by a developmentally induced protease called LonD, thus preventing its phosphorylation. This modified MrpC, termed MrpC2, activates the transcription of *mazF*, which in turn results in cell lysis (Nariya and Inouye, 2008). Importantly, the direct, or potentially indirect, mechanism by which MazF results in cell lysis remains unclear. Interestingly, MazF is not required for cell lysis in some strains *of M. xanthus* (Lee et al., 2012), thus suggesting alternative methods of lysis exist (Boynton et al., 2013). Not only are the exact mechanisms of cell lysis currently unknown, it remains unclear as to how immunity is conferred to sporulating cells, and at a more fundamental level, how cells differentiate into sporulating, lysing, and peripheral rod cells.

6.3 PCD helps build aerial hyphae in Streptomyces

The Streptomycetes family are unique in that they form mycelium in their natural soil habitat. Upon germination of a spore, *Streptomyces* sp. grow as a vegetative mycelium before entering into their reproductive lifecycle. During reproduction, aerial mycelium will protrude into the air. Spores will form in and eventually be expelled from aerial mycelium structures. Differentiation is tightly controlled by numerous *bld* regulators (reviewed in Claessen et al., 2006). BldD is the master regulator of differentiation as it regulates the expression of ~170 genes involved in development (Bush et al., 2015; Méndez et al., 1985). Similar to *M. xanthus*, PCD plays an important role during development in the reproductive stage. It is proposed that cell lysis releases important nutrients that help to build aerial hyphae, and in the formation of spores (Yagüe et al., 2013). There are two rounds of PCD in *Streptomyces*. The first round of PCD occurs prior to the formation of aerial hyphae. Under conditions of low nutrient availability, vegetative mycelium (called MI mycelia) undergo PCD. Some of the remaining viable cells form multinucleated cells (called MII mycelia), which facilitates the sporulation process and begins to form aerial hyphae. Just prior to sporulation of cells in the aerial hyphae, a second round of PCD occurs in some of the MII cells and any remaining MI cells forming the base (or substrate) of the aerial hyphae. Eventually, spores are formed in the aerial hyphae, which can propel the bacteria into new environments. We note that much of this is expanded upon in a previously published and thorough review (Claessen et al., 2006).

The signals and genetic elements that initiate and regulate this PCD process have yet to be determined. Similarly, the mechanisms that govern the development of subpopulations that do, and do not, undergo PCD are currently not known. More is known of the molecular players involved in causing cell lysis. Early studies, as well as proteomic analyses, have indicated that several hydrolyases, nucleases, proteases and additional enzymes are involved in the degradation of DNA, and a reduction in bacterial cell membrane integrity. More recently, however, three methods of PCD may have been identified in *Streptomyces*. First, it is possible that TA systems may play a role in PCD. For example, a TA system has been identified in two species of *Streptomyces* (*S. lividans* and *S. coelicolor*). This type II TA system called YefM/YoeB has been proposed to inhibit global translation through cleavage of mRNAs. Overexpression of the toxin protein, YoeB, increases the amount of cells undergoing PCD. However, deletion of *yefM* and *yoeB*, has no observable effect of PCD. Thus, there remain outstanding questions regarding the role of the YefM/YoeB TA system in PCD (Sevillano et al., 2012). An additional TA system has also been identified in *S. cattleya* (*relBE2sca*) (Li et al., 2016), but its role in PCD remains unknown. Second, there is evidence that prodiginines (a class of tripyrrole pigments) secreted by *S. coelicolor* cause cell lysis. The biosynthesis of these secondary metabolites appears to occur during PCD in the vegetative (M1) mycelium prior to the development of the aerial hyphae. Interestingly, cells that lack the ability to produce prodiginines through the deletion of the gene *redD* do not undergo PCD of vegetative (MI) mycelial cells (Tenconi et al., 2018). This suggests that prodiginines may be a key player in regulating PCD in *Streptomyces*. How their production is activated and regulated remains unknown. Third, recent work performed in *S. natalensis* has shown that PCD is dependent upon oxidative stress (Beites et al., 2015).

7. PCD releases bacterial DNA into the environment

It goes without saying that DNA is essential to all life, including bacteria. While most of the study of DNA has focused on its role inside the cell, extracellular DNA (eDNA) has critical roles in the ecology of bacteria. For example, some bacteria can enter into a state of competency where they can engulf, and subsequently incorporate, eDNA (Chen and Dubnau, 2004). Conjugative plasmids that are passed between bacteria through the extracellular milieu are responsible for the acquisition of novel traits, including antibiotic resistance (Lopatkin et al., 2016). DNA can serve as an important source of nutrients (Finkel and Kolter, 2001), and has roles in biofilm development and stability (Montanaro et al., 2011). As a result of PCD, bacterial DNA can be released into the environment. It has been hypothesized that some bacteria have evolved PCD systems for the purpose of providing a source of eDNA to the remaining viable population. Interestingly, as we will describe below, there is some evidence that the release of eDNA via PCD can increase the fitness of a bacterial population in certain instances. However, whether or not PCD evolved in these systems for the sole purposes of releasing DNA remains unresolved.

7.1 Microbial fratricide in *S. pneumoniae*

Fratricide is defined as the act of killing a sibling, and is apparently widespread throughout biological systems. Diverse organisms ranging from ants (Chapuisat et al., 1997), to eagles (Edwards and Collopy, 1983), to salamanders (Walls and Blaustein, 1995) commit fratricide. Fratricide is also observed in eukaryotic cells. For example, antigen-specific memory T-cells commit fratricide to regulate their own population density (Callard et al., 2003). Interestingly, bacteria can also commit fratricide. The most well studied example of fratricide in bacteria occurs in *S. pneumoniae*. However, due to the commonality of the genes involved in the fratricidal system, it is believed that additional, but not all, members of the *Streptococcus* genus may be able to commit fratricide (Claverys et al., 2007). In *S. pneumonia,* fratricide occurs when a fraction of a bacterial community reaches a natural state of competence for the purposes of up taking DNA from the environment, also known as genetic transformation (Popp and Mascher, 2019). Competent cells appear to kill their non-competent counterparts, thus releasing their internal contents, including DNA, into the environment. Competent cells can then acquire the released DNA through genetic transformation (Fig. 4A).

Competence is induced by the competence-stimulating peptide (CSP, Fig. 4B). Competency in *S. pneumoniae* occurs in nutrient rich conditions during logarithmic growth, where synthesis and accumulation of CSP occurs rapidly. The CSP autoinducer is encoded by the *comC* gene (Claverys and Håvarstein, 2007). Upon translation, the precursor peptide ComC is modified and subsequently secreted by ComA and ComB. Upon reaching a sufficiently high concentration, the secreted CSP interacts with a membrane-bound histidine-kinase receptor, ComD. Interestingly, recent evidence suggests that not all CSP is secreted and instead remains attached to the cell surface (Prudhomme et al., 2016). Nevertheless, CSP leads to the autophosphorylation of ComD and the transfer of a phosphoryl group to the ComE response regulator. Phosphorylated ComE then directly activates the expression of approximately 20 early competence (*com*) genes. A positive feedback loop, activated by the presence of the direct-repeat motif in the promoters of the *comAB* and *comCDE* operons, amplifies the signal toward inducing the competency state (Claverys and Håvarstein, 2007).

The specific mechanism that allows the development of two subpopulations, competent and non-competent cells, remains unknown to our knowledge. Recently, however, it has been shown that initiation of the

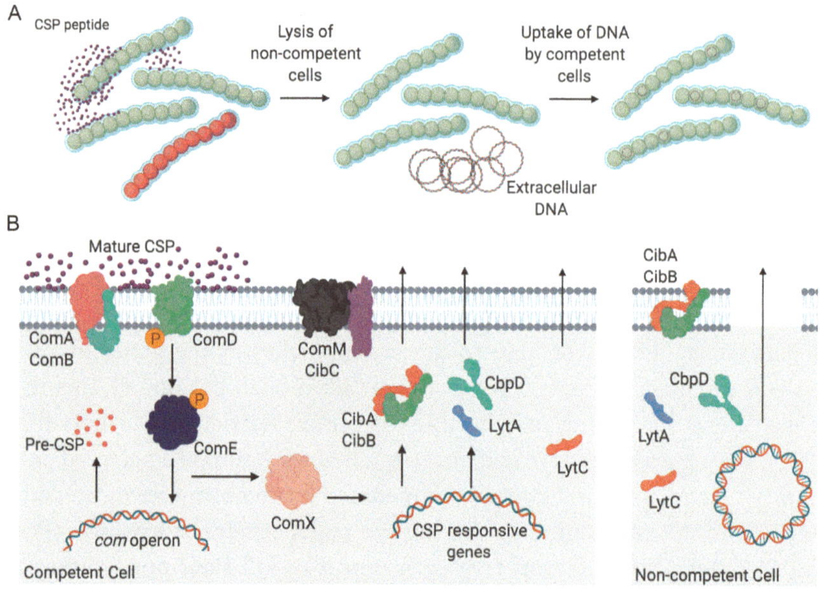

Fig. 4 Bacterial fratricide in *S. pneumoniae*. (A) Overview. The accumulation of CSP peptide on the cell surface and/or in the environment results in two subpopulations of cells: competent (green) and non-competent (red). The competent bacteria cause the non-competent bacteria to undergo lysis, which releases DNA. The competent bacteria internalize the DNA from the environment through genetic transformation. (B) In competent cells, pre-CSP is matured and exported from the bacteria by ComA/B. Accumulation of mature CSP in the environment activates ComD via phosphorylation, which passes its phosphate group to ComE. ComE activates expression of numerous *com*-responsive genes, including ComX. ComX activates CSP responsive genes. ComM/CibC confer self-immunity. Several lysis proteins are also produced and secreted from the bacteria. CibA/B embeds in the membrane of non-competent cells, resulting in lysis. LytA, CbpD and LytC (which is not regulated by CSP) also cause cell lysis through a yet to be discovered mechanism. Ultimately, DNA is released from the non-competent bacteria.

competency process is driven by a small number of cells in the population. In these cells, stochastic fluctuations in stress perception and response to stress activate the *comABCDE* operon. Initially, CSP produced as a result of *comABCDE* operon activation remains attached to the cell. Through cell-to-cell contacts, surface attached CSP activates the *comABCDE* operon in neighboring cells. This serves to amplify the production of CSP, and thus the number of cells entering into a state of competency (Prudhomme et al., 2016). It is possible that noise driven activation of competency could play a role in the development of competent and non-competent cells. This would be in line with competency development in *B. subtilis*, where stochastic fluctuations in gene expression allow a subpopulation of cells to enter into a state of competency (Süel et al., 2006).

One early *com* gene activated by ComE, which is critical for the development of competency and fratricidal behavior, is an alternative σ factor called ComX. Post-transcriptional activation of ComX by ComW leads to the expression of approximately 60 late *com* genes. The products of six late *com* genes activated by ComX—CboD, CibA, CibB, CibC, ComM and LytA—have been implicated in fratricide. Four of these proteins are directly involved in causing cell lysis: two murein hydrolases, CbpD and LytA, and a two-peptide bacteriocin composed of CibA and CibB (forming CibAB). In contrast, ComM and CibC are involved in immunity to fratricidal activity (Claverys and Håvarstein, 2007). However, the mechanisms by which they confer immunity have yet to be described (Popp and Mascher, 2019). Finally, LytC, a non-ComX regulated gene product, is also critical for fratricidal activity (Claverys and Håvarstein, 2007).

Cell lysis appears to involve two different methods. The bacteriocin CibAB produced by competent cells likely inserts itself into the membrane of non-competent cells, resulting in energy loss and promoting lysis. However, mutational studies have revealed that CibAB alone cannot trigger lysis. In contrast, the lytic enzymes LytA and LytC, are required for lysis (Claverys and Håvarstein, 2007). These enzymes are expressed in both competent and non-competent cells (Eldholm et al., 2009). CbpD appears to promote the release of DNA from cells undergoing lysis; however, its exact role and mechanism currently remain unclear. Moreover, there appears to be several unanswered questions with respect to the mechanism of cell lysis. What is the exact role and mechanism of CibAB? Why are both *lytA* and *lytC* expressed in both competent and non-competent cells? How is expression of *lytA* induced in non-competent cells? Finally, how is competency induced in a subset of an isogenic population of bacteria?

The functional role of fratricide is not yet fully understood, but is generally regarded as a mechanism to enhance the genetic diversity of the population. Here, eDNA released by non-competent cells could facilitate the acquisition of new beneficial genetic traits. For example, in biofilms, fratricide allows the efficient transfer of antibiotic resistant genes (Popp and Mascher, 2019). Indeed, the rapid acquisition of novel genetic traits under conditions of stress may be beneficial. Along this line, it appears that a state of competency may be induced in situations where *S. pneumonia* detects non-nutritional stress. Recall, competency in *S. pneumoniae* is induced in nutrient rich conditions. However, this species usually lives in the nasopharynx of its host, thus growing under nutrient-limited conditions. It has been speculated that injection of additional nutrients resulting in vigorous growth may signal *S. pneumoniae* that host immune defenses have become activated by an excessive bacterial load (Claverys and Håvarstein, 2007). Lysis of a subpopulation of bacteria could spur the acquisition of a novel genetic trait, which could be used to combat the immune system. Interestingly, two pherotypes of *S. pneumoniae* that differ in their structure of CSP (CSP-1 and CSP-2) are often isolated from clinical samples (Vestrheim et al., 2011). CSP released from one pherotype does not activate ComX in the other pherotype. If co-colonization of both pherotypes were to occur, the more abundant pherotype would likely enter a state of competency prior to the least abundant pherotype. Under this scenario, the less abundant pherotype would undergo lysis, and release its DNA, which would be subsequently absorbed by the competent population. Thus, each pherotype might serve as a reservoir of genetic diversity for the opposite pherotype (Claverys and Håvarstein, 2007).

7.2 DNA release during biofilm formation

One critical component of a biofilm in many bacteria, including *P. aeruginosa* and *S. aureus*, is eDNA. eDNA is thought to serve as a structural component that is used to facilitate intercellular connections, and is critical for the initiation and formation of biofilm structure. Given that eDNA shares many features with chromosomal DNA, it is thought that PCD facilitates the release and accumulation of eDNA. eDNA has several roles: it provides structural stability, enhances the ability of cells to attach to surfaces by promoting acid-base interactions, reduces repulsive intercellular forces, and can help guide cells into the biofilm (Okshevsky and Meyer, 2015).

There are different mechanisms by which the release of eDNA can be regulated in bacteria. The formation of biofilms in *S. aureus* is a tightly regulated process that has been recently reviewed in depth (Moormeier and Bayles, 2017). *In S. aureus*, PCD and subsequent release of eDNA is thought to be regulated by two operons: *cidABC* and *lrgAB*. CidA and LrgA serve as a holin-antiholin system, which behave very similarly to the TA systems discussed above (Ranjit et al., 2011). It is hypothesized that CidA forms pores in the cytoplasmic membrane, which results in membrane depolarization and subsequent cell lysis. LrgA is thought to complex with CidA, thus preventing it from depolarizing the membrane (Rice et al., 2003). Accordingly, the balance between CidA and LrgA regulates whether or not cell lysis occurs (Mann et al., 2009; Rice and Bayles, 2008). It has been proposed that cell-to-cell heterogeneity in the expression of the *cid* and *lrg* operons determines which cells are destined for lysis, and which are not, during biofilm development (Rice and Bayles, 2008). How this heterogeneity is established appears to be an unresolved question. Furthermore, there is evidence that expression of CidB increases the amount of reactive oxygen species (ROS) in the cell, which may also facilitate PCD (Windham et al., 2016). Finally, *cidC* encodes a pyruvate oxidase that can serve to acidify the growth medium (Patton et al., 2005) and facilitate the accumulation of ROS, which triggers PCD (Windham et al., 2016).

Expression of *cidABC* is regulated by a two component signaling system called SrrAB. Under conditions of excess glucose, SrrAB serves to repress the expression of *cidABC*, thus repressing PCD (Windham et al., 2016). In contrast, SrrAB activity is altered under conditions of impaired respiration, which leads to PCD. Recent additional evidence suggests that the SrrAB regulated gene *atlA*, which encodes a murein hydrolase, also facilitates cell lysis (Mashruwala et al., 2017). Overall, PCD in *S. aureus* is complex and appears to be caused by both a loss of cell membrane integrity and the generation of ROS.

While environmental conditions appear to regulate the release of eDNA in *S. aureus* biofilms, the release of eDNA can also be driven by quorum sensing. *P. aeruginosa* has multiple quorum sensing systems that are intertwined in a complex series of networks (Lee and Zhang, 2015). One receptor protein found in *P. aeruginosa*, called MvfR, is activated by a class of small diffusible molecules called hydroxyl-alkyl-quinolones (Déziel et al., 2005). Activation of MvfR results in the expression of virulence factors and proteins that produce autoinducers involved in quorum sensing. One of the many products regulated by MvfR is the production of 2-*n*-heptyl-4-hydroxyquinoline-*N*-oxide (HQNO), which is produced by the PQS quorum sensing system. Interestingly, the accumulation of HQNO has been shown to induce lysis of *P. aeruginosa* cells during the formation of biofilms, which subsequently results in the release of eDNA. Lysis appears to be regulated by an increased concentration of HQNO, as well as the reduction of intracellular glutathione, as cells enter stationary phase (Hazan et al., 2016). Currently, it is thought that HQNO causes cell lysis by interfering with the cytochrome bc_1 complex of the respiratory chain (Van Ark and Berden, 1977). This leads to the accumulation of ROS, which changes cellular membrane potential and compromises membrane integrity (Vercesi et al., 1997). It is currently unclear as to how lysing and non-lysing subpopulation are developed in *P. aeruginosa*. However, it appears that autolysis is not driven by stochastic processes (Hazan et al., 2016).

Prophages can also result in the release of eDNA. Indeed, the activity of prophages for the purposes of eDNA release during biofilm formation has been described in *P. aeruginosa* (Allesen-Holm et al., 2006) and *S. pneumoniae* (Carrolo et al., 2010). *Shewanella oneidensis* MR-1 is an aquatic bacterium that is unique in that it can reduce metal ions (Venkateswaran et al., 1999). During biofilm formation, activation of the λSo prophage leads to cell lysis, and the subsequent release of eDNA. *S. oneidensis* undergoes complex biofilm formation (McLean et al., 2008; Teal et al., 2006), part of which leads to the generation of a subpopulation of cells that are filamentous in nature. These filamentous cells undergo prophage-driven lysis and eDNA release as the biofilm is maturing. Why filamentous cells undergo lysis, whereas other cell forms do not, remains an open question. In the current paradigm to explain lysis and eDNA release, an increase in intracellular iron is thought to generate an abundance of ROS (Binnenkade et al., 2014). Increased ROS activates RecA of the SOS response pathway, which as noted above, has been previously implicated in facilitating the activation of λ-like phages. Here, activated RecA causes cleavage of the phage repressor protein, cI, thus allowing expression of the prophage. While a direct link between cI and RecA has yet to be observed in *S. oneidensis*, it is clear that prophage activation, and subsequent cellular lysis, is dependent upon the RecA protein. Moreover, deletion of the *fur* gene, which encodes a protein that represses genes involved in iron import from the environment, significantly increased λSo production. Finally, deletion of an oxidative stress defense regulator, *oxyR*, also increased λSo production. When taken together, in *S. oneidensis*, it appears that an influx of iron triggers oxidative damage in the cell via ROS. This, in turn, activates the RecA protein, which facilitates the expression of λSo resulting in cellular lysis and subsequent eDNA release (Binnenkade et al., 2014).

8. PCD to facilitate dispersal from biofilms

Dispersal into a new area is a critical behavior that allows for the survival and spread of a species. In the microbial world, dispersal has many benefits. Dispersal of pathogenic bacteria from location to location in a host may allow evasion of the immune system, or an intervention strategy, such as the application of antibiotics (Butler et al., 2010). Along this same line, bacteria can disperse from host to host, thus ensuring their spread within a population. In a less medical lens, the dispersal of bacteria is important to maintain biodiversity in environments as it prevents the dominance of one microbial species (Kerr et al., 2002). It also facilitates the acquisition of new nutrients (Blackburn et al., 1998), which can help ensure the survival of a microbial population.

Bacteria in biofilms have an advantage in that they are more resistant to environmental stressors, including antibiotics (Stewart and William Costerton, 2001), and the formation of biofilms can facilitate bacterial cooperation (Nadell et al., 2016). However, sessile bacteria that are housed in biofilms will eventually suffer from nutrient limitation, and increased intraspecific competition (Nadell et al., 2016). Thus, dispersal is eventually required. It is therefore unsurprising that dispersal from biofilms is a highly regulated process that is observed in diverse bacterial species. Interestingly, PCD plays an important role in facilitating active dispersal from biofilms.

Pseudoalteromonas tunicata is a marine dwelling bacterium that is commonly found attached to surfaces in a biofilm. It is a model system for host-microbe interactions in the marine environment. Members of the *Pseudoalteromonas* genera have been found to colonize a range of marine species including tunicates, algae, sponges, and pufferfish (Bowman, 2007). Like other bacteria that form biofilms, seeding dispersal is an important mechanism to promote the survival and spread of the species. Dispersal from the biofilm can occur through three mechanisms: erosion, sloughing, and seeding. While erosion (the near continual release of individual cells) and sloughing (spontaneous detachment of a large group of cells) are often passive, seeding dispersal is a tightly regulated, active process that involves a population of cells undergoing PCD (Kaplan, 2010). In *P. tunicata*, seeding dispersal involves the development of a large PCD driven cavity in the center of the biofilm. This results in the biofilm becoming detached from the substrate. The bacteria in the biofilm can now disperse to a new location. PCD is regulated by a lysine oxidase called AlpP (Mai-Prochnow et al., 2004, 2008). To cause PCD, AlpP produces hydrogen peroxide via the oxidation of L-lysine (Mai-Prochnow et al., 2008). Hydrogen peroxide is thought to result in death through a series of different mechanisms, including DNA damage (Imlay, 2013). Homologs of *alpP* have been identified in other biofilm forming bacteria suggesting this mechanism of PCD is not limited to *P. tunicata* (Mai-Prochnow et al., 2008).

In general, three hypotheses have been developed to explain the benefit of PCD and seeding dispersal in *P. tunicata*. First, lysis of cells through PCD would release both eDNA and nutrients as described above, which could help further the development and eventual dispersal of the biofilm. Second, it has been suggested that PCD and dispersal might provide a protective function to both the bacterial population and the hosts in which they reside. Overgrowth of biofilms and bacterial load would affect host quality, and would also serve to limit nutrient availability for *P. tunicata*. Regulated PCD and dispersal would limit the density of the bacterial population, increasing survivability of the host and the bacterial population alike (Mai-Prochnow et al., 2004). Third, after PCD, and upon seeding dispersal, viable *P. tunicata* cells have increased metabolic activity, and vary in their growth rate, motility, and ability to form biofilms. This suggests that, in addition to being critical in facilitating seeding dispersal, PCD also generates phenotypic diversity in *P. tunicata*, which may facilitate their ability to colonize a new area (Mai-Prochnow et al., 2006). It is important to note that seeding dispersal is widespread in bacteria, and additional PCD mechanisms to support such dispersal have been identified, including prophage mediated cell lysis in *P. aeruginosa* (Webb et al., 2003).

9. Inter-species activation of bacterial cell death

Given that resources are limiting in the microbial world, many bacteria often compete for these resources (Hibbing et al., 2010). Accordingly, bacteria have developed a number of different mechanisms to outcompete microbial competitors. Some of these mechanisms rely on triggering cell death in microbial competitors. For example, *P. aeruginosa* secretes membrane vesicles, which among other components, contain quinolones (Mashburn and Whiteley, 2005) and murein hydrolases (Li et al., 1996). Delivery of these membrane vesicles to microbial competitors results in cell death, likely through digestion of the cell wall and by interfering with DNA replication. Fungi secrete powerful antibiotic compounds to destroy a broad range of bacterial competitors (Brian, 1951). Finally, as noted above, some bacteria can produce antimicrobial peptides to eliminate their microbial competition (Cascales et al., 2007). Interestingly, some bacteria have developed the ability to trigger PCD in their microbial competitors. Below, we highlight examples of how bacteria can trigger PCD in bacterial competitors.

9.1 Cell death using TA systems

As described previously, the *mazEF* TA system has been implicated in causing PCD in diverse bacterial species (Engelberg-Kulka et al., 2006; Fu et al., 2007; Hayes, 2003; Kumar et al., 2013; Mittenhuber, 1999; Pandey and Gerdes, 2005; Ramage et al., 2009). Interestingly, *MazF*'s endoribonucleolytic activity can be triggered by extracellular death factors (EDFs), which are a set of diverse small peptides that are secreted by certain species of bacteria, including *E. coli, B. subtilis,* and *P. aeruginosa,* for the purposes of quorum sensing. These bacteria use EDFs to cause inter-species killing during stressful conditions. For example, the EDF found in *B. subtilis* called BsEDF, a hexapeptide with an amino acid sequence of RGQQNE, can lead to *mazF* driven cell death in *E. coli.* Specifically, experimental evidence suggests that BsEDF enhances the endoribonucleolytic activity of *E. coli*'s MazF by binding directly to it. This in turn leads to PCD in *E. coli.* Similarly, the three EDFs produced by *P. aeruginosa*, PaEDF-1, PaEDF-2, and PaEDF-3, also increase the endoribonucleolytic activity of MazF activity in *E. coli.* PCD mechanisms would be initiated inside *E. coli* as described above. It currently remains unclear as to how EDF traverse the cell membrane to interact with cytoplasmic MazF. Alternatively, there may exist a cell membrane receptor that facilitates the interaction between MazF and EDF (Kumar and Engelberg-Kulka, 2014).

9.2 PCD by interfering with cellular respiration

P. aeruginosa and *S. aureus* are important pathogens that are often found growing together during infections (Hotterbeekx et al., 2017). For instance, they are often found in co-culture in the lungs of individuals with cystic fibrosis (Mearns et al., 1972), and can be found together in chronic epithelial infections (Serra et al., 2015). Their propensity to encounter one another during infection has led to unique inter-species interactions including evasion, counter-inhibition, and eventually dominance in their battle for space and nutrients (Hotterbeekx et al., 2017). In general, *P. aeruginosa* strongly reduces the density of, or completely outcompetes, *S. aureus* during co-culture in many in vitro model systems (Baldan et al., 2014; DeLeon et al., 2014; Palmer et al., 2005).

The signaling interactions between *P. aeruginosa* and *S. aureus* are complex (Fig. 5). Initially, the cell wall component of *S. aureus*, *N*-acetylglucosamine (NAG), upregulates the PQS quorum sensing system in *P. aeruginosa* (Korgaonkar and Whiteley, 2011; Korgaonkar et al., 2013). The PQS quorum sensing system regulates the production of several virulence factors, including pyocyanin and HQNO (Machan et al., 1992). While these virulence factors are important for the pathogenesis of *P. aeruginosa,* they can also help it to compete with *S. aureus.* While HQNO can inhibit the growth of *P. aeruginosa* (as described above), it can also inhibit the growth

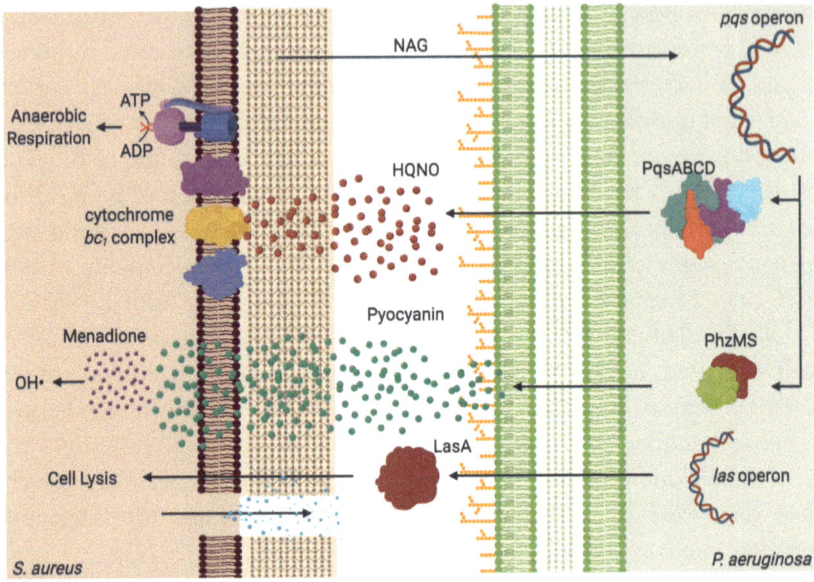

Fig. 5 Interspecies signaling between *S. aureus* and *P. aeruginosa* resulting in cell death. The cell wall component form *S. aureus*, *N*-acetylglucosamine (NAG), activate expression of the PQS operon in *P. aeruginosa*. This results in the production of HQNO via PqsABCD, and the production of pyocyanin via PhzMS. HQNO interferes with the cytochrome bc_1 complex in the electron transport chain of *S. aureus*, preventing the formation of new ATP and driving it into anaerobic respiration. Pyocyanin interacts with menadione to create hydroxyl radicles (OH•). Furthermore, LasA from *P. aeruginosa* compromises the cell wall of *S. aureus*, resulting in cell lysis, which may release nutrients. Overall, growth of *S. aureus* is significantly inhibited in the presence of *P. aeruginosa*.

of *S. aureus*. HQNO, which is produced by the PqsABCD proteins (Hazan et al., 2016), can cause cell death in *P. aeruginosa* by interfering with the activity of cytochrome bc_1. Similarly, HQNO reduces the growth of *S. aureus* by inhibiting oxidative respiration, when HQNO interferes with cytochrome bc_1 (Hoffman et al., 2006; Mitchell et al., 2010). Initial exposure to HQNO enhances the ability to *S. aureus* to form biofilms. However, prolonged exposure to HQNO significantly reduces the growth rate of *S. aureus* by attenuating ATP production (Hoffman et al., 2006). This activates anaerobic growth in *S. aureus*, and may also result in cell death (Filkins et al., 2015).

Pyocyanin is a blue-green pigmented virulence factor that is produced in part by the PhzMS proteins (Mavrodi et al., 2001). During the infection process, pyocyanin will interact with the host's cell. However, it also interacts with *S. aureus* to cause the generation of intercellular ROS. Experimental evidence suggests that interactions between intracellular menadione produced by *S. aureus*, and internalized pyocyanin produced by *P. aeruginosa*, force *S. aureus* into a fermentative state (Noto et al., 2017). While additional details have yet to be elucidated regarding this interaction between *P. aeruginosa* and *S. aureus*, interestingly, menadione has been implicated in ROS-dependent PCD in mammalian cells (Loor et al., 2010). While PCD involving menadione in *S. aureus* has yet to be described, it is worth nothing that a homolog involved in PCD in eukaryotic cells is affected by pyocyanin in *S. aureus*. Additional virulence factors produced by *P. aeruginosa* result in cell death of *S. aureus*. This includes elastase (LasA), which causes degradation of the cell wall of *S. aureus* (Kessler et al., 1993). It is important to note that additional interactions between *S. aureus* and *P. aeruginosa* occur during co-culture, but these do not necessarily result in cell death (Hotterbeekx et al., 2017).

10. Future outlook

Given that the mechanisms, and potential benefits, of PCD in bacteria have only recently been identified, there remain several philosophical and practical questions that have emerged. Several of the aforementioned bacterial behaviors appear to be altruistic in nature, where a subpopulation of bacteria sacrifice themselves for the benefit of the remaining population. However, it remains unclear as to if these behaviors evolved as true altruistic behaviors. That is, under selective pressures that would favor altruism. Alternatively, they may be reactions that just happen to result in cell death, and altruistic or group behavior was never under selective pressure. For example, the apparent altruistic nature of bacterial fratricide would only be beneficial if there was genetic diversity in the DNA released from the non-competent cells. Otherwise, lysing a subpopulation of clonal cells would not appear to have any benefit in the light of increasing genetic diversity of the remaining population. As an alternative hypothesis, DNA released by non-competent cells may serve as a source of nucleotides for competent cells. Along this line, it is more than likely that most traits acquired via genetic transformation are deleterious in nature, and thus a system to "augment" genetic transformation would not likely be under positive selective pressure (Redfield, 2001). It is also possible that these seemingly altruistic PCD behaviors are the by-product of systems meant to target other bacterial competitors. For example, it is possible that bacterial cannibalism and fratricide evolved to target competitor microbes (Claverys and Håvarstein, 2007). The majority of previous studies that have examined these populations have been performed in isogenic populations. Therefore, we attribute PCD in these systems to altruism when in fact the apparent "altruistic" nature of these behaviors is a by-product of experimental design. How would bacterial cannibalism and fratricide affect the composition of a diverse, non-clonal population of bacteria? On the practical side, how to harness the power of PCD for the rational manipulation of bacteria deserves to be explored. While some of the components involved in PCD have been harnessed for the use in areas such as synthetic biology (e.g., expression of TA systems to regulate cell growth; Wilson et al., 2017), the rational manipulation of PCD within the original host bacterium is somewhat unexplored (except for perhaps the manipulation of persister cells that are created using the *mazEF* TA system; Kim et al., 2011). For example, are there mechanisms that would allow us to prevent the PCD-driven release of virulence factors from hard to treat bacteria such as *C. difficile*? Now that we have identified and begun the characterization of these systems, answers to these questions are likely to be developed in the near future.

Acknowledgments

This research was partially sponsored by the Army Research Office and was accomplished under Grant Number W911NF-18-1-0443. The views and conclusions contained in this document are those of the authors and should not be interpreted as representing the official policies, either expressed or implied, of the Army Research Office or the U.S. Government. The U.S. Government is authorized to reproduce and distribute reprints for Government purposes notwithstanding any copyright notation herein. All figures were exported using a paid subscription and were created with Biorender.com.

References

Ackermann, M., Stecher, B., Freed, N.E., Songhet, P., Hardt, W.-D., Doebeli, M., 2008. Self-destructive cooperation mediated by phenotypic noise. Nature 454, 987–990.

Aizenman, E., Engelberg-Kulka, H., Glaser, G., 1996. An *Escherichia coli* chromosomal "addiction module" regulated by guanosine 3',5'-bispyrophosphate: a model for programmed bacterial cell death. Proc. Natl. Acad. Sci. U. S. A. 93, 6059–6063.

Allesen-Holm, M., Barken, K.B., Yang, L., Klausen, M., Webb, J.S., Kjelleberg, S., Molin, S., Givskov, M., Tolker-Nielsen, T., 2006. A characterization of DNA release in Pseudomonas aeruginosa cultures and biofilms. Mol. Microbiol. 59, 1114–1128.

Allocati, N., Masulli, M., Di Ilio, C., De Laurenzi, V., 2015. Die for the community: an overview of programmed cell death in bacteria. Cell Death Dis. 6, e1609.

Ameisen, J.C., 2002. On the origin, evolution, and nature of programmed cell death: a timeline of four billion years. Cell Death Differ. 9, 367.

Amitsur, M., Levitz, R., Kaufmann, G., 1987. Bacteriophage T4 anticodon nuclease, polynucleotide kinase and RNA ligase reprocess the host lysine tRNA. EMBO J. 6, 2499–2503.

Arias Del Angel, J.A., Escalante, A.E., Martínez-Castilla, L.P., Benítez, M., 2018. Cell-fate determination in Myxococcus xanthus development: network dynamics and novel predictions. Dev. Growth Differ. 60, 121–129.

Baldan, R., Cigana, C., Testa, F., Bianconi, I., De Simone, M., Pellin, D., Di Serio, C., Bragonzi, A., Cirillo, D.M., 2014. Adaptation of Pseudomonas aeruginosa in Cystic Fibrosis airways influences virulence of Staphylococcus aureus in vitro and murine models of co-infection. PLoS One 9, e89614.

Beites, T., Oliveira, P., Rioseras, B., Pires, S.D., Oliveira, R., Tamagnini, P., Moradas-Ferreira, P., Manteca, Á., Mendes, M.V., 2015. Streptomyces natalensis programmed cell death and morphological differentiation are dependent on oxidative stress. Sci. Rep. 5, 12887.

Bergsland, K.J., Kao, C., Yu, Y.-T.N., Gulati, R., Snyder, L., 1990. A site in the T4 bacteriophage major head protein gene that can promote the inhibition of all translation in *Escherichia coli*. J. Mol. Biol. 213, 477–494.

Berry, A.M., Paton, J.C., 2000. Additive attenuation of virulence of Streptococcus pneumoniae by mutation of the genes encoding pneumolysin and other putative pneumococcal virulence proteins. Infect. Immun. 68, 133–140.

Bingham, R., Ekunwe, S.I., Falk, S., Snyder, L., Kleanthous, C., 2000. The major head protein of bacteriophage T4 binds specifically to elongation factor Tu. J. Biol. Chem. 275, 23219–23226.

Binnenkade, L., Teichmann, L., Thormann, K.M., 2014. Iron triggers λSo prophage induction and release of extracellular DNA in Shewanella oneidensis MR-1 biofilms. Appl. Environ. Microbiol. 80, 5304–5316.

Blackburn, N., Fenchel, T., Mitchell, J., 1998. Microscale nutrient patches in planktonic habitats shown by chemotactic bacteria. Science 282, 2254–2256.

Bowman, J., 2007. Bioactive compound synthetic capacity and ecological significance of marine bacterial genus Pseudoalteromonas. Mar. Drugs 5, 220–241.

Boynton, T.O., McMurry, J.L., Shimkets, L.J., 2013. Characterization of Myxococcus xanthus MazF and implications for a new point of regulation. Mol. Microbiol. 87, 1267–1276.

Brian, P., 1951. Antibiotics produced by fungi. Bot. Rev. 17, 357–430.

Burbulys, D., Trach, K.A., Hoch, J.A., 1991. Initiation of sporulation in *B. subtilis* is controlled by a multicomponent phosphorelay. Cell 64, 545–552.

Bush, M.J., Tschowri, N., Schlimpert, S., Flärdh, K., Buttner, M.J., 2015. c-di-GMP signalling and the regulation of developmental transitions in streptomycetes. Nat. Rev. Microbiol. 13, 749–760.

Butler, M.T., Wang, Q., Harshey, R.M., 2010. Cell density and mobility protect swarming bacteria against antibiotics. Proc. Natl. Acad. Sci. U. S. A. 107, 3776–3781.

Callard, R.E., Stark, J., Yates, A.J., 2003. Fratricide: a mechanism for T memory-cell homeostasis. Trends Immunol. 24, 370–375.

Carrolo, M., Frias, M.J., Pinto, F.R., Melo-Cristino, J., Ramirez, M., 2010. Prophage spontaneous activation promotes DNA release enhancing biofilm formation in Streptococcus pneumoniae. PLoS One 5, e15678.

Cascales, E., Buchanan, S.K., Duché, D., Kleanthous, C., Lloubes, R., Postle, K., Riley, M., Slatin, S., Cavard, D., 2007. Colicin biology. Microbiol. Mol. Biol. Rev. 71, 158–229.

Chapuisat, M., Liselotte, S., Keller, L., 1997. Sex–ratio regulation: the economics of fratricide in ants. Proc. R. Soc. London, Ser. B 264, 1255–1260.

Chastanet, A., Vitkup, D., Yuan, G.-C., Norman, T.M., Liu, J.S., Losick, R.M., 2010. Broadly heterogeneous activation of the master regulator for sporulation in Bacillus subtilis. Proc. Natl. Acad. Sci. U. S. A. 107, 8486–8491.

Chen, I., Dubnau, D., 2004. DNA uptake during bacterial transformation. Nat. Rev. Microbiol. 2, 241.

Cheng, H.Y., Soo, V.W., Islam, S., McAnulty, M.J., Benedik, M.J., Wood, T.K., 2014. Toxin GhoT of the GhoT/GhoS toxin/antitoxin system damages the cell membrane to reduce adenosine triphosphate and to reduce growth under stress. Environ. Microbiol. 16, 1741–1754.

Claessen, D., De Jong, W., Dijkhuizen, L., Wösten, H.A., 2006. Regulation of Streptomyces development: reach for the sky! Trends Microbiol. 14, 313–319.

Claverys, J.-P., Håvarstein, L.S., 2007. Cannibalism and fratricide: mechanisms and raisons d'etre. Nature reviews. Microbiology 5, 219.

Claverys, J.-P., Martin, B., Håvarstein, L.S., 2007. Competence-induced fratricide in streptococci. Mol. Microbiol. 64, 1423–1433.

Connor, K.A., Zusman, D.R., 1991a. Behavior of peripheral rods and their role in the life cycle of Myxococcus xanthus. J. Bacteriol. 173, 3342.

Connor, K.A., Zusman, D.R., 1991b. Development in Myxococcus xanthus involves differentiation into two cell types, peripheral rods and spores. J. Bacteriol. 173, 3318.

Cummings, L.A., Wilkerson, W.D., Bergsbaken, T., Cookson, B.T., 2006. In vivo, fliC expression by Salmonella enterica serovar Typhimurium is heterogeneous, regulated by ClpX, and anatomically restricted. Mol. Microbiol. 61, 795–809.

D'Costa, V.M., King, C.E., Kalan, L., Morar, M., Sung, W.W., Schwarz, C., Froese, D., Zazula, G., Calmels, F., Debruyne, R., 2011. Antibiotic resistance is ancient. Nature 477, 457.

Dawkins, R., 1976. The Selfish Gene. Oxford University Press, Oxford, UK.

DeLeon, S., Clinton, A., Fowler, H., Everett, J., Horswill, A.R., Rumbaugh, K.P., 2014. Synergistic interactions of Pseudomonas aeruginosa and Staphylococcus aureus in an in vitro wound model. Infect. Immun. 82, 4718–4728.

Déziel, E., Gopalan, S., Tampakaki, A.P., Lépine, F., Padfield, K.E., Saucier, M., Xiao, G., Rahme, L.G., 2005. The contribution of MvfR to Pseudomonas aeruginosa pathogenesis and quorum sensing circuitry regulation: multiple quorum sensing-regulated genes are modulated without affecting lasRI, rhlRI or the production of N-acyl-l-homoserine lactones. Mol. Microbiol. 55, 998–1014.

Díaz-Orejas, R., Espinosa, M., Yeo, C.C., 2017. The importance of the expendable: toxin–antitoxin genes in plasmids and chromosomes. Front. Microbiol. 8, 1479.

Driks, A., 2002. Overview: development in bacteria: spore formation in Bacillus subtilis. Cell. Mol. Life Sci. 59, 389–391.

Edwards Jr., T.C., Collopy, M.W., 1983. Obligate and facultative brood reduction in eagles: an examination of factors that influence fratricide. Auk 100, 630–635.

Eldholm, V., Johnsborg, O., Haugen, K., Ohnstad, H.S., Håvarstein, L.S., 2009. Fratricide in Streptococcus pneumoniae: contributions and role of the cell wall hydrolases CbpD, LytA and LytC. Microbiology 155, 2223–2234.

Ellehauge, E., Nørregaard-Madsen, M., Søgaard-Andersen, L., 1998. The FruA signal transduction protein provides a checkpoint for the temporal co-ordination of inter-cellular signals in Myxococcus xanthus development. Mol. Microbiol. 30, 807–817.

Ellermeier, C.D., Hobbs, E.C., Gonzalez-Pastor, J.E., Losick, R., 2006. A three-protein signaling pathway governing immunity to a bacterial cannibalism toxin. Cell 124, 549–559.

Engelberg-Kulka, H., Hazan, R., 2003. Cannibals defy starvation and avoid sporulation. Science 301, 467–468.

Engelberg-Kulka, H., Hazan, R., Amitai, S., 2005. mazEF: a chromosomal toxin-antitoxin module that triggers programmed cell death in bacteria. J. Cell Sci. 118, 4327–4332.

Engelberg-Kulka, H., Amitai, S., Kolodkin-Gal, I., Hazan, R., 2006. Bacterial programmed cell death and multicellular behavior in bacteria. PLoS Genet. 2, e135.

Filkins, L.M., Graber, J.A., Olson, D.G., Dolben, E.L., Lynd, L.R., Bhuju, S., O'Toole, G.A., 2015. Coculture of Staphylococcus aureus with Pseudomonas aeruginosa drives S. aureus towards fermentative metabolism and reduced viability in a cystic fibrosis model. J. Bacteriol. 197, 2252–2264.

Finkel, S.E., Kolter, R., 2001. DNA as a nutrient: novel role for bacterial competence gene homologs. J. Bacteriol. 183, 6288–6293.
Fozo, E.M., Makarova, K.S., Shabalina, S.A., Yutin, N., Koonin, E.V., Storz, G., 2010. Abundance of type I toxin–antitoxin systems in bacteria: searches for new candidates and discovery of novel families. Nucleic Acids Res. 38, 3743–3759.
Fu, Z., Donegan, N.P., Memmi, G., Cheung, A.L., 2007. Characterization of MazFSa, an endoribonuclease from Staphylococcus aureus. J. Bacteriol. 189, 8871–8879.
Gardner, A., West, S.A., Buckling, A., 2004. Bacteriocins, spite and virulence. Proc. R. Soc. London, Ser. B 271, 1529–1535.
Gelvan, I., Varon, M., Rosenberg, E., 1987. Cell-density-dependent killing of Myxococcus xanthus by autocide AMV. J. Bacteriol. 169, 844.
Gerdes, K., 1988. The parB (hok/sok) locus of plasmid R1: a general purpose plasmid stabilization system. Biotechnology 6, 1402.
Gerdes, K., Rasmussen, P.B., Molin, S., 1986. Unique type of plasmid maintenance function: postsegregational killing of plasmid-free cells. Proc. Natl. Acad. Sci. U. S. A. 83, 3116–3120.
Gerdes, K., Thisted, T., Martinussen, J., 1990. Mechanism of post-segregational killing by the hoklsok system of plasmid R1: sok antisense RNA regulates formation of a hok mRNA species correlated with killing of plasmid-free cells. Mol. Microbiol. 4, 1807–1818.
Gerdes, K., Nielsen, A., Thorsted, P., Wagner, E.G.H., 1992. Mechanism of killer gene activation. Antisense RNA-dependent RNase III cleavage ensures rapid turnover of the stable hok, srnB and pndA effector messenger RNAs. J. Mol. Biol. 226, 637–649.
Germain-Amiot, N., Augagneur, Y., Camberlein, E., Nicolas, I., Lecureur, V., Rouillon, A., Felden, B., 2018. A novel Staphylococcus aureus cis–trans type I toxin–antitoxin module with dual effects on bacteria and host cells. Nucleic Acids Res. 47, 1759–1773.
Golubeva, Y.A., Sadik, A.Y., Ellermeier, J.R., Slauch, J.M., 2012. Integrating global regulatory input into the Salmonella pathogenicity island 1 type III secretion system. Genetics 190, 79–90.
Golubeva, Y.A., Ellermeier, J.R., Chubiz, J.E.C., Slauch, J.M., 2016. Intestinal long-chain fatty acids act as a direct signal to modulate expression of the Salmonella pathogenicity island 1 type III secretion system. MBio 7, e02170-02115.
González-Pastor, J.E., Hobbs, E.C., Losick, R., 2003. Cannibalism by sporulating bacteria. Science 301, 510–513.
Gorski, L., Gronewold, T., Kaiser, D., 2000. A ς54 activator protein necessary for spore differentiation within the fruiting body of Myxococcus xanthus. J. Bacteriol. 182, 2438–2444.
Götz, A., Lechner, M., Mader, A., von Bronk, B., Frey, E., Opitz, M., 2018. CsrA and its regulators control the time-point of ColicinE2 release in *Escherichia coli*. Sci. Rep. 8, 6537.
Govind, R., Dupuy, B., 2012. Secretion of Clostridium difficile toxins A and B requires the holin-like protein TcdE. PLoS Pathog. 8, e1002727.
Govind, R., Fitzwater, L., Nichols, R., 2015. Observations on the role of TcdE isoforms in Clostridium difficile toxin secretion. J. Bacteriol. 197, 2600–2609.
Guiboileau, A., Sormani, R., Meyer, C., Masclaux-Daubresse, C., 2010. Senescence and death of plant organs: nutrient recycling and developmental regulation. C. R. Biol. 333, 382–391.
Gutiérrez, P., Li, Y., Osborne, M.J., Pomerantseva, E., Liu, Q., Gehring, K., 2005. Solution structure of the carbon storage regulator protein CsrA from *Escherichia coli*. J. Bacteriol. 187, 3496–3501.
Hall, A.M., Gollan, B., Helaine, S., 2017. Toxin–antitoxin systems: reversible toxicity. Curr. Opin. Microbiol. 36, 102–110.

Harris, B.Z., Kaiser, D., Singer, M., 1998. The guanosine nucleotide (p) ppGpp initiates development and A-factor production in Myxococcus xanthus. Genes Dev. 12, 1022–1035.

Hayes, F., 2003. Toxins-antitoxins: plasmid maintenance, programmed cell death, and cell cycle arrest. Science 301, 1496–1499.

Hazan, R., Engelberg-Kulka, H., 2004. *Escherichia coli* mazEF-mediated cell death as a defense mechanism that inhibits the spread of phage P1. Mol. Genet. Genomics 272, 227–234.

Hazan, R., Sat, B., Engelberg-Kulka, H., 2004. *Escherichia coli* mazEF-mediated cell death is triggered by various stressful conditions. J. Bacteriol. 186, 3663–3669.

Hazan, R., Que, Y.A., Maura, D., Strobel, B., Majcherczyk, P.A., Hopper, L.R., Wilbur, D.J., Hreha, T.N., Barquera, B., Rahme, L.G., 2016. Auto poisoning of the respiratory chain by a quorum-sensing-regulated molecule favors biofilm formation and antibiotic tolerance. Curr. Biol. 26, 195–206.

Hibbing, M.E., Fuqua, C., Parsek, M.R., Peterson, S.B., 2010. Bacterial competition: surviving and thriving in the microbial jungle. Nat. Rev. Microbiol. 8, 15–25.

Hirst, R., Kadioglu, A., O'callaghan, C., Andrew, P., 2004. The role of pneumolysin in pneumococcal pneumonia and meningitis. Clin. Exp. Immunol. 138, 195–201.

Hoffman, L.R., Deziel, E., D'Argenio, D.A., Lepine, F., Emerson, J., McNamara, S., Gibson, R.L., Ramsey, B.W., Miller, S.I., 2006. Selection for Staphylococcus aureus small-colony variants due to growth in the presence of Pseudomonas aeruginosa. Proc. Natl. Acad. Sci. U. S. A. 103, 19890–19895.

Höfler, C., Heckmann, J., Fritsch, A., Popp, P., Gebhard, S., Fritz, G., Mascher, T., 2016. Cannibalism stress response in Bacillus subtilis. Microbiology 162, 164–176.

Hotterbeekx, A., Kumar-Singh, S., Goossens, H., Malhotra-Kumar, S., 2017. In vivo and In vitro Interactions between Pseudomonas aeruginosa and Staphylococcus spp. Front. Cell. Infect. Microbiol. 7, 106.

Howard, L.V., Gooder, H., 1974. Specificity of the autolysin of Streptococcus (Diplococcus) pneumoniae. J. Bacteriol. 117, 796–804.

Imlay, J.A., 2013. The molecular mechanisms and physiological consequences of oxidative stress: lessons from a model bacterium. Nat. Rev. Microbiol. 11, 443.

Jabbar, M.A., Snyder, L., 1984. Genetic and physiological studies of an *Escherichia coli* locus that restricts polynucleotide kinase-and RNA ligase-deficient mutants of bacteriophage T4. J. Virol. 51, 522–529.

Jandhyala, D.M., Thorpe, C.M., Magun, B., 2011. Ricin and Shiga toxins: effects on host cell signal transduction. In: Ricin and Shiga Toxins. Springer, pp. 41–65.

Jelsbak, L., Søgaard-Andersen, L., 2000. Pattern formation: fruiting body morphogenesis in Myxococcus xanthus. Curr. Opin. Microbiol. 3, 637–642.

Kamenšek, S., Podlesek, Z., Gillor, O., Zgur-Bertok, D., 2010. Genes regulated by the *Escherichia coli* SOS repressor LexA exhibit heterogeneous expression. BMC Microbiol. 10, 283.

Kaplan, J., 2010. Biofilm dispersal: mechanisms, clinical implications, and potential therapeutic uses. J. Dent. Res. 89, 205–218.

Kaplan, H.B., Plamann, L., 1996. A Myxococcus xanthus cell density-sensing system required for multicellular development. FEMS Microbiol. Lett. 139, 89–95.

Kerr, B., Riley, M.A., Feldman, M.W., Bohannan, B.J., 2002. Local dispersal promotes biodiversity in a real-life game of rock–paper–scissors. Nature 418, 171.

Kessler, E., Safrin, M., Olson, J.C., Ohman, D.E., 1993. Secreted LasA of Pseudomonas aeruginosa is a staphylolytic protease. J. Biol. Chem. 268, 7503–7508.

Kim, S.K., Kaiser, D., 1991. C-factor has distinct aggregation and sporulation thresholds during Myxococcus development. J. Bacteriol. 173, 1722–1728.

Kim, J.S., Heo, P., Yang, T.J., Lee, K.S., Cho, D.H., Kim, B.T., Suh, J.H., Lim, H.J., Shin, D., Kim, S.K., Kweon, D.H., 2011. Selective killing of bacterial persisters by a single chemical compound without affecting normal antibiotic-sensitive cells. Antimicrob. Agents Chemother. 55, 5380–5383.

Kim, J.-S., Schantz, A.B., Song, S., Kumar, M., Wood, T.K., 2018. GhoT of the GhoT/GhoS toxin/antitoxin system damages lipid membranes by forming transient pores. Biochem. Biophys. Res. Commun. 497, 467–472.

Kohanski, M.A., Dwyer, D.J., Wierzbowski, J., Cottarel, G., Collins, J.J., 2008. Mistranslation of membrane proteins and two-component system activation trigger antibiotic-mediated cell death. Cell 135, 679–690.

Korgaonkar, A.K., Whiteley, M., 2011. Pseudomonas aeruginosa enhances production of an antimicrobial in response to N-acetylglucosamine and peptidoglycan. J. Bacteriol. 193, 909–917.

Korgaonkar, A., Trivedi, U., Rumbaugh, K.P., Whiteley, M., 2013. Community surveillance enhances Pseudomonas aeruginosa virulence during polymicrobial infection. Proc. Natl. Acad. Sci. U. S. A. 110, 1059–1064.

Kostakioti, M., Hadjifrangiskou, M., Hultgren, S.J., 2013. Bacterial biofilms: development, dispersal, and therapeutic strategies in the dawn of the postantibiotic era. Cold Spring Harb. Perspect. Med. 3, a010306.

Kroos, L., 2017. Highly signal-responsive gene regulatory network governing Myxococcus development. Trends Genet. 33, 3–15.

Kumar, S., Engelberg-Kulka, H., 2014. Quorum sensing peptides mediating interspecies bacterial cell death as a novel class of antimicrobial agents. Curr. Opin. Microbiol. 21, 22–27.

Kumar, S., Kolodkin-Gal, I., Engelberg-Kulka, H., 2013. Novel quorum-sensing peptides mediating interspecies bacterial cell death. MBio 4, e00314-00313.

Labrie, S.J., Samson, J.E., Moineau, S., 2010. Bacteriophage resistance mechanisms. Nat. Rev. Microbiol. 8, 317.

Lamsa, A., Liu, W.T., Dorrestein, P.C., Pogliano, K., 2012. The Bacillus subtilis cannibalism toxin SDP collapses the proton motive force and induces autolysis. Mol. Microbiol. 84, 486–500.

Lechner, M., Schwarz, M., Opitz, M., Frey, E., 2016. Hierarchical post-transcriptional regulation of colicin E2 expression in *Escherichia coli*. PLoS Comput. Biol. 12, e1005243.

Lee, J., Zhang, L., 2015. The hierarchy quorum sensing network in Pseudomonas aeruginosa. Protein Cell 6, 26–41.

Lee, B., Holkenbrink, C., Treuner-Lange, A., Higgs, P.I., 2012. Myxococcus xanthus developmental cell fate production: heterogeneous accumulation of developmental regulatory proteins and reexamination of the role of MazF in developmental lysis. J. Bacteriol. 194, 3058–3068.

Li, Z., Clarke, A.J., Beveridge, T.J., 1996. A major autolysin of Pseudomonas aeruginosa: subcellular distribution, potential role in cell growth and division and secretion in surface membrane vesicles. J. Bacteriol. 178, 2479–2488.

Li, P., Tai, C., Deng, Z., Gan, J., Oggioni, M.R., Ou, H.-Y., 2016. Identification and characterization of chromosomal relBE toxin-antitoxin locus in Streptomyces cattleya DSM46488. Sci. Rep. 6, 32047.

Licking, E., Gorski, L., Kaiser, D., 2000. A common step for changing cell shape in fruiting body and starvation-independent sporulation of Myxococcus xanthus. J. Bacteriol. 182, 3553–3558.

Linde, D., Marischen, L., Müller, J.P., 2003. Characterisation of preYvaY export reveals differences in the substrate specificities of Bacillus subtilis and *Escherichia coli* leader peptidases. FEMS Microbiol. Lett. 227, 149–156.

Linder, P., Doelz, R., Gubler, M., Bickle, T., 1990. An anticodon nuclease gene inserted into a hsd region encoding a type I DNA restriction system. Nucleic Acids Res. 18, 7170.

Lindsten, T., Ross, A.J., King, A., Zong, W.-X., Rathmell, J.C., Shiels, H.A., Ulrich, E., Waymire, K.G., Mahar, P., Frauwirth, K., 2000. The combined functions of proapoptotic Bcl-2 family members bak and bax are essential for normal development of multiple tissues. Mol. Cell 6, 1389–1399.

Little, J., 1993. LexA cleavage and other self-processing reactions. J. Bacteriol. 175, 4943.

Liu, M.Y., Yang, H., Romeo, T., 1995. The product of the pleiotropic *Escherichia coli* gene csrA modulates glycogen biosynthesis via effects on mRNA stability. J. Bacteriol. 177, 2663–2672.

Loor, G., Kondapalli, J., Schriewer, J.M., Chandel, N.S., Hoek, T.L.V., Schumacker, P.T., 2010. Menadione triggers cell death through ROS-dependent mechanisms involving PARP activation without requiring apoptosis. Free Radic. Biol. Med. 49, 1925–1936.

Lopatkin, A.J., Huang, S., Smith, R.P., Srimani, J.K., Sysoeva, T.A., Bewick, S., Karig, D.K., You, L., 2016. Antibiotics as a selective driver for conjugation dynamics. Nat. Microbiol. 1, 16044.

López, D., Vlamakis, H., Losick, R., Kolter, R., 2009. Cannibalism enhances biofilm development in Bacillus subtilis. Mol. Microbiol. 74, 609–618.

López-Garrido, J., Casadesús, J., 2010. Regulation of Salmonella enterica pathogenicity island 1 by DNA adenine methylation. Genetics 184, 637–649.

Los, J.M., Los, M., Wegrzyn, A., Wegrzyn, G., 2013. Altruism of Shiga toxin-producing *Escherichia coli*: recent hypothesis versus experimental results. Front. Cell. Infect. Microbiol. 2, 166.

Machan, Z.A., Taylor, G.W., Pitt, T.L., Cole, P.J., Wilson, R., 1992. 2-Heptyl-4-hydroxyquinoline N-oxide, an antistaphylococcal agent produced by Pseudomonas aeruginosa. J. Antimicrob. Chemother. 30, 615–623.

Mader, A., von Bronk, B., Ewald, B., Kesel, S., Schnetz, K., Frey, E., Opitz, M., 2015. Amount of colicin release in *Escherichia coli* is regulated by lysis gene expression of the colicin E2 operon. PLoS One 10, e0119124.

Mai-Prochnow, A., Evans, F., Dalisay-Saludes, D., Stelzer, S., Egan, S., James, S., Webb, J.S., Kjelleberg, S., 2004. Biofilm development and cell death in the marine bacterium Pseudoalteromonas tunicata. Appl. Environ. Microbiol. 70, 3232–3238.

Mai-Prochnow, A., Webb, J.S., Ferrari, B.C., Kjelleberg, S., 2006. Ecological advantages of autolysis during the development and dispersal of Pseudoalteromonas tunicata biofilms. Appl. Environ. Microbiol. 72, 5414–5420.

Mai-Prochnow, A., Lucas-Elio, P., Egan, S., Thomas, T., Webb, J.S., Sanchez-Amat, A., Kjelleberg, S., 2008. Hydrogen peroxide linked to lysine oxidase activity facilitates biofilm differentiation and dispersal in several gram-negative bacteria. J. Bacteriol. 190, 5493–5501.

Mann, E.E., Rice, K.C., Boles, B.R., Endres, J.L., Ranjit, D., Chandramohan, L., Tsang, L.H., Smeltzer, M.S., Horswill, A.R., Bayles, K.W., 2009. Modulation of eDNA release and degradation affects Staphylococcus aureus biofilm maturation. PLoS One 4, e5822.

Martner, A., Skovbjerg, S., Paton, J.C., Wold, A.E., 2009. Streptococcus pneumoniae autolysis prevents phagocytosis and production of phagocyte-activating cytokines. Infect. Immun. 77, 3826–3837.

Mashburn, L.M., Whiteley, M., 2005. Membrane vesicles traffic signals and facilitate group activities in a prokaryote. Nature 437, 422.

Mashruwala, A.A., van de Guchte, A., Boyd, J.M., 2017. Impaired respiration elicits SrrAB-dependent programmed cell lysis and biofilm formation in Staphylococcus aureus. Elife 6, e23845.

Matz, K., Schmandt, M., Gussin, G.N., 1982. The rex gene of bacteriophage λ is really two genes. Genetics 102, 319–327.
Mavridou, D.A., Gonzalez, D., Kim, W., West, S.A., Foster, K.R., 2018. Bacteria use collective behavior to generate diverse combat strategies. Curr. Biol. 28, 345-355.e344.
Mavrodi, D.V., Bonsall, R.F., Delaney, S.M., Soule, M.J., Phillips, G., Thomashow, L.S., 2001. Functional analysis of genes for biosynthesis of pyocyanin and phenazine-1-carboxamide from Pseudomonas aeruginosa PAO1. J. Bacteriol. 183, 6454–6465.
McLean, J.S., Majors, P.D., Reardon, C.L., Bilskis, C.L., Reed, S.B., Romine, M.F., Fredrickson, J.K., 2008. Investigations of structure and metabolism within Shewanella oneidensis MR-1 biofilms. J. Microbiol. Methods 74, 47–56.
Mearns, M.B., Hunt, G., Rushworth, R.J.A., 1972. Bacterial flora of respiratory tract in patients with cystic fibrosis, 1950-1971. Arch. Dis. Child. 47, 902–907.
Mellroth, P., Daniels, R., Eberhardt, A., Rönnlund, D., Blom, H., Widengren, J., Normark, S., Henriques-Normark, B., 2012. LytA, major autolysin of Streptococcus pneumoniae, requires access to nascent peptidoglycan. J. Biol. Chem. 287, 11018–11029.
Méndez, C., Brana, A.F., Manzanal, M.B., Hardisson, C., 1985. Role of substrate mycelium in colony development in Streptomyces. Can. J. Microbiol. 31, 446–450.
Miller, M.B., Bassler, B.L., 2001. Quorum sensing in bacteria. Annu. Rev. Microbiol. 55, 165–199.
Mitchell, G., Séguin, D.L., Asselin, A.-E., Déziel, E., Cantin, A.M., Frost, E.H., Michaud, S., Malouin, F., 2010. Staphylococcus aureus sigma B-dependent emergence of small-colony variants and biofilm production following exposure to Pseudomonas aeruginosa 4-hydroxy-2-heptylquinoline-N-oxide. BMC Microbiol. 10, 33.
Mittenhuber, G., 1999. Occurrence of mazEF-like antitoxin/toxin systems in bacteria. J. Mol. Microbiol. Biotechnol. 1, 295–302.
Montanaro, L., Poggi, A., Visai, L., Ravaioli, S., Campoccia, D., Speziale, P., Arciola, C.R., 2011. Extracellular DNA in biofilms. Int. J. Artif. Organs 34, 824–831.
Moormeier, D.E., Bayles, K.W., 2017. Staphylococcus aureus biofilm: a complex developmental organism. Mol. Microbiol. 104, 365–376.
Mrak, P., Podlesek, Z., van Putten, J.P., Žgur-Bertok, D., 2007. Heterogeneity in expression of the *Escherichia coli* colicin K activity gene cka is controlled by the SOS system and stochastic factors. Mol. Genet. Genomics 277, 391–401.
Muñoz-Dorado, J., Marcos-Torres, F.J., García-Bravo, E., Moraleda-Muñoz, A., Pérez, J., 2016. Myxobacteria: moving, killing, feeding, and surviving together. Front. Microbiol. 7, 781.
Nadell, C.D., Drescher, K., Foster, K.R., 2016. Spatial structure, cooperation and competition in biofilms. Nat. Rev. Microbiol. 14, 589–600.
Nagata, S., Tanaka, M., 2017. Programmed cell death and the immune system. Nat. Rev. Immunol. 17, 333.
Nariya, H., Inouye, M., 2008. MazF, an mRNA interferase, mediates programmed cell death during multicellular Myxococcus development. Cell 132, 55–66.
Nöllmann, M., Gilbert, R., Mitchell, T., Sferrazza, M., Byron, O., 2004. The role of cholesterol in the activity of pneumolysin, a bacterial protein toxin. Biophys. J. 86, 3141–3151.
Norbury, C.J., Hickson, I.D., 2001. Cellular responses to DNA damage. Annu. Rev. Pharmacol. Toxicol. 41, 367–401.
Nordström, K., Molin, S., Aagaard-Hansen, H., 1980. Partitioning of plasmid R1 in *Escherichia coli*: I. Kinetics of loss of plasmid derivatives deleted of the par region. Plasmid 4, 215–227.

Noto, M.J., Burns, W.J., Beavers, W.N., Skaar, E.P., 2017. Mechanisms of pyocyanin toxicity and genetic determinants of resistance in Staphylococcus aureus. J. Bacteriol. 199, e00221-00217.
Okshevsky, M., Meyer, R.L., 2015. The role of extracellular DNA in the establishment, maintenance and perpetuation of bacterial biofilms. Crit. Rev. Microbiol. 41, 341–352.
Page, R., Peti, W., 2016. Toxin-antitoxin systems in bacterial growth arrest and persistence. Nat. Chem. Biol. 12, 208.
Palmer, K.L., Mashburn, L.M., Singh, P.K., Whiteley, M., 2005. Cystic fibrosis sputum supports growth and cues key aspects of Pseudomonas aeruginosa physiology. J. Bacteriol. 187, 5267–5277.
Pandey, D.P., Gerdes, K., 2005. Toxin-antitoxin loci are highly abundant in free-living but lost from host-associated prokaryotes. Nucleic Acids Res. 33, 966–976.
Parma, D.H., Snyder, M., Sobolevski, S., Nawroz, M., Brody, E., Gold, L., 1992. The Rex system of bacteriophage lambda: tolerance and altruistic cell death. Genes Dev. 6, 497–510.
Patton, T.G., Rice, K.C., Foster, M.K., Bayles, K.W., 2005. The Staphylococcus aureus cidC gene encodes a pyruvate oxidase that affects acetate metabolism and cell death in stationary phase. Mol. Microbiol. 56, 1664–1674.
Pecota, D.C., Osapay, G., Selsted, M.E., Wood, T.K., 2003. Antimicrobial properties of the *Escherichia coli* R1 plasmid host killing peptide. J. Biotechnol. 100, 1–12.
Penner, M., Morad, I., Snyder, L., Kaufmann, G., 1995. Phage T4-coded Stp: double-edged effector of coupled DNA and tRNA-restriction systems. J. Mol. Biol. 249, 857–868.
Perry, J., Waglechner, N., Wright, G., 2016. The prehistory of antibiotic resistance. Cold Spring Harb. Perspect. Med. 6, a025197.
Polage, C.R., Solnick, J.V., Cohen, S.H., 2012. Nosocomial diarrhea: evaluation and treatment of causes other than Clostridium difficile. Clin. Infect. Dis. 55, 982–989.
Popp, P.F., Mascher, T., 2019. Coordinated cell death in isogenic bacterial populations: sacrificing some for the benefit of many? J. Mol. Biol. 431 (23), 4656–4669.
Prudhomme, M., Berge, M., Martin, B., Polard, P., 2016. Pneumococcal competence coordination relies on a cell-contact sensing mechanism. PLoS Genet. 12, e1006113.
Pucci, B., Kasten, M., Giordano, A., 2000. Cell cycle and apoptosis. Neoplasia 2, 291–299.
Ramage, H.R., Connolly, L.E., Cox, J.S., 2009. Comprehensive functional analysis of Mycobacterium tuberculosis toxin-antitoxin systems: implications for pathogenesis, stress responses, and evolution. PLoS Genet. 5, e1000767.
Ranjit, D.K., Endres, J.L., Bayles, K.W., 2011. Staphylococcus aureus CidA and LrgA proteins exhibit holin-like properties. J. Bacteriol. 193, 2468–2476.
Redfield, R.J., 2001. Do bacteria have sex? Nat. Rev. Genet. 2, 634.
Rice, K.C., Bayles, K.W., 2008. Molecular control of bacterial death and lysis. Microbiol. Mol. Biol. Rev. 72, 85–109.
Rice, K.C., Firek, B.A., Nelson, J.B., Yang, S.-J., Patton, T.G., Bayles, K.W., 2003. The Staphylococcus aureus cidAB operon: evaluation of its role in regulation of murein hydrolase activity and penicillin tolerance. J. Bacteriol. 185, 2635–2643.
Riley, M.A., Chavan, M.A., 2007. Bacteriocins. Springer.
Roberts, J.W., Roberts, C.W., Craig, N.L., 1978. *Escherichia coli* recA gene product inactivates phage lambda repressor. Proc. Natl. Acad. Sci. U. S. A. 75, 4714–4718.
Ryan, G.L., Rutenberg, A.D., 2007. Clocking out: modeling phage-induced lysis of *Escherichia coli*. J. Bacteriol. 189, 4749–4755.
Saier, M.H., Reddy, B.L., 2015. Holins in bacteria, eukaryotes, and archaea: multifunctional xenologues with potential biotechnological and biomedical applications. J. Bacteriol. 197, 7–17.
Saini, S., Ellermeier, J.R., Slauch, J.M., Rao, C.V., 2010. The role of coupled positive feedback in the expression of the SPI1 type three secretion system in Salmonella. PLoS Pathog. 6, e1001025.

Sayed, N., Nonin-Lecomte, S., Réty, S., Felden, B., 2012. Functional and structural insights of a Staphylococcus aureus apoptotic-like membrane peptide from a toxin-antitoxin module. J. Biol. Chem. 287, 43454–43463.
Schlumberger, M.C., Müller, A.J., Ehrbar, K., Winnen, B., Duss, I., Stecher, B., Hardt, W.-D., 2005. Real-time imaging of type III secretion: Salmonella SipA injection into host cells. Proc. Natl. Acad. Sci. U. S. A. 102, 12548–12553.
Serra, R., Grande, R., Butrico, L., Rossi, A., Settimio, U.F., Caroleo, B., Amato, B., Gallelli, L., de Franciscis, S., 2015. Chronic wound infections: the role of Pseudomonas aeruginosa and Staphylococcus aureus. Expert Rev. Anti Infect. Ther. 13, 605–613.
Sevillano, L., Díaz, M., Yamaguchi, Y., Inouye, M., Santamaría, R.I., 2012. Identification of the first functional toxin-antitoxin system in Streptomyces. PLoS One 7, e32977.
Shimkets, L.J., Gill, R.E., Kaiser, D., 1983. Developmental cell interactions in Myxococcus xanthus and the spoC locus. Proc. Natl. Acad. Sci. U. S. A. 80, 1406–1410.
Short, F.L., Akusobi, C., Broadhurst, W.R., Salmond, G.P., 2018. The bacterial Type III toxin-antitoxin system, ToxIN, is a dynamic protein-RNA complex with stability-dependent antiviral abortive infection activity. Sci. Rep. 8, 1013.
Snyder, L., 1995. Phage-exclusion enzymes: a bonanza of biochemical and cell biology reagents? Mol. Microbiol. 15, 415–420.
Søgaard-Andersen, L., Kaiser, D., 1996. C factor, a cell-surface-associated intercellular signaling protein, stimulates the cytoplasmic Frz signal transduction system in Myxococcus xanthus. Proc. Natl. Acad. Sci. U. S. A. 93, 2675–2679.
Song, S., Wood, T.K., 2018. Post-segregational killing and phage inhibition are not mediated by cell death through toxin/antitoxin systems. Front. Microbiol. 9, 814.
Stecher, B., Robbiani, R., Walker, A.W., Westendorf, A.M., Barthel, M., Kremer, M., Chaffron, S., Macpherson, A.J., Buer, J., Parkhill, J., 2007. Salmonella enterica serovar typhimurium exploits inflammation to compete with the intestinal microbiota. PLoS Biol. 5, e244.
Stewart, M.K., Cookson, B.T., 2014. Mutually repressing repressor functions and multi-layered cellular heterogeneity regulate the bistable S almonella fliC census. Mol. Microbiol. 94, 1272–1284.
Stewart, P.S., William Costerton, J., 2001. Antibiotic resistance of bacteria in biofilms. Lancet 358, 135–138.
Stewart, M.K., Cummings, L.A., Johnson, M.L., Berezow, A.B., Cookson, B.T., 2011. Regulation of phenotypic heterogeneity permits Salmonella evasion of the host caspase-1 inflammatory response. Proc. Natl. Acad. Sci. U. S. A. 108, 20742–20747.
Süel, G.M., Garcia-Ojalvo, J., Liberman, L.M., Elowitz, M.B., 2006. An excitable gene regulatory circuit induces transient cellular differentiation. Nature 440, 545.
Summers, D.K., 1991. The kinetics of plasmid loss. Trends Biotechnol. 9, 273–278.
Tan, K.S., Wee, B.Y., Song, K.P., 2001. Evidence for holin function of tcdE gene in the pathogenicity of Clostridium difficile. J. Med. Microbiol. 50, 613–619.
Tanouchi, Y., Pai, A., Buchler, N.E., You, L., 2012. Programming stress-induced altruistic death in engineered bacteria. Mol. Syst. Biol. 8, 626.
Tanouchi, Y., Lee, A.J., Meredith, H., You, L., 2013. Programmed cell death in bacteria and implications for antibiotic therapy. Trends Microbiol. 21, 265–270.
Teal, T.K., Lies, D.P., Wold, B.J., Newman, D.K., 2006. Spatiometabolic stratification of Shewanella oneidensis biofilms. Appl. Environ. Microbiol. 72, 7324–7330.
Temme, K., Salis, H., Tullman-Ercek, D., Levskaya, A., Hong, S.-H., Voigt, C.A., 2008. Induction and relaxation dynamics of the regulatory network controlling the type III secretion system encoded within Salmonella pathogenicity island 1. J. Mol. Biol. 377, 47–61.
Tenconi, E., Traxler, M.F., Hoebreck, C., Van Wezel, G.P., Rigali, S., 2018. Production of prodiginines is part of a programmed cell death process in Streptomyces coelicolor. Front. Microbiol. 9, 1742.

Thisted, T., Gerdes, K., 1992. Mechanism of post-segregational killing by the hok/sok system of plasmid R1: Sok antisense RNA regulates hok gene expression indirectly through the overlapping mok gene. J. Mol. Biol. 223, 41–54.
Tyndall, C., Meister, J., Bickle, T.A., 1994. The *Escherichia coli* prr region encodes a functional type IC DNA restriction system closely integrated with an anticodon nutlease gene. J. Mol. Biol. 237, 266–274.
Van Ark, G., Berden, J.A., 1977. Binding of HQNO to beef-heart sub-mitochondrial particles. Biochim. Biophys. Acta, Bioenerg. 459, 119–137.
Varon, M., Cohen, S., Rosenberg, E., 1984. Autocides produced by Myxococcus xanthus. J. Bacteriol. 160, 1146–1150.
Venkateswaran, K., Moser, D.P., Dollhopf, M.E., Lies, D.P., Saffarini, D.A., MacGregor, B.J., Ringelberg, D.B., White, D.C., Nishijima, M., Sano, H., 1999. Polyphasic taxonomy of the genus Shewanella and description of Shewanella oneidensis sp. nov. Int. J. Syst. Evol. Microbiol. 49, 705–724.
Vercesi, A.E., Kowaltowski, A.J., Grijalba, M.T., Meinicke, A.R., Castilho, R.F., 1997. The role of reactive oxygen species in mitochondrial permeability transition. Biosci. Rep. 17, 43–52.
Vesper, O., Amitai, S., Belitsky, M., Byrgazov, K., Kaberdina, A.C., Engelberg-Kulka, H., Moll, I., 2011. Selective translation of leaderless mRNAs by specialized ribosomes generated by MazF in *Escherichia coli*. Cell 147, 147–157.
Vestrheim, D.F., Gaustad, P., Aaberge, I.S., Caugant, D.A., 2011. Pherotypes of pneumococcal strains co-existing in healthy children. Infect. Genet. Evol. 11, 1703–1708.
Wagner, P.L., Acheson, D.W., Waldor, M.K., 2001. Human neutrophils and their products induce Shiga toxin production by enterohemorrhagic *Escherichia coli*. Infect. Immun. 69, 1934–1937.
Walls, S.C., Blaustein, A.R., 1995. Larval marbled salamanders, Ambystoma opacum, eat their kin. Anim. Behav. 50, 537–545.
Wang, X., Lord, D.M., Cheng, H.-Y., Osbourne, D.O., Hong, S.H., Sanchez-Torres, V., Quiroga, C., Zheng, K., Herrmann, T., Peti, W., 2012. A new type V toxin-antitoxin system where mRNA for toxin GhoT is cleaved by antitoxin GhoS. Nat. Chem. Biol. 8, 855.
Ward, M.J., Zusman, D.R., 2000. Developmental aggregation and fruiting body formation in the gliding bacterium Myxococcus xanthus. In: Prokaryotic Development. American Society of Microbiology, pp. 243–262.
Webb, J.S., Thompson, L.S., James, S., Charlton, T., Tolker-Nielsen, T., Koch, B., Givskov, M., Kjelleberg, S., 2003. Cell death in Pseudomonas aeruginosa biofilm development. J. Bacteriol. 185, 4585–4592.
Wein, T., Hülter, N.F., Mizrahi, I., Dagan, T., 2019. Emergence of plasmid stability under non-selective conditions maintains antibiotic resistance. Nat. Commun. 10, 2595.
Wen, Y., Behiels, E., Devreese, B., 2014. Toxin–antitoxin systems: their role in persistence, biofilm formation, and pathogenicity. Pathog. Dis. 70, 240–249.
Wilson, C., Lopatkin, A., Craddock, T., Eldakar, O., Driscoll, W., Lopez, J., Smith, R., 2017. Cooperation and competition shape ecological resistance during periodic spatial disturbance of engineered bacteria. Sci. Rep. 7, 440.
Windham, I.H., Chaudhari, S.S., Bose, J.L., Thomas, V.C., Bayles, K.W., 2016. SrrAB modulates Staphylococcus aureus cell death through regulation of cidABC transcription. J. Bacteriol. 198, 1114–1122.
Wireman, J.W., Dworkin, M., 1977. Developmentally induced autolysis during fruiting body formation by Myxococcus xanthus. J. Bacteriol. 129, 798–802.
Wu, F., Tan, C., 2019. Dead bacterial absorption of antimicrobial peptides underlies collective tolerance. J. R. Soc. Interface 16, 20180701.

Wydau-Dematteis, S., El Meouche, I., Courtin, P., Hamiot, A., Lai-Kuen, R., Saubaméa, B., Fenaille, F., Butel, M.-J., Pons, J.-L., Dupuy, B., 2018. Cwp19 is a novel lytic transglycosylase involved in stationary-phase autolysis resulting in toxin release in Clostridium difficile. MBio 9, e00648-00618.

Yagüe, P., López-García, M.T., Rioseras, B., Sánchez, J., Manteca, A., 2013. Pre-sporulation stages of Streptomyces differentiation: state-of-the-art and future perspectives. FEMS Microbiol. Lett. 342, 79–88.

Yamaguchi, Y., Inouye, M., 2016. Toxin–antitoxin systems in bacteria and archaea. In: Stress and Environmental Regulation of Gene Expression and Adaptation in Bacteria. Wiley, pp. 97–107.

Yang, J.H., Wright, S.N., Hamblin, M., McCloskey, D., Alcantar, M.A., Schrübbers, L., Lopatkin, A.J., Satish, S., Nili, A., Palsson, B.O., 2019. A white-box machine learning approach for revealing antibiotic mechanisms of action. Cell 177, 1649-1661.e1649.

Yu, Y., Snyder, L., 1994. Translation elongation factor Tu cleaved by a phage-exclusion system. Proc. Natl. Acad. Sci. U. S. A. 91, 802–806.

Zhang, Y., Zhang, J., Hoeflich, K.P., Ikura, M., Qing, G., Inouye, M., 2003. MazF cleaves cellular mRNAs specifically at ACA to block protein synthesis in *Escherichia coli*. Mol. Cell 12, 913–923.

CHAPTER TWO

Transcriptional and epigenetic control of regulated cell death in yeast

Andreas Zimmermann[a], Jelena Tadic[a,b], Katharina Kainz[a], Sebastian J. Hofer[a], Maria A. Bauer[a], Didac Carmona-Gutierrez[a,]*, Frank Madeo[a,c,]*

[a]Institute of Molecular Biosciences, NAWI Graz, University of Graz, Graz, Austria
[b]Division of Immunology and Pathophysiology, Otto Loewi Research Center, Medical University of Graz, Graz, Austria
[c]BioTechMed Graz, Graz, Austria
*Corresponding authors: e-mail address: carmonag@uni-graz.at; frank.madeo@uni-graz.at

Contents

Abstract

Unicellular organisms like yeast can undergo controlled demise in a manner that is partly reminiscent of mammalian cell death. This is true at the levels of both mechanistic and functional conservation. Yeast offers the combination of unparalleled genetic amenability and a comparatively simple biology to understand both the regulation and evolution of cell death. In this minireview, we address the capacity of the nucleus as a regulatory hub during yeast regulated cell death (RCD), which is becoming an increasingly central question in yeast RCD research. In particular, we explore and critically discuss the available data on stressors and signals that specifically impinge on the nucleus. Moreover, we also analyze the current knowledge on nuclear factors as well as on transcriptional control and epigenetic events that orchestrate yeast RCD. Altogether we conclude that the functional significance of the nucleus for yeast RCD in undisputable, but that further exploration beyond correlative work is necessary to disentangle the role of nuclear events in the regulatory network.

International Review of Cell and Molecular Biology, Volume 352
ISSN 1937-6448
https://doi.org/10.1016/bs.ircmb.2019.12.003

1. Introduction

Throughout their life, all organisms strive for cellular homeostasis and thus quickly adapt to changing endogenous and environmental conditions. Yet, under specific physiological trigger events or excessive stress that cannot be mitigated through stress response mechanisms, cells may undergo either unregulated "accidental cell death" (ACD, also called "necrosis") or commit to a suicide program termed "regulated cell death" (RCD). RCD involves a sophisticated network of molecular events, which orchestrates and enforces cellular demise in a controlled fashion. In multicellular organisms, several subroutines of RCD have been discovered, which feature specific triggers and downstream execution pathways. Among them are, for instance, mitochondrial permeability transition pore-driven necrosis or lysosomal-dependent cell death (Galluzzi et al., 2018), although the best described RCD program is apoptosis. Apoptosis can be triggered extrinsically through death receptor signaling or intrinsically through the mitochondrial apoptosis pathway. Morphological hallmarks of apoptosis are cellular blebbing, nuclear DNA fragmentation, phosphatidylserine (PdSer) exposure at the plasma membrane and the activation of specific effector proteins such as the cysteinyl-aspartate specific proteases (caspases) (Galluzzi et al., 2016). Since multicellular organisms rely on the collective cooperation of up to billions of cells, RCD plays a pivotal role in controlling compliance and maintaining organismal homeostasis, e.g., by removing superfluous, damaged or malignant individual cells. In humans, this removal may amount to several billion cells each day (Renehan, 2001). In addition, a specialized form of RCD, termed "programmed cell death" (PCD), which carries out a stipulated cellular date of expiry, is essential for embryonic development and tissue homeostasis. Thus, the existence of RCD in multicellular organisms appears coherent: it participates in shaping an organism and keeping malignant transformation at bay.

Intriguingly, it has become evident that conserved RCD routines do also exist in unicellular organisms. About 20 years ago, morphological markers of apoptosis were described in baker's yeast (*Saccharomyces cerevisiae*), specifically in a strain harboring a mutated form of the AAA-ATPase Cdc48p[a]/VCP (Madeo et al., 1997). Thereafter, a number of yeast orthologs to metazoan RCD regulators have been identified (Carmona-Gutierrez et al., 2010a).

[a] Lowercase 'p' indicates yeast proteins.

Important examples include the mitochondrial effectors cytochrome *c* (Ludovico et al., 2002; Manon et al., 1997), apoptosis-inducing factor (Aif1p) (Wissing et al., 2004), the yeast homolog of endonuclease-G (Nuc1p) (Büttner et al., 2007) or the BH3-domain-containing protein Ybh3p (Büttner et al., 2011, p. 3) among others. Importantly, yeast also harbors a metacaspase, Yca1p/Mca1p, which is homologous to mammalian caspases and participates in cell death execution (Madeo et al., 2002). Importantly, the genetic amenability and comparatively simple biology make yeast a highly effective tool to understand both the regulation (Munoz et al., 2012; Teng et al., 2011) and evolution of cell death (Carmona-Gutierrez and Madeo, 2009; Gao et al., 2019). In fact, several apoptotic regulators such as Cdc48p/VCP (Braun and Zischka, 2008; Madeo et al., 1997; Shirogane et al., 1999), or Bax inhibitor-1 (TMBIM6) (Xu and Reed, 1998) as well as the protective role of Pep4p/cathepsin D in autophagy-independent mitochondrial degradation (Oliveira et al., 2015; Pereira et al., 2010) were originally studied in yeast and then confirmed in multicellular models.

Emerging evidence suggests that these yeast RCD factors do not act autonomously and randomly, but are part of a regulated network that responds to external and internal stressor factors (Carmona-Gutierrez et al., 2018; Falcone and Mazzoni, 2016; Strich, 2015). Here, the regulatory role of the nucleus is becoming a critical question. Which signals reach the nucleus to ignite an RCD routine? Which nuclear agents execute RCD? Are RCD executors transcriptionally regulated? Do posttranslational histone modifications influence RCD? In this minireview, we explore and critically discuss the currently available data on transcriptional and epigenetic events in the regulation of yeast RCD.

2. The role and hallmarks of RCD in yeast

The mere existence of RCD in a unicellular organism immediately poses a teleological question—what is the purpose of suicide in a single cell that represents a discrete organism? Importantly, a population of yeast cells does not behave as a group of isolated cells but rather represents a multicellular collective of communicating individuals (Carmona-Gutierrez et al., 2018). In fact, yeast colonies grown on solid media exhibit a rudimentary degree of differentiation, exhibiting hallmarks of RCD in sub-surface cell-layers (Váchová and Palková, 2018). Notably, this may be regarded as an ancient example of PCD, in which the development of a colony is driven by unicellular death events. Since a population of yeast cells usually originates from a single clone, the genetic conformity between yeast cells originating from the same clone provides an evolutionary basis for altruistic kin selection (Fabrizio et al., 2004; Herker et al., 2004).

In metazoans, the elimination of apoptotic cells traverses different stages, including cell membrane blebbing, followed by the formation of membrane protrusions that finally result in vesicular apoptotic bodies, which are removed via phagocytosis. In contrast, it is commonly assumed that dying yeast cells (maybe owing to the fungal cell wall) do not form apoptotic bodies but can undergo a biphasic cell death execution mechanism, with apoptotic hallmarks such as PdSer exposure and accumulation of reactive oxygen species (ROS) preceding necrotic hallmarks such as plasma membrane rupture (Fig. 1) (Carmona-Gutierrez et al., 2018). The subsequent release of cellular content may promote the survival of surrounding cells in times of dwindling nutrient supply, for example, in aging batch cultures (Herker et al., 2004). In addition, the RCD machinery in yeast might have been selected for through additional mechanisms. For example, during yeast spore formation, mitochondrial nucleases that also participate in RCD are released to keep viral nucleic acids in check, which otherwise might replicate unhamperedly (Gao et al., 2019). Other physiological scenarios, in which yeast RCD plays a regulatory role include the elimination of infertile or otherwise damaged cells during failed mating (Severin and Hyman, 2002; Zhang et al., 2006) and genetic recombinants not fit for adaptation to the environment as it occurs during meiosis and sporulation (Knorre et al., 2005). Importantly, unicellular RCD is not restricted to *S. cerevisiae*, as many other fungi and even bacteria and protozoan parasites incorporate similar regulatory processes (Debrabant et al., 2003; Dwyer et al., 2012).

Both ACD and RCD can be found in yeast (Box 1), with the latter mainly manifesting as either regulated necrosis or apoptosis. A third RCD subroutine that has been discussed to be present in yeast is autophagy-dependent cell death (ADCD), which in higher eukaryotes causally requires the autophagic machinery for lethal execution (Denton and Kumar, 2019). Although single reports on yeast ADCD do exist (Singh et al., 2019), this type of RCD remains rather unexplored and a matter of debate for *S. cerevisiae*. Regulated necrosis is a cell death subroutine exhibiting necrotic features, which however can be decelerated through pharmacological or genetic interference (Carmona-Gutierrez et al., 2018). For instance, during yeast chronological aging, the time a yeast population can remain viable (Fabrizio, 2001; Longo and Fabrizio, 2012), the necrotic population can be specifically reduced by external application of natural compounds like spermidine

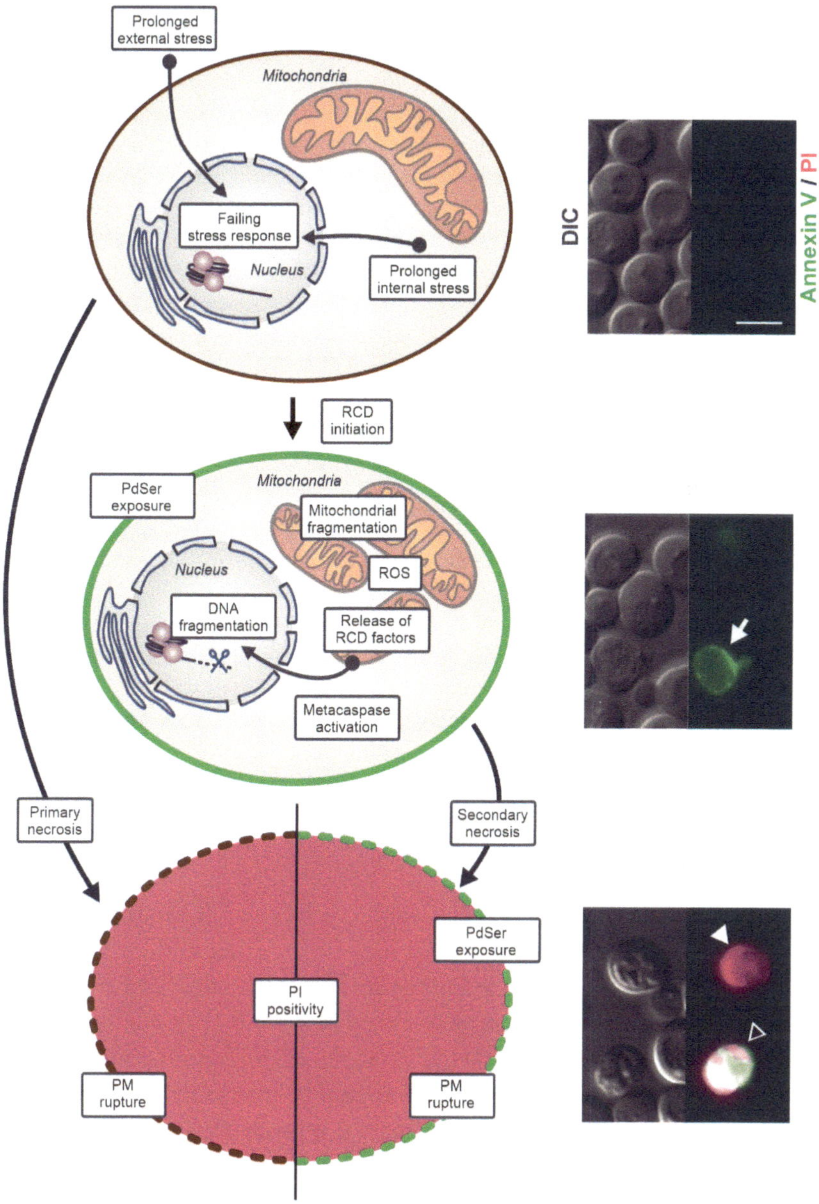

Fig. 1 Hallmarks and kinetics of yeast RCD. Annexin V/PI staining assays exposure of PdSer to the outer plasma membrane evaluated by binding of PdSer to Annexin V and plasma membrane rupture monitored by PI staining. This double staining allows discrimination between yeast subpopulations going through different stages of cell death.

BOX 1 General principles and hallmarks of yeast cell death.
Accidental Cell Death (ACD) represents uncontrolled cellular death induced by severe external stress factors.

Regulated Cell Death (RCD) describes cellular demise in a controlled fashion and is involved in various developmental stages of an organism, as well as in tissue maintenance and repair, shaping and protecting the organism as a whole. Yeast RCD can occur with characteristic morphological hallmarks of mammalian apoptosis, including PdSer externalization, nDNA degradation, mitochondrial fragmentation, and ROS accumulation in a metacaspase-dependent or -independent fashion. Yet, RCD scenarios without these typical hallmarks (see Regulated necrosis) have been described.

Programmed cell death (PCD) represents a form of RCD exclusively initiated by physiological conditions (e.g., aging, development).

Yeast RCD triggers include external (e.g., acetic acid, H_2O_2) and internal (e.g., aging, ROS, DNA damage) triggers.

Necrosis defines a cell death category primarily characterized by plasma membrane rupturing, which can occur as primary necrosis (immediate) or secondary necrosis (final stage of apoptosis).

Regulated necrosis is an RCD subtype exhibiting necrosis morphology (ruptured plasma membrane but no PdSer exposure) that can be prevented by specific genetic and pharmacological interventions.

Autophagy-dependent cell death (ADCD) describes an RCD subroutine driven by autophagic degradation or its machinery.

(Eisenberg et al., 2009) or 4,4′-dimethoxychalcone (Carmona-Gutierrez et al., 2019). Some of the factors that may be involved in the regulation of yeast regulated necrosis include hydrolases like the homolog of mammalian cathepsin D (Pep4p) (Carmona-Gutiérrez et al., 2011) or the calpain-like protease Rim13p/Cpl1p (Rockenfeller et al., 2018). Yeast apoptosis is currently characterized mainly by morphological hallmarks shared by mammalian apoptosis (Box 1), while in mammals, it is more strictly defined as a cell death subroutine precipitated by executioner caspases (Galluzzi et al., 2018). Yeast does not encode caspases, but at least one metacaspase (Yca1p) (Madeo et al., 2002). While caspases and metacaspases diverge in their proteolytic characteristics (cleavage site after arginine or lysine for metacaspases, but after aspartate for caspases), they do share natural, death-related substrates (Silva et al., 2011; Strobel and Osiewacz, 2013; Sundström et al., 2009; Tsiatsiani et al., 2013) and can both induce cell death, clearly suggesting functional homology (Carmona-Gutierrez et al., 2010b). Indeed, some scenarios in yeast RCD do follow Yca1p-dependent lethal execution (Almeida et al., 2008; Bink et al., 2010; Guaragnella et al., 2010; Johansson et al., 2016; Kajiwara et al., 2012; Khan et al., 2005; Madeo et al., 2002; Mazzoni et al., 2005; Rinnerthaler et al., 2012; Walter et al., 2014), while others are characterized by metacaspase-independent pathways (Carmona-Gutierrez et al., 2018).

In general, RCD in yeast depends on a designated molecular machinery and can be pharmacologically or genetically modulated (Carmona-Gutierrez et al., 2018). In contrast to stress-induced RCD, which, in yeast, can typically be provoked by external triggers such as H_2O_2 (Madeo et al., 1999) or acetic acid (Ludovico et al., 2001), PCD is triggered by physiological scenarios like development or aging (Eastwood and Meneghini, 2015; Fabrizio et al., 2004; Herker et al., 2004) (Box 1). Consensus nomenclature guidelines and an extensive palette of methods facilitate to differentiate between cell death subroutines in yeast (Carmona-Gutierrez et al., 2018).

3. Nuclear effectors of yeast RCD

Besides transcription factors, which may regulate RCD indirectly by driving stress responses or changing the expression of RCD effectors, a number of factors that either translocate from or to the nucleus have been directly associated with controlled demise in yeast (Fig. 2). A typical hallmark of yeast RCD is the fragmentation of DNA, which can be mediated by nuclear nucleases, such as the dsDNA endo−/exonuclease Tat-D (encoded by the gene *YBL055C*) in situ. Consistently, overexpression of Tat-D leads to increased cell death rates (Qiu et al., 2005), while deletion of this nuclease confers resistance to both H_2O_2 (Qiu et al., 2005) and heat stress-induced cell death (Teng et al., 2011). Another nuclear-resident factor is the serine protease and HtrA2/Omi homolog nuclear mediator of apoptosis (Nma111p). Similar to Tat-D, *NMA111* disruption reduces apoptosis hallmarks, whereas its overexpression exacerbates cell death upon H_2O_2 treatment or hyperthermia (Fahrenkrog et al., 2004). In contrast to the mammalian HtrA2/Omi that resides in mitochondria, Nma111p executes its pro-death function dependent on the nuclear localization sequence as well as the serine protease activity (Belanger et al., 2009). Nma111p processes the nucleo-cytosolic protein Bir1p, the only known inhibitor of apoptosis (IAP) protein in yeast. *BIR1* deletion sensitizes cells to oxidative stress, and in turn, Bir1p overexpression reduces apoptosis markers, an effect

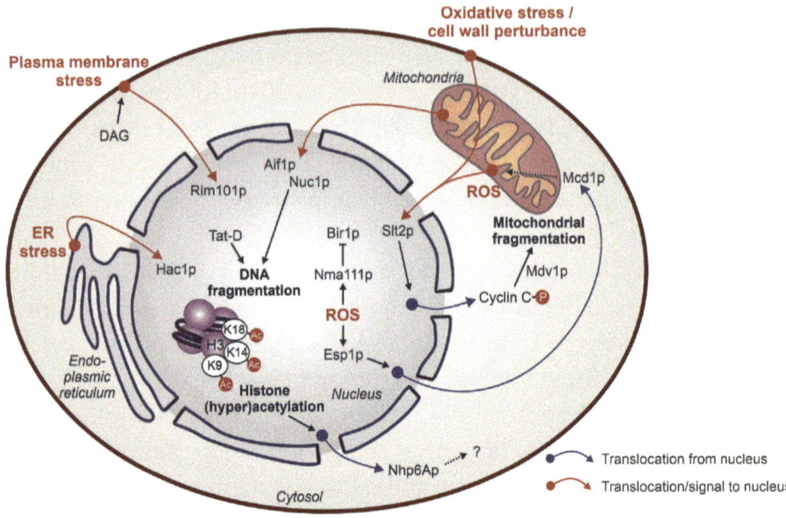

Fig. 2 Nuclear regulation of yeast RCD. Cell death signals originating from diverse intra- or extracellular stress sources may reach the nucleus either via activation of nuclear-resident factors or by promoting the translocation of lethal effectors to the nucleus. Upon RCD activation, various cell death effectors may translocate from the nucleus to other compartments to ignite cellular deterioration.

that can be antagonized by co-overexpression of Nma111p (Walter et al., 2006). Interestingly, Bir1p is cleaved by the yeast metacaspase Yca1p in vitro, similar to the cleavage of IAPs by caspases in higher eukaryotes (Wong et al., 2012).

In addition to these nuclear factors, multiple proteins translocate either from other organelles to the nucleus or vice versa during yeast RCD, suggesting inter-organelle communication when yeast cells succumb to controlled demise. The mitochondrial nuclease Nuc1p and the mitochondrial oxidoreductase Aif1p are released from the mitochondrial intermembrane space upon cytotoxic insults and enter the nucleus, where they initiate DNA degradation (Büttner et al., 2007; Wissing et al., 2004). In contrast, nuclear Mcd1p, the homolog of human cohesin RAD21, is processed *in nucleo* from the caspase-like protease Esp1p during H_2O_2 stress and subsequently the C-terminal fragment leaves the nucleus and translocates to mitochondria, where it leads to decreased mitochondrial membrane potential and cell death in a cytochrome *c*-dependent fashion (Yang et al., 2008). Oxidative stress can also trigger the nuclear release of cyclin C (also called Ssn8p), which can then initiate mitochondrial hyperfission, a common hallmark of yeast RCD (Cooper et al., 2014; Willis et al., 2018).

This direct cell death regulation by RCD executors can be accompanied by indirect regulation through extensive communication between the nucleus and mitochondria via anterograde (nucleus to mitochondria) and retrograde (mitochondria to nucleus) signaling (Guaragnella et al., 2018).

Aging-induced regulated necrotic cell death in yeast is accompanied by nuclear release of Nhp6Ap, the yeast homolog of human high-mobility group (HMG) protein 1 (Eisenberg et al., 2009; Scaffidi et al., 2002). This non-histone protein is associated with the chromatin and released during necrotic cell death in mammals (Galluzzi et al., 2011). In yeast, similar effects have been observed upon inhibition of target of rapamycin complex 1 (TORC1) with rapamycin, which leads to reduced chromatin binding and increased cytoplasmic localization of the HMG proteins Hmo1p and Nhp10p. Overexpression of these HMG proteins was followed by markers for necrosis (Chen et al., 2013). Despite these similarities between yeast and humans, it remains to be determined whether nuclear release of yeast HMG proteins has a functional role in the execution of regulated necrosis or if it is merely a correlative event.

4. Transcriptional regulation of yeast RCD

In metazoans, transcriptional RCD regulation is one of the main mechanisms of PCD regulation during development (Kumar and Cakouros, 2004). In the eukaryotic model organism *Caenorhabditis elegans*, the transcription factors HLH-2 and HLH-3 promote the expression of the pro-apoptotic Bcl-2 homology 3 (BH3) domain-only protein EGL-1, which triggers the death of obsolete cells during embryonal neurogenesis (Thellmann et al., 2003). During *Drosophila melanogaster* neuronal development, Hox genes mediate apoptosis of neural stem cells, e.g., through activation of the pro-apoptotic protein *reaper*, which, in turn, inactivates the caspase inhibitor death-associated inhibitor of apoptosis 1 (Diap1) (Khandelwal et al., 2017; White et al., 1994). Another transcriptional circuit couples genome stability and DNA integrity to the RCD machinery in multicellular organisms, most prominently through the tumor suppressor p53, which is stabilized by different stress stimuli and subsequently transcriptionally activates pro-apoptotic factors such as the BH3-only proteins PUMA and NOXA (Aubrey et al., 2018). These examples illustrate the straightforward "trigger > transcription factor > effector > RCD" pathways in metazoans that transmit pro-death signals, like DNA damage to the RCD machinery. In yeast, the situation is more convoluted and rather follows the scheme "stress > adaptation > failure to adapt > RCD" without obvious, isolated transcriptional events that directly trigger RCD. Thus, in such cases, yeast RCD unfolds as a consequence of failed cellular response to a given stress. In the following sub-sections we (i) describe how various upstream stress signaling pathways may feed into triggering RCD and (ii) discuss evidence for transcriptional regulation of RCD executors that may act downstream of stress signaling.

4.1 Stress response pathways upstream of RCD

Several signal transduction pathways in yeast orchestrate stress responses that aim at adaptation to external or internal noxa and restoration of cellular homeostasis (Ana Sanz et al., 2017; Deprez et al., 2018; Jin et al., 2015; Postnikoff et al., 2017; Rodrigues-Pousada et al., 2019). Yet, the same pathways are also implicated in yeast RCD, as interference with stress signaling often protects from cell death under excessive or prolonged cellular damage (Fig. 2). A prominent example is the integrated stress response (ISR), which senses nutrient depletion, ER stress and the accumulation of uncharged tRNAs through the kinase Gcn2p, and as a response initiates an adaptive transcriptional program via the transcription factor Gcn4p (Postnikoff et al., 2017). Accordingly, Gcn4p is generally considered to have pro-survival effects (Mittal et al., 2017, p. 4). At the same time, though, *GCN4* deletion mutants are more resistant to acetic acid-induced cell death (Almeida et al., 2009), suggesting that the ISR may have additional, maladaptive functions. The ISR is remarkably well conserved between yeast and humans, where it is controlled by the Gcn4p homolog ATF4 (Pakos-Zebrucka et al., 2016; Postnikoff et al., 2017). ATF4 can ignite RCD programs if the adaptive response is unable to restore homeostasis, for example, by TP53-independent transcriptional activation of PUMA and NOXA in glutamine-starved neuroblastoma cells (Qing et al., 2012, p. 4).

Along these lines, cellular damage induces mitochondrial fragmentation, also termed "hyperfission" which is a typical morphological hallmark of adaptation to stress but at the same time accompanies cell death in yeast (Strich, 2015). Mitochondrial fission is controlled through both transcriptional and post-translational mechanisms: under non-stressed conditions, a protein kinase consisting of cyclin C and Cdk8 (Ssn3p) represses stress-inducible genes through interaction with the RNA polymerase II holoenzyme (Bourbon, 2008; Strich, 2015). Stress signaling proteins, for example, the MAP kinase Slt2p, which can be activated by disturbed cell wall homeostasis through the cell wall integrity pathway, can phosphorylate cyclin C, leading to its translocation to the cytoplasm and de-repression of stress gene transcription (Cooper, 1997; Jin et al., 2014). The upstream signaling leading to cyclin C release involves the yeast homolog of AMP-activated kinase (AMPK, in yeast encoded by *SNF1* and *SNF4*), which senses the cellular energy status and glucose limitation stress (Coccetti et al., 2018; Willis et al., 2018), suggesting that cyclin C responds both to the cell wall integrity pathway and to AMPK signaling. Cytoplasmic cyclin C can interact with the mitochondrial fission factor Mdv1p to initiate fission (Cooper et al., 2014). Importantly, cells lacking cyclin C are protected from oxidative stress-induced RCD, while cells with constitutively cytoplasmic cyclin C show increased cell death upon oxidative damage (Khakhina et al., 2014). Cyclin C also responds to the so-called "cell wall integrity pathway," which senses oxidative stress or structural damage at the fungal cell wall and initiates a transcriptional response (Ana Sanz et al., 2017; García et al., 2017;Jin et al., 2014 ; Levin, 2011). The membrane proteins Mid2p, Mlt1p and Wsc1p can be stimulated by external stressors such as H_2O_2 and subsequently activate the GTPase Rho1p, which activates Slt2p and cytoplasmic translocation of cyclin C (Jin et al., 2014; Levin, 2011).

Nuclear signaling also has an important function in cellular proteostasis. For instance, ER stress is sensed by a conserved signal transduction cascade called "unfolded protein response" (UPR). Accumulation of unfolded or underglycosylated proteins in the ER trigger the oligomerization of the ER transmembrane protein Ire1p, which leads to a non-conventional cytoplasmic splicing of the *HAC1* mRNA, encoding the transcription factor Hac1p (Gardner et al., 2013; Karagöz et al., 2019). The following transcriptional response includes the expression of chaperons and heat-shock proteins which mediate oxidative protein folding processes in the ER (Haynes et al., 2004). Consistently, excessive ER stress, e.g., by inhibiting protein N-glycosylation is accompanied by elevated ROS production and cell death with apoptotic features (Hauptmann et al., 2006; Hauptmann and Lehle, 2008). This parallels findings from mammalian cells, where persistent ER stress and UPR activation induces apoptosis by activating an intracellular pool of death receptor 5 (DR5), which subsequently activates caspase-8 (Lu et al., 2014).

Besides ER Stress, lipid stress is also able to prompt RCD by triggering a transcriptional feedback. Excess build-up of diacylglycerol in yeast causes necrotic cell death through a pathway that involves the Rim21p-dependent protein complex at the plasma membrane (Rockenfeller et al., 2018). This sentinel complex usually senses pH stress-induced changes in the plasma membrane phospholipid asymmetry, but in this case, it responds to diacylglycerol-elicited lipid disturbance and subsequently activates the protease Rim13p, a homolog of the mammalian necrosis-regulating calpains. Rim13p cleaves and activates the transcriptional repressor Rim101p, which translocates to the nucleus where a complex transcriptional response, likely involving the transcriptional repressor Nrg1p, culminates in a maladaptive response to diacylglycerol stress (Rockenfeller et al., 2018; Rockenfeller and Gourlay, 2018).

Finally, transcriptional responses have also been implicated in the evasion from RCD. Importantly, crosstalk between the nucleus and mitochondria has a crucial role in responding to intracellular and extracellular stress and can serve as a decision point between maintaining cellular homeostasis or activating cell death (Guaragnella et al., 2018). For example, yeast retrograde signaling, a mitochondria-to-nucleus communication pathway that is activated by the sensory protein Rtg2p and executed by the heteromeric Rtg1/Rtg3 transcription factor, protects from acetic acid-induced RCD (Laera et al., 2016). Interestingly, retrograde signaling only mediates RCD evasion when cells are grown on raffinose, but not when cells grow on glucose. This carbon source-dependency is also reflected by the fact that the carbon catabolite repression transcription factors Adr1p and Cat8p are required for retrograde signaling and RCD inhibition under acetic acid treatment in raffinose media. Yet, when cells are grown on glucose, Rtg2p may even promote necrotic cell death (Laera et al., 2016). Target genes of retrograde signaling are also activated by the stress-activated protein kinase Hog1p, which can mitigate cellular stress such as ROS by increased catalase expression (Guaragnella et al., 2018, 2019). At the same time, sustained activation of Hog1p promotes ROS accumulation and cell death in an Nma111p-dependent manner (Vendrell et al., 2011, p. 2). In sum, a common theme crystallizes, where transcriptional stress responses initially mitigate damage but switch to a pro-death role upon excessive or prolonged cellular disbalance. Yet, the precise decision points, timing and stress thresholds that decide between cell survival and cell death remain to be determined.

4.2 Direct regulation of RCD executors

Given the conservation of many RCD factors between humans and yeast, a transcriptional regulation of RCD through modulating the expression of pro- or anti-apoptotic factors seems likely. However, in yeast, there is little manually curated experimental data on transcriptional regulation and thus, high-throughput data provide the only evidence for a direct contribution of transcription to RCD execution. This comes with obvious limitations: (i) large-scale data seldom provides causal links but rather allows correlative analyses; (ii) transcriptomics data without concomitant chromatin immunoprecipitation sequencing cannot differentiate between modulation of mRNA transcription and altered mRNA stability; (iii) since many (if not all) RCD regulators have dual functions—under non-lethal conditions they execute a vital role—any transcriptional regulation might simply reflect a cell's need to modulate the respective protein's non-death activity.

Despite these limitations, there is accumulating evidence that yeast RCD genes do indeed respond to stress conditions. For instance, on the one hand, global transcriptional responses to DNA damage under various genotoxic stresses, including H_2O_2-mediated oxidative stress have been described (Fu et al., 2008; Shapira et al., 2004). On the other hand, excessive DNA damage by H_2O_2 treatment can reportedly trigger RCD (Ribeiro et al., 2006). In addition, transcriptome microarray datasets from yeast indicate that pro-death factors are subjected to transcriptional regulation under diverse environmental stress conditions. For instance, exposure to H_2O_2 in exponentially growing cells leads to a transient upregulation of *AIF1* and *YBH3*, while the mRNAs of other RCD regulators like the proposed inhibitor of apoptosis (IAP) *BIR1* (Walter et al., 2006) or the pro-apoptotic protein *NUC1* are downregulated under similar stress conditions (Berry and Gasch, 2008; Gasch et al., 2001; Shapira et al., 2004). Likewise, ethanol stress, which reportedly induces cell death with apoptotic features (Kitagaki et al., 2007), upregulates the pro-apoptotic factors *NDI1*, one of two cytochrome *c* paralogs (CYC7), and the mitochondrial fission regulator *DNM1*, while, again, *BIR1* and *NUC1* are downregulated (Lewis et al., 2014). Dong et al. reported that yeast exposed to 150 mM acetic acid upregulates several RCD-related genes, as assessed by RNA-Seq-based transcriptomics (Dong et al., 2017), although only two of them, *CYC7* and *AAC3*, have been directly implicated in yeast RCD: analogous to human cells, cytochrome *c* is released from mitochondria upon heterologous expression of BAX in yeast (Manon et al., 1997), and deletion of the ADP/ATP carrier *AAC3* (along with the two isoforms *AAC1* and *AAC2*) protects yeast from acetic acid-induced cell death (Trindade et al., 2016). *NUC1* counterintuitively appears to be less expressed upon RCD-inducing stress, yet, it is reportedly upregulated when yeast spores or cells that are just undergoing sporulation return to growth (Friedlander et al., 2006; Geijer et al., 2012). Intriguingly, mitochondrial release of Nuc1p has recently been shown to contribute to the constriction of RNA mycoviruses, which try to replicate during yeast sporulation (Gao et al., 2019). Future studies should clarify whether *NUC1* fulfills a similar role during germination and the return to vegetative growth. Surprisingly, mRNA levels of the yeast metacaspase *YCA1* appear to be largely unaffected, at least by typical RCD-inducing environmental stress conditions (Berry and Gasch, 2008; Gasch et al., 2001; Lewis et al., 2014; Shapira et al., 2004), except for an induction after treatment with the ER-stress agent DTT (Travers et al., 2000).

Nevertheless, there is a considerable body of data indicating a general responsiveness of the yeast RCD genes to environmental stress. But does increased expression of pro-apoptotic genes necessarily initiate RCD in yeast? Most studies that led to the identification of yeast RCD factors used overexpression experiments to demonstrate pro-apoptotic properties. Yet, overexpression of pro-apoptotic genes alone does not suffice to induce cell death in most cases but requires additional environmental stressors like H_2O_2 or acetic acid (Büttner et al., 2007, 2011; Li et al., 2006; Madeo et al., 2002; Wissing et al., 2004). Accordingly, it is tempting to speculate that stress-induced expression of RCD regulators does not represent a point-of-no-return but rather primes yeast for cell death in case of prolonged stress (Strich, 2015).

This motif also extends to internal noxa such as DNA damage. DNA damage-inducing agents can upregulate the large subunit ribonucleotide-diphosphate reductase (consisting of the paralogs Rnr1p or Rnr3p) up to 100-fold, and to a minor degree the small subunit (consisting of the paralogs Rnr2p or Rnr4p) (Elledge and Davis, 1990, 1989), resulting in elevated dNTP levels, which is essential for yeast survival (Chabes et al., 2003). Removing the feedback inhibition of Rnr3p by dATP through mutation of Asp-57 to Asn results in an even more pronounced dNTP accumulation. This buildup of dNTPs is suspected to be causal for the subsequent cell cycle arrest and elevated DNA damage sensitivity (Chabes and Stillman, 2007), both states, which have been linked to yeast cell death (Burhans et al., 2003; Palermo et al., 2010; Wysocki and Kron, 2004). Indeed, imbalance of NTP/dNTP levels has been suggested to contribute to RCD in cells with disturbed mRNA stability (Palermo et al., 2010). One hypothetical scenario is that moderate DNA damage is effectively remedied by Rnr3p but that prolonged DNA damage might result in chronically elevated Rnr3p levels and increased dNTP pools which could drive the cell into an RCD program. In contrast, Almeida et al. reported decreased protein expression of Rnr2p and Rnr4p in cells that underwent cell death in response to acetic acid, although the expression level of Rnr1p or Rnr3p was not assessed (Almeida et al., 2009). Of note, while acetic acid can provoke DNA damage at cell-death inducing concentrations, the phenotypical DNA damage markers occur with much slower kinetics compared to H_2O_2 stress even though the survival rate is comparable (Ribeiro et al., 2006), which could indicate that DNA damage stress is not the main cause for acetic acid-mediated RCD.

5. Epigenetic regulation of yeast RCD

RCD is not only modulated by activation of transcriptional responses but also via epigenetic modifications (Fig. 3). The genetic material in eukaryotes is tightly packed into nucleosomes, which consist of the core histone proteins H2A, H2B, H3 and H4. In mammalian cells, histone tails can be post-translationally (epigenetically) modified during RCD, for example, by the Ste20p kinase homolog Mst1, which phosphorylates histone variant H2AX at Ser-139 (Wen et al., 2010, p. 1). Phosphorylated H2AX sequesters the guanine nucleotide exchange factor for the Ran GTPase, RCC1, at the chromatin, reducing nuclear RanGTP levels, which subsequently allows an apoptosis program to proceed (Wong et al., 2009). In yeast, Ste20p phosphorylates histone H2B at Ser-10, which is essential for the induction of cell death with apoptotic hallmarks in cells treated with H_2O_2 (Ahn et al., 2005), as well as cell death induced by Nuc1p overexpression (Büttner et al., 2007).

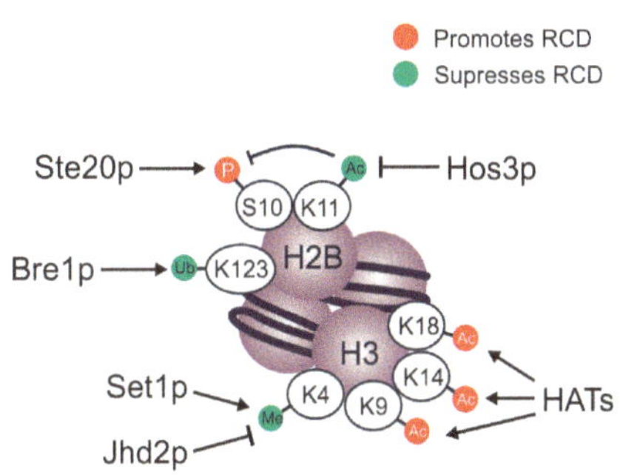

Fig. 3 Epigenetic regulation of yeast RCD. The histones H2B and H3 are subject to different posttranslational modifications that control yeast RCD, including acetylation, phosphorylation, methylation and ubiquitination. Notably, some of these epigenetic markers show a regulatory cross-talk between them.

Ste20p is also involved in cell death evoked by prolonged exposure to mating pheromones as a downstream effector of pheromone-induced MAPK signaling (Severin and Hyman, 2002). Of note, mammalian Mst1 contributes to beta-cell apoptosis under diabetogenic conditions by upregulating the BH3-only protein BIM and by phosphorylating and thus promoting rapid proteasomal degradation of the beta-cell-specific transcription factor PDX1 (Ardestani et al., 2014, p. 1). It remains to be investigated to what extent Ste20p and Mst1 share common regulatory functions in evolutionary distant cell death routines. Interestingly, H2B Lys-11 deacetylation by the histone deacetylase (HDAC) Hos3p has been shown to precede H2B Ser-10 phosphorylation. Accordingly, introducing an acetylation mimicking H2B Lys-11 to Gln mutant protects from H_2O_2-induced cell death. Vice versa, mutation of Lys-11 to Arg and thus mimicking constitutive deacetylation, promotes cell death (Ahn et al., 2006, p. 3). During budding, Hos3p is asymmetrically enriched at the daughter cell spindle pole body (SPB, equivalent to centrosomes)—a unique feature among all other yeast HDACs—and might be responsible for mitigating spindle orientation defects by stalling mitotic exit (Wang and Collins, 2014, p. 3). It will be interesting to investigate whether Hos3p represents a potential link between SPB misalignment, failure to exit mitosis and RCD. Defective mitotic exit, e.g., by deletion of the cell cycle initiator Cdc6p, has been linked to apoptotic phenotypes in yeast (Burhans et al., 2003) and mammalian Cdc6 has been shown to be inactivated during RCD (Blanchard et al., 2002).

H2B can also be modified at Lys-123 by ubiquitination via the ubiquitin ligase Bre1p, which is conserved in human cells and is important for preventing aberrant RNA-DNA structures (R-loops) (Chernikova et al., 2012). *BRE1*-deficient cells show increased cell death rates with apoptotic hallmarks, which depend on the metacaspase Yca1p. In turn, increased Bre1p levels protect cells from H_2O_2 stress (Walter et al., 2010). Bre1p has also been implicated in cyclin regulation, and similar to observation in old mother cells, *BRE1* deletion leads to increased nuclear localization of the transcription co-factor Whi5p and reduced expression of the G1 cyclins Cln1p and Cln2p, while the expression of the (presumably) pro-apoptotic cyclin Cln3p remained unaffected (Albrecht et al., 2016). Of note, H2B Lys-123 monoubiquitination enhances H3 Lys-4 methylation via the methyltransferase Set1p, and loss of this histone modification sensitizes yeast to RCD triggers, again in a Yca1p-dependent manner (Walter et al., 2014). In sum, epigenetic events have important regulatory roles in RCD, although mechanistic details remain to be investigated (Fahrenkrog, 2015). It is possible that histone-modifying enzymes have a sentinel role and translate nucleus-intrinsic stress (e.g., DNA damage or replication stress) into a signal to the RCD machinery. However, no such signal, which could also be realized via increased transcription of pro-death factors, has reliably been identified so far. One report showed that reducing H3 Lys-36 methylation by disrupting the methyltransferase *SET2* displayed significantly reduced expression of Aif1p (Tompa and Madhani, 2007). However, a formal investigation of Set2p activity and RCD induction has not been conducted so far.

Histone hyperacetylation, especially at H3 Lys-9, Lys-14 and Lys-18, has been implicated in regulated necrosis in yeast. Restoring cellular levels of spermidine, a naturally occurring polyamine, during aging, reduces H3 acetylation and markers of necrosis such as PI positivity and translocation of Nhp6Ap (Eisenberg et al., 2009). Mechanistically, this action of spermidine was originally attributed to the inhibition of histone acetyltransferases (Eisenberg et al., 2009), which has later been confirmed in human cell culture (Pietrocola et al., 2015). Recent studies in mammals suggest that supplementing spermidine also reverts the age-dependent decline of hypusination of the translation initiation factor eIF5A, a unique posttranslational modification which requires spermidine as a substrate (Zhang et al., 2019). Importantly, both pathways culminate in the induction of autophagy, an intracellular, cytoprotective degradation process. Interference with autophagy by disrupting essential autophagy components precludes spermidine's anti-necrotic effects (Eisenberg et al., 2009). This suggests that increased histone acetylation during aging blocks cellular repair mechanisms, which normally keep necrosis-inducing cellular damage at bay. Consistently, H3 hyperacetylation by disrupting mitochondrial consumption of the sole cellular donor for acetylation reactions, acetyl-CoA, leads to increased cell death. However, the phenotypical hallmarks slightly differ: while aging-induced H3 acetylation increases both primary and secondary necrotic cell populations, H3 hyperacetylation in *ACH1* deletion mutants, which are unable to use up intracellular acetate and accumulate nucleo-cytosolic acetyl-CoA, correlates only with apoptotic and secondary necrotic cells (Eisenberg et al., 2014, 2009).

Finally, epigenetic modifications also encompass direct modification of DNA, in particular cytosine methylation at CpG dinucleotides. In contrast to mammalian cells, where CpG methylation has been implicated in RCD regulation (Meng et al., 2011), understanding the role of yeast cytosine methylation is at its infancy. This is probably due to the fact that this modification is reportedly below detection limits (Capuano et al., 2014) or, according to others, occurs considerably (about 1–2 orders of magnitude) less frequently than in mammalian cells (Tang et al., 2012).

6. Conclusion

The nucleus represents a central hub for the execution of yeast RCD, (i) through the destruction of the genetic material by resident or incoming hydrolases, (ii) as a lair for pro-apoptotic proteins that can be released in a regulated manner, and (iii) as a pivot for the regulation of the RCD machinery. This regulatory role has largely been overlooked in yeast RCD research and thus, the available data is mainly correlative. Hence, additional omics data of different RCD scenarios in yeast showing the expression kinetics of RCD-related proteins are urgently needed. In addition, the available correlative data should be revisited and the causal relationship of expression changes tested, e.g., by means of controllable, conditional expression systems. While RCD in yeast is certainly less complex than mammalian RCD in mammalian cells, the power of yeast genetics and technical possibilities for systems biology approaches could yield novel insight into transcriptional and epigenetic control of cell death beyond the fungal kingdom.

Acknowledgments

This work was supported by the Austrian Science Fund FWF [F3012, P23490-B20, P24381, P27893, P29203, P29262, W1226], the Austrian Federal Ministry of Science, Research and Economy [BMWFW-80.109/0001-WF/V/3b/2015], BioTechMedGraz [EPIAge], and the University of Graz [Unkonventionelle Forschung]. We acknowledge support from NAWI Graz.

Conflict of interest

No competing financial interests exist.

References

Ahn, S.-H., Cheung, W.L., Hsu, J.-Y., Diaz, R.L., Smith, M.M., Allis, C.D., 2005. Sterile 20 kinase phosphorylates histone H2B at serine 10 during hydrogen peroxide-induced apoptosis in S. cerevisiae. Cell 120, 25–36. https://doi.org/10.1016/j.cell.2004.11.016.

Ahn, S.-H., Diaz, R.L., Grunstein, M., Allis, C.D., 2006. Histone H2B deacetylation at lysine 11 is required for yeast apoptosis induced by phosphorylation of H2B at serine 10. Mol. Cell 24, 211–220. https://doi.org/10.1016/j.molcel.2006.09.008.

Albrecht, D., Ceschin, J., Dompierre, J., Gueniot, F., Pinson, B., Daignan-Fornier, B., 2016. Chemo-genetic interactions between histone modification and the antiproliferation drug AICAR are conserved in yeast and humans. Genetics 204, 1447–1460. https://doi.org/10.1534/genetics.116.192518.

Almeida, T., Marques, M., Mojzita, D., Amorim, M.A., Silva, R.D., Almeida, B., Rodrigues, P., Ludovico, P., Hohmann, S., Moradas-Ferreira, P., Côrte-Real, M., Costa, V., 2008. Isc 1p plays a key role in hydrogen peroxide resistance and chronological lifespan through modulation of iron levels and apoptosis. Mol. Biol. Cell 19, 865–876. https://doi.org/10.1091/mbc.e07-06-0604.

Almeida, B., Ohlmeier, S., Almeida, A.J., Madeo, F., Leão, C., Rodrigues, F., Ludovico, P., 2009. Yeast protein expression profile during acetic acid-induced apoptosis indicates causal involvement of the TOR pathway. Proteomics 9, 720–732. https://doi.org/10.1002/pmic.200700816.

Ardestani, A., Paroni, F., Azizi, Z., Kaur, S., Khobragade, V., Yuan, T., Frogne, T., Tao, W., Oberholzer, J., Pattou, F., Kerr Conte, J., Maedler, K., 2014. MST1 is a key regulator of beta cell apoptosis and dysfunction in diabetes. Nat. Med. 20, 385–397. https://doi.org/10.1038/nm.3482.

Aubrey, B.J., Kelly, G.L., Janic, A., Herold, M.J., Strasser, A., 2018. How does p53 induce apoptosis and how does this relate to p53-mediated tumour suppression? Cell. Death Differ. 25, 104–113. https://doi.org/10.1038/cdd.2017.169.

Belanger, K.D., Walter, D., Henderson, T.A., Yelton, A.L., O'Brien, T.G., Belanger, K.G., Geier, S.J., Fahrenkrog, B., 2009. Nuclear localisation is crucial for the proapoptotic activity of the HtrA-like serine protease Nma111p. J. Cell Sci. 122, 3931–3941. https://doi.org/10.1242/jcs.056887.

Berry, D.B., Gasch, A.P., 2008. Stress-activated genomic expression changes serve a preparative role for impending stress in yeast. Mol. Biol. Cell 19, 4580–4587. https://doi.org/10.1091/mbc.e07-07-0680.

Bink, A., Govaert, G., François, I.E.J.A., Pellens, K., Meerpoel, L., Borgers, M., Van Minnebruggen, G., Vroome, V., Cammue, B.P.A., Thevissen, K., 2010. A fungicidal piperazine-1-carboxamidine induces mitochondrial fission-dependent apoptosis in yeast: piperazine-1-carboxamidine and apoptosis. FEMS Yeast Res. 10, 812–818. https://doi.org/10.1111/j.1567-1364.2010.00663.x.

Blanchard, F., Rusiniak, M.E., Sharma, K., Sun, X., Todorov, I., Castellano, M.M., Gutierrez, C., Baumann, H., Burhans, W.C., 2002. Targeted destruction of DNA replication protein Cdc6 by cell death pathways in mammals and yeast. Mol. Biol. Cell 13, 1536–1549. https://doi.org/10.1091/mbc.02-02-0010.

Bourbon, H.-M., 2008. Comparative genomics supports a deep evolutionary origin for the large, four-module transcriptional mediator complex. Nucleic Acids Res. 36, 3993–4008. https://doi.org/10.1093/nar/gkn349.

Braun, R.J., Zischka, H., 2008. Mechanisms of Cdc48/VCP-mediated cell death—from yeast apoptosis to human disease. Biochim. Biophys. Acta, Mol. Cell Res. 1783, 1418–1435. https://doi.org/10.1016/j.bbamcr.2008.01.015.

Burhans, W.C., Weinberger, M., Marchetti, M.A., Ramachandran, L., D'Urso, G., Huberman, J.A., 2003. Apoptosis-like yeast cell death in response to DNA damage and replication defects. Mutat. Res. 532, 227–243.

Büttner, S., Eisenberg, T., Carmona-Gutierrez, D., Ruli, D., Knauer, H., Ruckenstuhl, C., Sigrist, C., Wissing, S., Kollroser, M., Fröhlich, K.-U., Sigrist, S., Madeo, F., 2007. Endonuclease G regulates budding yeast life and death. Mol. Cell 25, 233–246. https://doi.org/10.1016/j.molcel.2006.12.021.

Büttner, S., Ruli, D., Vögtle, F.-N., Galluzzi, L., Moitzi, B., Eisenberg, T., Kepp, O., Habernig, L., Carmona-Gutierrez, D., Rockenfeller, P., Laun, P., Breitenbach, M., Khoury, C., Fröhlich, K.-U., Rechberger, G., Meisinger, C., Kroemer, G., Madeo, F., 2011. A yeast BH3-only protein mediates the mitochondrial pathway of apoptosis. EMBO J. 30, 2779–2792. https://doi.org/10.1038/emboj.2011.197.

Capuano, F., Mülleder, M., Kok, R., Blom, H.J., Ralser, M., 2014. Cytosine DNA methylation is found in *Drosophila melanogaster* but absent in *Saccharomyces cerevisiae, Schizosaccharomyces pombe*, and other yeast species. Anal. Chem. 86, 3697–3702. https://doi.org/10.1021/ac500447w.

Carmona-Gutierrez, D., Madeo, F., 2009. Tracing the roots of death: apoptosis in Saccharomyces cerevisiae. In: Dong, Z., Yin, X.-M. (Eds.), Essentials of Apoptosis. Humana Press, Totowa, NJ, pp. 325–354. https://doi.org/10.1007/978-1-60327-381-7_14.
Carmona-Gutierrez, D., Eisenberg, T., Büttner, S., Meisinger, C., Kroemer, G., Madeo, F., 2010a. Apoptosis in yeast: triggers, pathways, subroutines. Cell Death Differ. 17, 763–773. https://doi.org/10.1038/cdd.2009.219.
Carmona-Gutierrez, D., Fröhlich, K.-U., Kroemer, G., Madeo, F., 2010b. Metacaspases are caspases. Doubt no more. Cell Death Differ. 17, 377–378. https://doi.org/10.1038/cdd.2009.198.
Carmona-Gutiérrez, D., Bauer, M.A., Ring, J., Knauer, H., Eisenberg, T., Büttner, S., Ruckenstuhl, C., Reisenbichler, A., Magnes, C., Rechberger, G.N., Birner-Gruenberger, R., Jungwirth, H., Fröhlich, K.-U., Sinner, F., Kroemer, G., Madeo, F., 2011. The propeptide of yeast cathepsin D inhibits programmed necrosis. Cell Death Dis. 2, e161. https://doi.org/10.1038/cddis.2011.43.
Carmona-Gutierrez, D., Bauer, M.A., Zimmermann, A., Aguilera, A., Austriaco, N., Ayscough, K., Balzan, R., Bar-Nun, S., Barrientos, A., Belenky, P., Blondel, M., Braun, R.J., Breitenbach, M., Burhans, W.C., Büttner, S., Cavalieri, D., Chang, M., Cooper, K.F., Côrte-Real, M., Costa, V., Cullin, C., Dawes, I., Dengjel, J., Dickman, M.B., Eisenberg, T., Fahrenkrog, B., Fasel, N., Fröhlich, K.-U., Gargouri, A., Giannattasio, S., et al., 2018. Guidelines and recommendations on yeast cell death nomenclature. Microb. Cell 5, 4–31. https://doi.org/10.15698/mic2018.01.607.
Carmona-Gutierrez, D., Zimmermann, A., Kainz, K., Pietrocola, F., Chen, G., Maglioni, S., Schiavi, A., Nah, J., Mertel, S., Beuschel, C.B., Castoldi, F., Sica, V., Trausinger, G., Raml, R., Sommer, C., Schroeder, S., Hofer, S.J., Bauer, M.A., Pendl, T., Tadic, J., Dammbrueck, C., Hu, Z., Ruckenstuhl, C., Eisenberg, T., Durand, S., Bossut, N., Aprahamian, F., Abdellatif, M., Sedej, S., Enot, D.P., Wolinski, H., Dengjel, J., Kepp, O., Magnes, C., Sinner, F., Pieber, T.R., Sadoshima, J., Ventura, N., Sigrist, S.J., Kroemer, G., Madeo, F., 2019. The flavonoid 4,4′-dimethoxychalcone promotes autophagy-dependent longevity across species. Nat. Commun. 10 651https://doi.org/10.1038/s41467-019-08555-w.
Chabes, A., Stillman, B., 2007. Constitutively high dNTP concentration inhibits cell cycle progression and the DNA damage checkpoint in yeast Saccharomyces cerevisiae. Proc. Natl. Acad. Sci. U. S. A. 104, 1183–1188. https://doi.org/10.1073/pnas.0610585104.
Chabes, A., Georgieva, B., Domkin, V., Zhao, X., Rothstein, R., Thelander, L., 2003. Survival of DNA damage in yeast directly depends on increased dNTP levels allowed by relaxed feedback inhibition of ribonucleotide reductase. Cell 112, 391–401. https://doi.org/10.1016/S0092-8674(03)00075-8.
Chen, H., Workman, J.J., Tenga, A., Laribee, R., 2013. Target of rapamycin signaling regulates high mobility group protein association to chromatin, which functions to suppress necrotic cell death. Epigenetics Chromatin 6, 29. https://doi.org/10.1186/1756-8935-6-29.
Chernikova, S.B., Razorenova, O.V., Higgins, J.P., Sishc, B.J., Nicolau, M., Dorth, J.A., Chernikova, D.A., Kwok, S., Brooks, J.D., Bailey, S.M., Game, J.C., Brown, J.M., 2012. Deficiency in mammalian histone H2B ubiquitin ligase Bre1 (Rnf20/Rnf40) leads to replication stress and chromosomal instability. Cancer Res. 72, 2111–2119. https://doi.org/10.1158/0008-5472.CAN-11-2209.
Coccetti, P., Nicastro, R., Tripodi, F., 2018. Conventional and emerging roles of the energy sensor Snf1/AMPK in Saccharomyces cerevisiae. Microb. Cell 5, 482–494. https://doi.org/10.15698/mic2018.11.655.

Cooper, K.F., 1997. Stress and developmental regulation of the yeast C-type cyclin Ume3p (Srb11p/Ssn8p). EMBO J. 16, 4665–4675. https://doi.org/10.1093/emboj/16.15.4665.
Cooper, K.F., Khakhina, S., Kim, S.K., Strich, R., 2014. Stress-induced nuclear-to-cytoplasmic translocation of cyclin C promotes mitochondrial fission in yeast. Dev. Cell 28, 161–173. https://doi.org/10.1016/j.devcel.2013.12.009.
Debrabant, A., Lee, N., Bertholet, S., Duncan, R., Nakhasi, H.L., 2003. Programmed cell death in trypanosomatids and other unicellular organisms. Int. J. Parasitol. 33, 257–267.
Denton, D., Kumar, S., 2019. Autophagy-dependent cell death. Cell Death Differ. 26, 605–616. https://doi.org/10.1038/s41418-018-0252-y.
Deprez, M.-A., Eskes, E., Wilms, T., Ludovico, P., Winderickx, J., 2018. pH homeostasis links the nutrient sensing PKA/TORC1/Sch9 ménage-à-trois to stress tolerance and longevity. Microb. Cell Graz Austria 5, 119–136. https://doi.org/10.15698/mic2018.03.618.
Dong, Y., Hu, J., Fan, L., Chen, Q., 2017. RNA-Seq-based transcriptomic and metabolomic analysis reveal stress responses and programmed cell death induced by acetic acid in Saccharomyces cerevisiae. Sci. Rep. 7 42659https://doi.org/10.1038/srep42659.
Dwyer, D.J., Camacho, D.M., Kohanski, M.A., Callura, J.M., Collins, J.J., 2012. Antibiotic-induced bacterial cell death exhibits physiological and biochemical hallmarks of apoptosis. Mol. Cell 46, 561–572. https://doi.org/10.1016/j.molcel.2012.04.027.
Eastwood, M.D., Meneghini, M.D., 2015. Developmental coordination of gamete differentiation with programmed cell death in sporulating yeast. Eukaryot. Cell 14, 858–867. https://doi.org/10.1128/EC.00068-15.
Eisenberg, T., Knauer, H., Schauer, A., Büttner, S., Ruckenstuhl, C., Carmona-Gutierrez, D., Ring, J., Schroeder, S., Magnes, C., Antonacci, L., Fussi, H., Deszcz, L., Hartl, R., Schraml, E., Criollo, A., Megalou, E., Weiskopf, D., Laun, P., Heeren, G., Breitenbach, M., Grubeck-Loebenstein, B., Herker, E., Fahrenkrog, B., Fröhlich, K.-U., Sinner, F., Tavernarakis, N., Minois, N., Kroemer, G., Madeo, F., 2009. Induction of autophagy by spermidine promotes longevity. Nat. Cell Biol. 11, 1305–1314. https://doi.org/10.1038/ncb1975.
Eisenberg, T., Schroeder, S., Andryushkova, A., Pendl, T., Küttner, V., Bhukel, A., Mariño, G., Pietrocola, F., Harger, A., Zimmermann, A., Moustafa, T., Sprenger, A., Jany, E., Büttner, S., Carmona-Gutierrez, D., Ruckenstuhl, C., Ring, J., Reichelt, W., Schimmel, K., Leeb, T., Moser, C., Schatz, S., Kamolz, L.-P., Magnes, C., Sinner, F., Sedej, S., Fröhlich, K.-U., Juhasz, G., Pieber, T.R., Dengjel, J., Sigrist, S.J., Kroemer, G., Madeo, F., 2014. Nucleocytosolic depletion of the energy metabolite acetyl-coenzyme a stimulates autophagy and prolongs lifespan. Cell Metab. 19, 431–444. https://doi.org/10.1016/j.cmet.2014.02.010.
Elledge, S.J., Davis, R.W., 1989. Identification of the DNA damage-responsive element of RNR2 and evidence that four distinct cellular factors bind it. Mol. Cell. Biol. 9, 5373–5386. https://doi.org/10.1128/MCB.9.12.5373.
Elledge, S.J., Davis, R.W., 1990. Two genes differentially regulated in the cell cycle and by DNA-damaging agents encode alternative regulatory subunits of ribonucleotide reductase. Genes Dev. 4, 740–751. https://doi.org/10.1101/gad.4.5.740.
Fabrizio, P., 2001. Regulation of longevity and stress resistance by Sch9 in yeast. Science 292, 288–290. https://doi.org/10.1126/science.1059497.
Fabrizio, P., Battistella, L., Vardavas, R., Gattazzo, C., Liou, L.-L., Diaspro, A., Dossen, J.W., Gralla, E.B., Longo, V.D., 2004. Superoxide is a mediator of an altruistic aging program in Saccharomyces cerevisiae. J. Cell Biol. 166, 1055–1067. https://doi.org/10.1083/jcb.200404002.
Fahrenkrog, B., 2015. Histone modifications as regulators of life and death in *Saccharomyces cerevisiae*. Microb. Cell 3, 1–13. https://doi.org/10.15698/mic2016.01.472.

Fahrenkrog, B., Sauder, U., Aebi, U., 2004. The S. cerevisiae HtrA-like protein Nma111p is a nuclear serine protease that mediates yeast apoptosis. J. Cell Sci. 117, 115–126. https://doi.org/10.1242/jcs.00848.

Falcone, C., Mazzoni, C., 2016. External and internal triggers of cell death in yeast. Cell. Mol. Life Sci. 73, 2237–2250. https://doi.org/10.1007/s00018-016-2197-y.

Friedlander, G., Joseph-Strauss, D., Carmi, M., Zenvirth, D., Simchen, G., Barkai, N., 2006. Modulation of the transcription regulatory program in yeast cells committed to sporulation. Genome Biol. 7, R20. https://doi.org/10.1186/gb-2006-7-3-r20.

Fu, Y., Pastushok, L., Xiao, W., 2008. DNA damage-induced gene expression in *Saccharomyces cerevisiae*. FEMS Microbiol. Rev. 32, 908–926. https://doi.org/10.1111/j.1574-6976.2008.00126.x.

Galluzzi, L., Vanden Berghe, T., Vanlangenakker, N., Buettner, S., Eisenberg, T., Vandenabeele, P., Madeo, F., Kroemer, G., 2011. Programmed necrosis from molecules to health and disease. Int. Rev. Cell Mol. Biol. 289, 1–35. https://doi.org/10.1016/B978-0-12-386039-2.00001-8.

Galluzzi, L., Kepp, O., Kroemer, G., 2016. Mitochondrial regulation of cell death: a phylogenetically conserved control. Microb. Cell 3, 101–108. https://doi.org/10.15698/mic2016.03.483.

Galluzzi, L., Vitale, I., Aaronson, S.A., Abrams, J.M., Adam, D., Agostinis, P., Alnemri, E.S., Altucci, L., Amelio, I., Andrews, D.W., Annicchiarico-Petruzzelli, M., Antonov, A.V., Arama, E., Baehrecke, E.H., Barlev, N.A., Bazan, N.G., Bernassola, F., Bertrand, M.J.M., Bianchi, K., Blagosklonny, M.V., Blomgren, K., Borner, C., Boya, P., Brenner, C., Campanella, M., Candi, E., Carmona-Gutierrez, D., Cecconi, F., Chan, F.K.-M., Chandel, N.S., Cheng, E.H., Chipuk, J.E., Cidlowski, J.A., Ciechanover, A., Cohen, G.M., Conrad, M., Cubillos-Ruiz, J.R., Czabotar, P.E., D'Angiolella, V., Dawson, T.M., Dawson, V.L., De Laurenzi, V., De Maria, R., Debatin, K.-M., DeBerardinis, R.J., Deshmukh, M., Di Daniele, N., Di Virgilio, F., Dixit, V.M., Dixon, S.J., Duckett, C.S., Dynlacht, B.D., El-Deiry, W.S., Elrod, J.W., Fimia, G.M., Fulda, S., García-Sáez, A.J., Garg, A.D., Garrido, C., Gavathiotis, E., Golstein, P., Gottlieb, E., Green, D.R., Greene, L.A., Gronemeyer, H., Gross, A., Hajnoczky, G., Hardwick, J.M., Harris, I.S., Hengartner, M.O., Hetz, C., Ichijo, H., Jäättelä, M., Joseph, B., Jost, P.J., Juin, P.P., Kaiser, W.J., Karin, M., Kaufmann, T., Kepp, O., Kimchi, A., Kitsis, R.N., Klionsky, D.J., Knight, R.A., Kumar, S., Lee, S.W., Lemasters, J.J., Levine, B., Linkermann, A., Lipton, S.A., Lockshin, R.A., López-Otín, C., Lowe, S.W., Luedde, T., Lugli, E., MacFarlane, M., Madeo, F., Malewicz, M., Malorni, W., Manic, G., Marine, J.-C., Martin, S.J., Martinou, J.-C., Medema, J.P., Mehlen, P., Meier, P., Melino, S., Miao, E.A., Molkentin, J.D., Moll, U.M., Muñoz-Pinedo, C., Nagata, S., Nuñez, G., Oberst, A., Oren, M., Overholtzer, M., Pagano, M., Panaretakis, T., Pasparakis, M., Penninger, J.M., Pereira, D.M., Pervaiz, S., Peter, M.E., Piacentini, M., Pinton, P., Prehn, J.H.M., Puthalakath, H., Rabinovich, G.A., Rehm, M., Rizzuto, R., Rodrigues, C.M.P., Rubinsztein, D.C., Rudel, T., Ryan, K.M., Sayan, E., Scorrano, L., Shao, F., Shi, Y., Silke, J., Simon, H.-U., Sistigu, A., Stockwell, B.R., Strasser, A., Szabadkai, G., Tait, S.W.G., Tang, D., Tavernarakis, N., Thorburn, A., Tsujimoto, Y., Turk, B., Vanden Berghe, T., Vandenabeele, P., Vander Heiden, M.G., Villunger, A., Virgin, H.W., Vousden, K.H., Vucic, D., Wagner, E.F., Walczak, H., Wallach, D., Wang, Y., Wells, J.A., Wood, W., Yuan, J., Zakeri, Z., Zhivotovsky, B., Zitvogel, L., Melino, G., Kroemer, G., 2018. Molecular mechanisms of cell death: recommendations of the Nomenclature Committee on Cell Death 2018. Cell Death Differ. 25, 486–541. https://doi.org/10.1038/s41418-017-0012-4.

Gao, J., Chau, S., Chowdhury, F., Zhou, T., Hossain, S., McQuibban, G.A., Meneghini, M.D., 2019. Meiotic viral attenuation through an ancestral apoptotic pathway. Proc. Natl. Acad. Sci. 116, 16454–16462. 201900751. https://doi.org/10.1073/pnas.1900751116.

García, R., Bravo, E., Diez-Muñiz, S., Nombela, C., Rodríguez-Peña, J.M., Arroyo, J., 2017. A novel connection between the Cell Wall Integrity and the PKA pathways regulates cell wall stress response in yeast. Sci. Rep. 7, 5703. https://doi.org/10.1038/s41598-017-06001-9.

Gardner, B.M., Pincus, D., Gotthardt, K., Gallagher, C.M., Walter, P., 2013. Endoplasmic reticulum stress sensing in the unfolded protein response. Cold Spring Harb. Perspect. Biol. 5, a013169. https://doi.org/10.1101/cshperspect.a013169.

Gasch, A.P., Huang, M., Metzner, S., Botstein, D., Elledge, S.J., Brown, P.O., 2001. Genomic expression responses to DNA-damaging agents and the regulatory role of the yeast ATR homolog Mec1p. Mol. Biol. Cell 12, 2987–3003. https://doi.org/10.1091/mbc.12.10.2987.

Geijer, C., Pirkov, I., Vongsangnak, W., Ericsson, A., Nielsen, J., Krantz, M., Hohmann, S., 2012. Time course gene expression profiling of yeast spore germination reveals a network of transcription factors orchestrating the global response. BMC Genomics 13, 554. https://doi.org/10.1186/1471-2164-13-554.

Guaragnella, N., Bobba, A., Passarella, S., Marra, E., Giannattasio, S., 2010. Yeast acetic acid-induced programmed cell death can occur without cytochrome c release which requires metacaspase YCA1. FEBS Lett. 584, 224–228. https://doi.org/10.1016/j.febslet.2009.11.072.

Guaragnella, N., Coyne, L.P., Chen, X.J., Giannattasio, S., 2018. Mitochondria–cytosol–nucleus crosstalk: learning from Saccharomyces cerevisiae. FEMS Yeast Res. 18, foy088. https://doi.org/10.1093/femsyr/foy088.

Guaragnella, N., Stirpe, M., Marzulli, D., Mazzoni, C., Giannattasio, S., 2019. Acid stress triggers resistance to acetic acid-induced regulated cell death through *Hog 1* activation which requires *RTG2* in Yeast. Oxid. Med. Cell. Longev. 2019, 1–9. https://doi.org/10.1155/2019/4651062.

Hauptmann, P., Lehle, L., 2008. Kex1 protease is involved in yeast cell death induced by defective N-glycosylation, acetic acid, and chronological aging. J. Biol. Chem. 283, 19151–19163. https://doi.org/10.1074/jbc.M801303200.

Hauptmann, P., Riel, C., Kunz-Schughart, L.A., Frohlich, K.-U., Madeo, F., Lehle, L., 2006. Defects in N-glycosylation induce apoptosis in yeast. Mol. Microbiol. 59, 765–778. https://doi.org/10.1111/j.1365-2958.2005.04981.x.

Haynes, C.M., Titus, E.A., Cooper, A.A., 2004. Degradation of misfolded proteins prevents ER-derived oxidative stress and cell death. Mol. Cell 15, 767–776. https://doi.org/10.1016/j.molcel.2004.08.025.

Herker, E., Jungwirth, H., Lehmann, K.A., Maldener, C., Fröhlich, K.-U., Wissing, S., Büttner, S., Fehr, M., Sigrist, S., Madeo, F., 2004. Chronological aging leads to apoptosis in yeast. J. Cell Biol. 164, 501–507. https://doi.org/10.1083/jcb.200310014.

Jin, C., Strich, R., Cooper, K.F., 2014. Slt2p phosphorylation induces cyclin C nuclear-to-cytoplasmic translocation in response to oxidative stress. Mol. Biol. Cell 25, 1396–1407. https://doi.org/10.1091/mbc.e13-09-0550.

Jin, C., Kim, S., Willis, S., Cooper, K., 2015. The MAPKKKs Ste11 and Bck1 jointly transduce the high oxidative stress signal through the cell wall integrity MAP kinase pathway. Microb. Cell 2, 329–342. https://doi.org/10.15698/mic2015.09.226.

Johansson, M., Chen, X., Milanova, S., Santos, C., Petranovic, D., 2016. PUFA-induced cell death is mediated by Yca1p-dependent and -independent pathways, and is reduced by vitamin C in yeast. FEMS Yeast Res. 16 fow007. https://doi.org/10.1093/femsyr/fow007.

Kajiwara, K., Muneoka, T., Watanabe, Y., Karashima, T., Kitagaki, H., Funato, K., 2012. Perturbation of sphingolipid metabolism induces endoplasmic reticulum stress-mediated mitochondrial apoptosis in budding yeast. Mol. Microbiol. 86, 1246–1261. https://doi.org/10.1111/mmi.12056.

Karagöz, G.E., Acosta-Alvear, D., Walter, P., 2019. The unfolded protein response: detecting and responding to fluctuations in the protein-folding capacity of the endoplasmic reticulum. Cold Spring Harb. Perspect. Biol. 11 a033886. https://doi.org/10.1101/cshperspect.a033886.

Khakhina, S., Cooper, K.F., Strich, R., 2014. Med13p prevents mitochondrial fission and programmed cell death in yeast through nuclear retention of cyclin C. Mol. Biol. Cell 25, 2807–2816. https://doi.org/10.1091/mbc.e14-05-0953.

Khan, M.A.S., Chock, P.B., Stadtman, E.R., 2005. Knockout of caspase-like gene, YCA1, abrogates apoptosis and elevates oxidized proteins in Saccharomyces cerevisiae. Proc. Natl. Acad. Sci. U. S. A. 102, 17326–17331. https://doi.org/10.1073/pnas.0508120102.

Khandelwal, R., Sipani, R., Govinda Rajan, S., Kumar, R., Joshi, R., 2017. Combinatorial action of Grainyhead, Extradenticle and Notch in regulating Hox mediated apoptosis in Drosophila larval CNS. PLOS Genet. 13 e1007043. https://doi.org/10.1371/journal.pgen.1007043.

Kitagaki, H., Araki, Y., Funato, K., Shimoi, H., 2007. Ethanol-induced death in yeast exhibits features of apoptosis mediated by mitochondrial fission pathway. FEBS Lett. 581, 2935–2942. https://doi.org/10.1016/j.febslet.2007.05.048.

Knorre, D.A., Smirnova, E.A., Severin, F.F., 2005. Natural conditions inducing programmed cell death in the yeast Saccharomyces cerevisiae. Biochem. Biokhimiia 70, 264–266. https://doi.org/10.1007/s10541-005-0110-7.

Kumar, S., Cakouros, D., 2004. Transcriptional control of the core cell-death machinery. Trends Biochem. Sci. 29, 193–199. https://doi.org/10.1016/j.tibs.2004.02.001.

Laera, L., Guaragnella, N., Zdralevic, M., Marzulli, D., Liu, Z., Giannattasio, S., 2016. The transcription factors ADR1 or CAT8 are required for RTG pathway activation and evasion from yeast acetic acid-induced programmed cell death in raffinose. Microb. Cell 3, 621–631. https://doi.org/10.15698/mic2016.12.549.

Levin, D.E., 2011. Regulation of cell wall biogenesis in *Saccharomyces cerevisiae* : the cell wall integrity signaling pathway. Genetics 189, 1145–1175. https://doi.org/10.1534/genetics.111.128264.

Lewis, J.A., Broman, A.T., Will, J., Gasch, A.P., 2014. Genetic architecture of ethanol-responsive transcriptome variation in Saccharomyces cerevisiae strains. Genetics 198, 369–382. https://doi.org/10.1534/genetics.114.167429.

Li, W., Sun, L., Liang, Q., Wang, J., Mo, W., Zhou, B., 2006. Yeast AMID homologue Ndi1p displays respiration-restricted apoptotic activity and is involved in chronological aging. Mol. Biol. Cell 17, 1802–1811. https://doi.org/10.1091/mbc.e05-04-0333.

Longo, V.D., Fabrizio, P., 2012. Chronological aging in Saccharomyces cerevisiae. Subcell. Biochem. 57, 101–121. https://doi.org/10.1007/978-94-007-2561-4_5.

Lu, M., Lawrence, D.A., Marsters, S., Acosta-Alvear, D., Kimmig, P., Mendez, A.S., Paton, A.W., Paton, J.C., Walter, P., Ashkenazi, A., 2014. Opposing unfolded-protein-response signals converge on death receptor 5 to control apoptosis. Science 345, 98–101. https://doi.org/10.1126/science.1254312.

Ludovico, P., Sousa, M.J., Silva, M.T., Leão, C., Côrte-Real, M., 2001. Saccharomyces cerevisiae commits to a programmed cell death process in response to acetic acid. Microbiol. Read. Engl. 147, 2409–2415. https://doi.org/10.1099/00221287-147-9-2409.

Ludovico, P., Rodrigues, F., Almeida, A., Silva, M.T., Barrientos, A., Côrte-Real, M., 2002. Cytochrome c release and mitochondria involvement in programmed cell death induced by acetic acid in Saccharomyces cerevisiae. Mol. Biol. Cell 13, 2598–2606. https://doi.org/10.1091/mbc.E01-12-0161.

Madeo, F., Fröhlich, E., Fröhlich, K.-U., 1997. A yeast mutant showing diagnostic markers of early and late apoptosis. J. Cell Biol. 139, 729–734. https://doi.org/10.1083/jcb.139.3.729.

Madeo, F., Fröhlich, E., Ligr, M., Grey, M., Sigrist, S.J., Wolf, D.H., Fröhlich, K.-U., 1999. Oxygen stress: a regulator of apoptosis in yeast. J. Cell Biol. 145, 757–767. https://doi.org/10.1083/jcb.145.4.757.

Madeo, F., Herker, E., Maldener, C., Wissing, S., Lächelt, S., Herlan, M., Fehr, M., Lauber, K., Sigrist, S.J., Wesselborg, S., Fröhlich, K.-U., 2002. A caspase-related protease regulates apoptosis in yeast. Mol. Cell 9, 911–917. https://doi.org/10.1016/S1097-2765(02)00501-4.

Manon, S., Chaudhuri, B., Guérin, M., 1997. Release of cytochrome c and decrease of cytochrome c oxidase in Bax-expressing yeast cells, and prevention of these effects by coexpression of Bcl-x_L. FEBS Lett. 415, 29–32. https://doi.org/10.1016/S0014-5793(97)01087-9.

Mazzoni, C., Herker, E., Palermo, V., Jungwirth, H., Eisenberg, T., Madeo, F., Falcone, C., 2005. Yeast caspase 1 links messenger RNA stability to apoptosis in yeast. EMBO Rep. 6, 1076–1081. https://doi.org/10.1038/sj.embor.7400514.

Meng, H.X., Hackett, J.A., Nestor, C., Dunican, D.S., Madej, M., Reddington, J.P., Pennings, S., Harrison, D.J., Meehan, R.R., 2011. Apoptosis and DNA methylation. Cancers 3, 1798–1820. https://doi.org/10.3390/cancers3021798.

Mittal, N., Guimaraes, J.C., Gross, T., Schmidt, A., Vina-Vilaseca, A., Nedialkova, D.D., Aeschimann, F., Leidel, S.A., Spang, A., Zavolan, M., 2017. The Gcn4 transcription factor reduces protein synthesis capacity and extends yeast lifespan. Nat. Commun. 8, 457. https://doi.org/10.1038/s41467-017-00539-y.

Munoz, A.J., Wanichthanarak, K., Meza, E., Petranovic, D., 2012. Systems biology of yeast cell death. FEMS Yeast Res. 12, 249–265. https://doi.org/10.1111/j.1567-1364.2011.00781.x.

Oliveira, C.S.F., Pereira, H., Alves, S., Castro, L., Baltazar, F., Chaves, S.R., Preto, A., Côrte-Real, M., 2015. Cathepsin D protects colorectal cancer cells from acetate-induced apoptosis through autophagy-independent degradation of damaged mitochondria. Cell Death Dis. 6, e1788. https://doi.org/10.1038/cddis.2015.157.

Pakos-Zebrucka, K., Koryga, I., Mnich, K., Ljujic, M., Samali, A., Gorman, A.M., 2016. The integrated stress response. EMBO Rep. 17, 1374–1395. https://doi.org/10.15252/embr.201642195.

Palermo, V., Cundari, E., Mangiapelo, E., Falcone, C., Mazzoni, C., 2010. Yeast lsm pro-apoptotic mutants show defects in S-phase entry and progression. Cell Cycle 9, 3991–3996. https://doi.org/10.4161/cc.9.19.13210.

Pereira, C., Chaves, S., Alves, S., Salin, B., Camougrand, N., Manon, S., Sousa, M.J., Côrte-Real, M., 2010. Mitochondrial degradation in acetic acid-induced yeast apoptosis: the role of Pep4 and the ADP/ATP carrier: mitochondria degradation in apoptosis. Mol. Microbiol. 76, 1398–1410. https://doi.org/10.1111/j.1365-2958.2010.07122.x.

Pietrocola, F., Lachkar, S., Enot, D.P., Niso-Santano, M., Bravo-San Pedro, J.M., Sica, V., Izzo, V., Maiuri, M.C., Madeo, F., Mariño, G., Kroemer, G., 2015. Spermidine induces autophagy by inhibiting the acetyltransferase EP300. Cell Death Differ. 22, 509–516. https://doi.org/10.1038/cdd.2014.215.

Postnikoff, S.D.L., Johnson, J.E., Tyler, J.K., 2017. The integrated stress response in budding yeast lifespan extension. Microb. Cell 4, 368–375. https://doi.org/10.15698/mic2017.11.597.

Qing, G., Li, B., Vu, A., Skuli, N., Walton, Z.E., Liu, X., Mayes, P.A., Wise, D.R., Thompson, C.B., Maris, J.M., Hogarty, M.D., Simon, M.C., 2012. ATF4 regulates MYC-mediated neuroblastoma cell death upon glutamine deprivation. Cancer Cell 22, 631–644. https://doi.org/10.1016/j.ccr.2012.09.021.

Qiu, J., Yoon, J.-H., Shen, B., 2005. Search for apoptotic nucleases in yeast: role of Tat-D nuclease in apoptotic DNA degradation. J. Biol. Chem. 280, 15370–15379. https://doi.org/10.1074/jbc.M413547200.

Renehan, A.G., 2001. What is apoptosis, and why is it important? BMJ 322, 1536–1538. https://doi.org/10.1136/bmj.322.7301.1536.

Ribeiro, G.F., Côrte-Real, M., Johansson, B., 2006. Characterization of DNA damage in yeast apoptosis induced by hydrogen peroxide, acetic acid, and hyperosmotic shock. Mol. Biol. Cell 17, 4584–4591. https://doi.org/10.1091/mbc.E06-05-0475.

Rinnerthaler, M., Büttner, S., Laun, P., Heeren, G., Felder, T.K., Klinger, H., Weinberger, M., Stolze, K., Grousl, T., Hasek, J., Benada, O., Frydlova, I., Klocker, A., Simon-Nobbe, B., Jansko, B., Breitenbach-Koller, H., Eisenberg, T., Gourlay, C.W., Madeo, F., Burhans, W.C., Breitenbach, M., 2012. Yno1p/Aim14p, a NADPH-oxidase ortholog, controls extramitochondrial reactive oxygen species generation, apoptosis, and actin cable formation in yeast. Proc. Natl. Acad. Sci. U. S. A. 109, 8658–8663. https://doi.org/10.1073/pnas.1201629109.

Rockenfeller, P., Gourlay, C.W., 2018. Lipotoxicty in yeast: a focus on plasma membrane signalling and membrane contact sites. FEMS Yeast Res. 18, foy034. https://doi.org/10.1093/femsyr/foy034.

Rockenfeller, P., Smolnig, M., Diessl, J., Bashir, M., Schmiedhofer, V., Knittelfelder, O., Ring, J., Franz, J., Foessl, I., Khan, M.J., Rost, R., Graier, W.F., Kroemer, G., Zimmermann, A., Carmona-Gutierrez, D., Eisenberg, T., Büttner, S., Sigrist, S.J., Kühnlein, R.P., Kohlwein, S.D., Gourlay, C.W., Madeo, F., 2018. Diacylglycerol triggers Rim101 pathway–dependent necrosis in yeast: a model for lipotoxicity. Cell Death Differ. 25, 767–783. https://doi.org/10.1038/s41418-017-0014-2.

Rodrigues-Pousada, C., Devaux, F., Caetano, S.M., Pimentel, C., da Silva, S., Cordeiro, A.C., Amaral, C., 2019. Yeast AP-1 like transcription factors (Yap) and stress response: a current overview. Microb. Cell 6, 267–285. https://doi.org/10.15698/mic2019.06.679.

Sanz, A., García, R., Rodríguez-Peña, J., Arroyo, J., 2017. The CWI pathway: regulation of the transcriptional adaptive response to cell wall stress in yeast. J. Fungi 4, 1. https://doi.org/10.3390/jof4010001.

Scaffidi, P., Misteli, T., Bianchi, M.E., 2002. Release of chromatin protein HMGB1 by necrotic cells triggers inflammation. Nature 418, 191–195. https://doi.org/10.1038/nature00858.

Severin, F.F., Hyman, A.A., 2002. Pheromone induces programmed cell death in S. cerevisiae. Curr. Biol. 12, R233–R235. https://doi.org/10.1016/S0960-9822(02)00776-5.

Shapira, M., Segal, E., Botstein, D., 2004. Disruption of yeast forkhead-associated cell cycle transcription by oxidative stress. Mol. Biol. Cell 15, 5659–5669. https://doi.org/10.1091/mbc.e04-04-0340.

Shirogane, T., Fukada, T., Muller, J.M.M., Shima, D.T., Hibi, M., Hirano, T., 1999. Synergistic roles for Pim-1 and c-Myc in STAT3-mediated cell cycle progression and antiapoptosis. Immunity 11, 709–719. https://doi.org/10.1016/S1074-7613(00)80145-4.

Silva, A., Almeida, B., Sampaio-Marques, B., Reis, M.I.R., Ohlmeier, S., Rodrigues, F., do Vale, A., Ludovico, P., 2011. Glyceraldehyde-3-phosphate dehydrogenase (GAPDH) is a specific substrate of yeast metacaspase. Biochim. Biophys. Acta, Mol. Cell Res. 1813, 2044–2049. https://doi.org/10.1016/j.bbamcr.2011.09.010.

Singh, K., Lee, M.E., Entezari, M., Jung, C.-H., Kim, Y., Park, Y., Fioretti, J.D., Huh, W.-K., Park, H.-O., Kang, P.J., 2019. Genome-wide studies of Rho5-interacting proteins that are involved in oxidant-induced cell death in dudding yeast. G3 Bethesda 9, 921–931. https://doi.org/10.1534/g3.118.200887.

Strich, R., 2015. Programmed cell death initiation and execution in budding yeast. Genetics 200, 1003–1014. https://doi.org/10.1534/genetics.115.179150.

Strobel, I., Osiewacz, H.D., 2013. Poly(ADP-ribose) polymerase is a substrate recognized by two metacaspases of Podospora anserina. Eukaryot. Cell 12, 900–912. https://doi.org/10.1128/EC.00337-12.

Sundström, J.F., Vaculova, A., Smertenko, A.P., Savenkov, E.I., Golovko, A., Minina, E., Tiwari, B.S., Rodriguez-Nieto, S., Zamyatnin, A.A., Välineva, T., Saarikettu, J., Frilander, M.J., Suarez, M.F., Zavialov, A., Ståhl, U., Hussey, P.J., Silvennoinen, O., Sundberg, E., Zhivotovsky, B., Bozhkov, P.V., 2009. Tudor staphylococcal nuclease is an evolutionarily conserved component of the programmed cell death degradome. Nat. Cell Biol. 11, 1347–1354. https://doi.org/10.1038/ncb1979.

Tang, Y., Gao, X.-D., Wang, Y., Yuan, B.-F., Feng, Y.-Q., 2012. Widespread existence of cytosine methylation in yeast DNA measured by gas chromatography/mass spectrometry. Anal. Chem. 84, 7249–7255. https://doi.org/10.1021/ac301727c.

Teng, X., Cheng, W.-C., Qi, B., Yu, T.-X., Ramachandran, K., Boersma, M.D., Hattier, T., Lehmann, P.V., Pineda, F.J., Hardwick, J.M., 2011. Gene-dependent cell death in yeast. Cell Death Dis. 2, e188. https://doi.org/10.1038/cddis.2011.72.

Thellmann, M., Hatzold, J., Conradt, B., 2003. The Snail-like CES-1 protein of C. elegans can block the expression of the BH3-only cell-death activator gene egl-1 by antagonizing the function of bHLH proteins. Dev. Camb. Engl. 130, 4057–4071. https://doi.org/10.1242/dev.00597.

Tompa, R., Madhani, H.D., 2007. Histone H3 lysine 36 methylation antagonizes silencing in *Saccharomyces cerevisiae* independently of the Rpd3S histone deacetylase complex. Genetics 175, 585–593. https://doi.org/10.1534/genetics.106.067751.

Travers, K.J., Patil, C.K., Wodicka, L., Lockhart, D.J., Weissman, J.S., Walter, P., 2000. Functional and genomic analyses reveal an essential coordination between the unfolded protein response and ER-associated degradation. Cell 101, 249–258. https://doi.org/10.1016/s0092-8674(00)80835-1.

Trindade, D., Pereira, C., Chaves, S., Manon, S., Corte-Real, M., Joao Sousa, M., 2016. VDAC regulates AAC-mediated apoptosis and cytochrome c release in yeast. Microb. Cell 3, 500–510. https://doi.org/10.15698/mic2016.10.533.

Tsiatsiani, L., Timmerman, E., De Bock, P.-J., Vercammen, D., Stael, S., van de Cotte, B., Staes, A., Goethals, M., Beunens, T., Van Damme, P., Gevaert, K., Van Breusegem, F., 2013. The *Arabidopsis* metacaspase9 degradome. Plant Cell 25, 2831–2847. https://doi.org/10.1105/tpc.113.115287.

Váchová, L., Palková, Z., 2018. How structured yeast multicellular communities live, age and die? FEMS Yeast Res. 18, foy033. https://doi.org/10.1093/femsyr/foy033.

Vendrell, A., Martínez-Pastor, M., González-Novo, A., Pascual-Ahuir, A., Sinclair, D.A., Proft, M., Posas, F., 2011. Sir2 histone deacetylase prevents programmed cell death caused by sustained activation of the Hog1 stress-activated protein kinase. EMBO Rep. 12, 1062–1068. https://doi.org/10.1038/embor.2011.154.

Walter, D., Wissing, S., Madeo, F., Fahrenkrog, B., 2006. The inhibitor-of-apoptosis protein Bir1p protects against apoptosis in S. cerevisiae and is a substrate for the yeast homologue of Omi/HtrA2. J. Cell Sci. 119, 1843–1851. https://doi.org/10.1242/jcs.02902.

Walter, D., Matter, A., Fahrenkrog, B., 2010. Bre1p-mediated histone H2B ubiquitylation regulates apoptosis in Saccharomyces cerevisiae. J. Cell Sci. 123, 1931–1939. https://doi.org/10.1242/jcs.065938.

Walter, D., Matter, A., Fahrenkrog, B., 2014. Loss of histone H3 methylation at lysine 4 triggers apoptosis in Saccharomyces cerevisiae. PLoS Genet. 10, e1004095. https://doi.org/10.1371/journal.pgen.1004095.

Wang, M., Collins, R.N., 2014. A lysine deacetylase Hos3 is targeted to the bud neck and involved in the spindle position checkpoint. Mol. Biol. Cell 25, 2720–2734. https://doi.org/10.1091/mbc.e13-10-0619.

Wen, W., Zhu, F., Zhang, J., Keum, Y.-S., Zykova, T., Yao, K., Peng, C., Zheng, D., Cho, Y.-Y., Ma, W., Bode, A.M., Dong, Z., 2010. MST1 promotes apoptosis through phosphorylation of histone H2AX. J. Biol. Chem. 285, 39108–39116. https://doi.org/10.1074/jbc.M110.151753.

White, K., Grether, M.E., Abrams, J.M., Young, L., Farrell, K., Steller, H., 1994. Genetic control of programmed cell death in Drosophila. Science 264, 677–683. https://doi.org/10.1126/science.8171319.

Willis, S.D., Stieg, D.C., Ong, K.L., Shah, R., Strich, A.K., Grose, J.H., Cooper, K.F., 2018. Snf1 cooperates with the CWI MAPK pathway to mediate the degradation of Med13 following oxidative stress. Microb. Cell 5, 357–370. https://doi.org/10.15698/mic2018.08.641.

Wissing, S., Ludovico, P., Herker, E., Büttner, S., Engelhardt, S.M., Decker, T., Link, A., Proksch, A., Rodrigues, F., Corte-Real, M., Fröhlich, K.-U., Manns, J., Candé, C., Sigrist, S.J., Kroemer, G., Madeo, F., 2004. An AIF orthologue regulates apoptosis in yeast. J. Cell Biol. 166, 969–974. https://doi.org/10.1083/jcb.200404138.

Wong, C.-H., Chan, H., Ho, C.-Y., Lai, S.-K., Chan, K.-S., Koh, C.-G., Li, H.-Y., 2009. Apoptotic histone modification inhibits nuclear transport by regulating RCC1. Nat. Cell Biol. 11, 36–45. https://doi.org/10.1038/ncb1810.

Wong, A.H.-H., Yan, C., Shi, Y., 2012. Crystal structure of the yeast metacaspase Yca1. J. Biol. Chem. 287, 29251–29259. https://doi.org/10.1074/jbc.M112.381806.

Wysocki, R., Kron, S.J., 2004. Yeast cell death during DNA damage arrest is independent of caspase or reactive oxygen species. J. Cell Biol. 166, 311–316. https://doi.org/10.1083/jcb.200405016.

Xu, Q., Reed, J.C., 1998. Bax inhibitor-1, a mammalian apoptosis suppressor identified by functional screening in yeast. Mol. Cell 1, 337–346.

Yang, H., Ren, Q., Zhang, Z., 2008. Cleavage of Mcd1 by caspase-like protease Esp1 promotes apoptosis in budding yeast. Mol. Biol. Cell 19, 2127–2134. https://doi.org/10.1091/mbc.E07-11-1113.

Zhang, N.-N., Dudgeon, D.D., Paliwal, S., Levchenko, A., Grote, E., Cunningham, K.W., 2006. Multiple signaling pathways regulate yeast cell death during the response to mating pheromones. Mol. Biol. Cell 17, 3409–3422. https://doi.org/10.1091/mbc.e06-03-0177.

Zhang, H., Alsaleh, G., Feltham, J., Sun, Y., Napolitano, G., Riffelmacher, T., Charles, P., Frau, L., Hublitz, P., Yu, Z., Mohammed, S., Ballabio, A., Balabanov, S., Mellor, J., Simon, A.K., 2019. Polyamines control eIF5A hypusination, TFEB translation, and autophagy to reverse B cell senescence. Mol. Cell 76, 110–125.e9. https://doi.org/10.1016/j.molcel.2019.08.005.

CHAPTER THREE

Cell death in the avian brain with emphasis on the development and plasticity of the song control system

Tracy A. Larson*
Department of Biology, University of Virginia, Charlottesville, VA, United States
*Corresponding author: e-mail address: larson.tracy@virginia.edu

Contents

Abstract

Programmed cell death is a fundamental feature of brain development, homeostasis, and adult plasticity. One model system, in which the role of cell death in establishment, maintenance and plasticity of neural tissues is evident throughout both early development and in the adult, is the neural circuitry underlying the learning and production of singing behavior in songbirds. The dramatic sexual dimorphism and natural, cyclical growth and regression of the song control system provides a useful environment for studying programmed cell death. Especially valuable and unique to songbirds, the occurrence of cell death in the song control system is correlated to quantifiable changes in a biologically relevant and learned sensorimotor behavior—that is singing. Within this review I explore the topic of cell death in the avian brain primarily within the context of the song circuits. I first establish why songbirds are a useful model for studying cell death and provide a brief overview of the organization of the circuitry underlying song learning and production. I then discuss the processes and mechanisms of cell death during early development and sexual differentiation of the song control system. I present the classic and recent work exploring cell death in the adult avian brain by covering topics of homeostasis and neuronal turnover, seasonal plasticity, and neural

International Review of Cell and Molecular Biology, Volume 352
ISSN 1937-6448
https://doi.org/10.1016/bs.ircmb.2020.01.001

injury and insult. Finally, I propose several outstanding questions in the field of cell death biology in the avian brain, which when addressed have great potential to provide unique insight into the role of cell death in the organization and maintenance of neural tissues, the plasticity of developmentally organized neural circuits in the adult, and the mechanisms underlying functional recovery from both natural and injury-induced neurodegeneration.

1. Introduction

Cell death is critical for proper development of neural tissues and for subsequent homeostasis and plasticity in the adult central nervous system. During development, the programmed cell death of neurons that where produced in overabundance facilitates proper circuit refinement and function. In the adult central nervous system, cell death was initially assumed to be limited to disease and damage until the discovery that new neurons are born and functionally incorporated in the adult brain across vertebrate species (Altman and Das, 1965; Kaplan and Hinds, 1977; Paton and Nottebohm, 1984). The discovery of functional adult neurogenesis provided the first evidence that cell death in the adult brain does not always negatively impact neural function. In fact, neuronal turnover, or replacement of old neurons with adult-born neurons, likely contributes to the maintenance of healthy, efficient neural circuits through the replacement of weak or damaged neurons (reviewed in (Brenowitz and Larson, 2015)). The birth and death of neurons in the adult brain also supports natural plasticity of neuronal number and circuit architecture. Perhaps one of the most striking examples in which neuronal programmed cell death occurs not only during developmental circuit refinement, but also throughout adulthood for circuit maintenance and plasticity is that which occurs within the neural circuits underlying singing behavior in songbirds.

Herein, I aim to provide a comprehensive view of the processes and mechanisms of programmed cell death in both the developing and adult avian brain. After introducing the neural circuits that are responsible for the audition, learning, and production of songs, I detail the involvement of cell death in establishing these circuits. I then present the foundational work in songbirds, in addition to recent discoveries, that led to the general hypothesis that adult neurogenesis, or the birth of new neurons in the adult brain, is functionally linked to cell death. Under the umbrella of this hypothesis I will cover topics including homeostatic turnover of neurons, seasonal plasticity, and selective neuronal-type cell death. Much, but certainly not all, of the work in the area of cell death that I will discuss has focused at the phenomena level as opposed to mechanistic level. Until recently, the songbird model has lacked the genetic and genomic tools necessary delve into the cellular and molecular interactions underlying cell death in the song control system. I will present an update on the recent emergence of a suite of tools that will markedly facilitate the mechanistic examination of several standing, important questions regarding programmed cell death in the song control system.

2. Why study cell death in songbirds?

The birdsong system offers several advantages as a model for studying neuronal death and replacement by new neurons:

(1) Unlike mammals, in which cell death in the brain is primarily limited to early development, damage and disease, cell death in the brain of adult song birds occurs naturally and cyclically within the neural circuits controlling singing behavior. The circuits that are responsible for the learning and production of song not only undergo dramatic morphological changes in cell number and connectivity during development, but also exhibit profound changes in cell death and neurogenesis during each annual breeding season. In both wild and domesticated seasonally breeding songbirds, nearly 30% of the projection neurons in the sensorimotor nucleus that integrates auditory input with motor output of song, called HVC (proper name; (Reiner et al., 2004)), are added anew and die via caspase-dependent mechanisms each breeding and nonreeding season, respectively (Kirn and Nottebohm, 1993; Thompson and Brenowitz, 2008; Tramontin and Brenowitz, 1999).

(2) Cell death in the adult brain of songbirds is robust during both regression of the song control circuits and periods of stable homeostasis. In the songbird white-crowned sparrow (*Zonotrichia leucophrys*), the 30% of projection neurons that die back each non-breeding season account for over 50,000 total neurons that die over the course of just a few days (Thompson et al., 2007). By as soon as 1 day following transition into non-breeding conditions, apoptosis is quantifiably elevated under laboratory conditions (Larson et al., 2014). Cell death within the sensorimotor nucleus HVC peaks around 4 days and returns to baseline levels by 7 days (Larson et al., 2014). Once HVC neuronal number returns to baseline numbers, neuronal turnover continues at quantifiable levels (Larson et al., 2019).

(3) Cell death is causally linked to neurogenesis. As HVC regresses in neuronal number, proliferation of cells in the nearby neural stem cell niche increases (Larson et al., 2014). This increase in stem cell proliferation is dependent on cell death in HVC (Larson et al., 2014) and presumably contributes to an increase in newly added neurons to HVC following seasonal regression (Thompson and Brenowitz, 2008). Likewise, during periods of stable neurogenesis, neuronal death and replacement in the adult avian brain continue at levels substantially higher than those seen in mammalian brains. In songbirds an estimated 0.1–0.7% of all HVC neurons turnover (i.e., die and are replaced by new neurons) each day (Gahr et al., 2002; Vellema et al., 2010), whereas in mammals approximately 0.02% of hippocampal neurons are generated each day (Jabes et al., 2010).

(4) Cell death in the song system correlates to changes in a biologically relevant behavior—singing (Catchpole and Slater, 1995). As neurons die in HVC, seasonally breeding birds discontinue singing highly stereotyped song and nearly discontinue singing even "low quality" song altogether (Kirn et al., 1994; Larson et al., 2014).

(5) Rich species diversity within the order of Passerine perching birds and suborder of Oscine songbirds provides a unique opportunity to test both ultimate and proximate mechanisms of cell death in the contexts of development, homeostasis, and adult neural plasticity. The approximate 10,000 species of extant Aves, 6600 species of Passerines, and 4000 species songbirds exhibit an immense variety of innate and learned behavior. Thus, putative and established diversity in neural organization and neural resource allocation provides opportunities to utilize comparative studies across bird species for exploring outstanding questions in the area of cell death biology.

3. The song system

Vocal learning, and in the case of Passerine Oscine birds, singing, is a complex behavior unique to just a few vertebrates including humans, cetaceans, bats, parrots, and hummingbirds. Just as with human vocal learning and production, singing in birds relies upon a developmentally programed neural circuit that is capable of a high degree of developmental and adult plasticity, auditory perception, and social context for performing the behavior (Wilbrecht and Nottebohm, 2003). In songbirds, singing is primarily used to attract mates, defend territories and for aggregating in flocks (Catchpole and Slater, 1995).

Although the schedule of song learning can be quite variable in length and frequency across species of songbirds, all songbirds learn to sing their song through a common series of processes. Songbirds begin vocal learning by listening to and memorizing one or more tutor song(s)—most typically those of their father or neighboring birds (reviewed in (Mooney, 2009; Woolley, 2012)). After this sensory phase, young songbirds enter the sensorimotor phase of learning (Mooney, 2009), in which the birds, or pupils, "babble" or produce highly variable and poorly structured "subsong" (reviewed in (Mooney, 2009; Wilbrecht and Nottebohm, 2003)). For weeks to months, depending on the species, the pupil utilizes auditory feedback and error correction to gradually modify its motor output until the "plastic" song nearly matches the memorized tutor song in structure. Once the pupil has achieved stereotypy of its song(s), the song(s) becomes "crystalized" or fixed in spectral structure and syllable content. In all songbirds, each phase of learning occurs during sensitive periods; however, these sensitive periods vary in timing and duration across songbird species. Many species engage in song learning just once prior to its first breeding season. These birds are referred to as close-ended learners. Alternatively, other species including the canary (*Serinus canaria domestica*; (Nottebohm et al., 1986)) and the European starling (*Sturnus vulgaris*; (Van Hout et al., 2009)), among others, are open-ended learners and re-enter the sensory and sensorimotor phases of learning prior to each breeding season.

The neural circuits underlying the acquisition and production of learned song are comprised of several discrete, interconnected brain regions, or nuclei, organized into three major pathways. These pathways include the (1) ascending auditory pathway, which allows for the perception of conspecific (same species) song, heterospecific (different species) song, and even a bird's own song (Mooney et al., 2001), and the (2) song control pathway (Fig. 1). The song control pathway is further delineated into the (2a) anterior forebrain pathway, which is extensively involved in the sensorimotor stage of song learning, and the (2b) descending motor pathway, which precisely encodes the motor pattern of song output (reviewed in (Fee et al., 2004; Mooney, 2009); Fig. 1). All three pathways converge on the telencephalon sensorimotor nucleus HVC.

Auditory information integrates with the song control pathway in HVC (Bauer et al., 2008; Fortune and Margoliash, 1995) after ascending through a

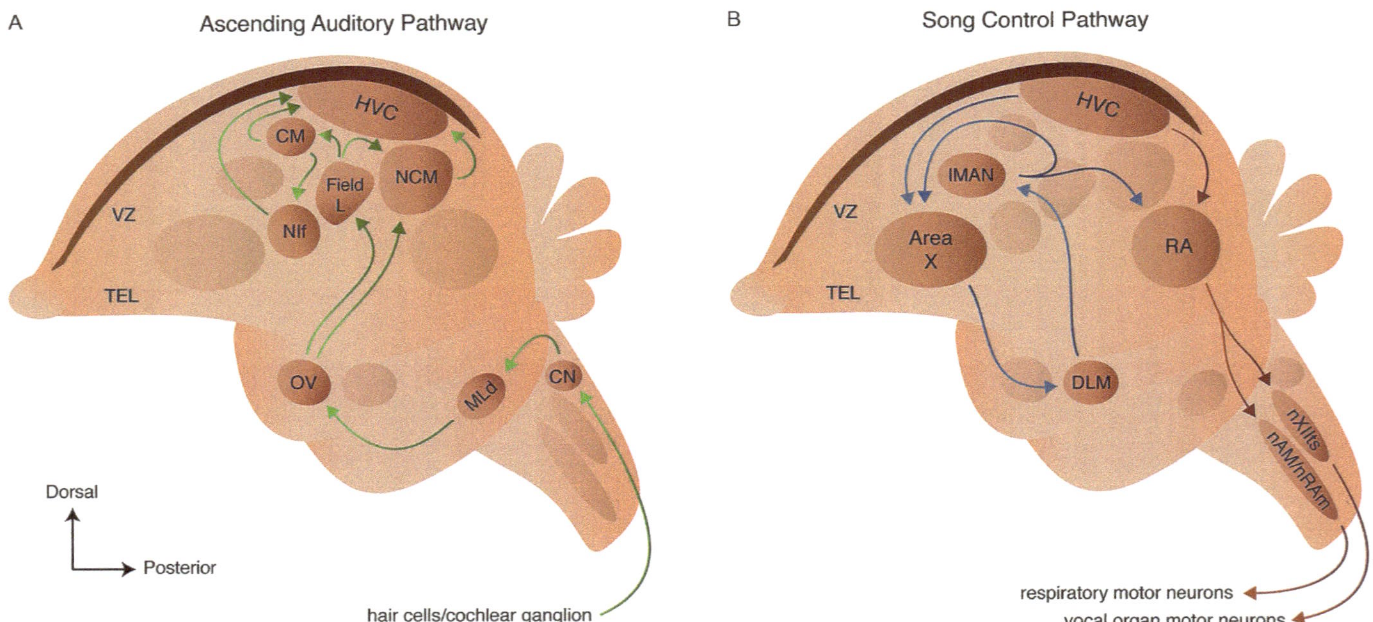

Fig. 1 Simplified schematics of the song control system. (A) The nuclei and projections (green arrows) of the ascending auditory pathway. (B) The nuclei and projections of the anterior forebrain pathway (blue arrows) and the descending motor pathway (red arrows). Faded nuclei in both A and B represent the major nuclei of the alternative pathway for ease of reference to relative location across pathways.

series of auditory regions beginning with the eighth cranial nerve (Fig. 1A). Auditory input continues through several mid- and hindbrain auditory nuclei, which ultimately terminate on neurons to the thalamic nucleus ovoidalis (Ov). Auditory information from Ov is sent to Field L of the telencephalon and then on to secondary auditory regions including the caudal medial nidopallium (NCM), the caudal medial mesopallium (CM) and the interfacial nucleus of nidopallium (Nif). Neurons of CM and Nif project into HVC to complete the ascending auditory pathway (reviewed in (Mooney, 2009)).

From HVC the anterior forebrain pathway and descending motor pathway diverge with HVC neurons projecting to Area X and the robust nucleus of archopallium (RA), respectively (Fig. 1B; (Fortune and Margoliash, 1995; Kirn et al., 1991)). Area X of the anterior forebrain pathway receives input from HVC, and in turn projects to the dorsal lateral nucleus of the medial thalamus (DLM). DLM sends axonal projections to the lateral magnocellular nucleus of the anterior nidopallium (lMAN). LMAN projection neurons are bifurcating, with one bifurcation terminating in RA and the other back in Area X to complete the recurrent anterior forebrain pathway loop (Vates and Nottebohm, 1995). In addition to the indirect projections from HVC to RA through the anterior forebrain pathway, RA also receives over half of all HVC neuronal projections directly as part of the descending motor pathway (Fig. 1B; (Kirn et al., 1991)). To complete the motor pathway, RA sends projections to brain stem regions including the tracheosyringeal portion of the hypoglossal nucleus (nXIIts) and respiratory premotor neurons (Mooney, 2009; Nottebohm et al., 1976; Wild, 1993).

The distinct functions of the anterior forebrain and descending motor pathway nuclei have been identified using lesion studies. In the zebra finch (*Taeniopygia guttata*) lesions in the anterior forebrain pathway, specifically lMAN and RA, before post-hatch day 45, disrupt juvenile subsong, suggesting that early song learning and subsong are controlled by the anterior forebrain pathway (Aronov et al., 2008; Bottjer et al., 1984). Alternatively, lesions to HVC before post-hatch day 45 have little effect on subsong, but after day 45 severely impair song crystallization and production (Aronov et al., 2008). These results indicate that subsong is dependent on activity in lMAN and RA, but not HVC, whereas crystalized song relies upon HVC and RA.

4. Early development of the song system

Developmental neurogenesis, herein distinguished from adult neurogenesis as the birth and addition of neurons in the pre-sexually mature bird, in nuclei of the song circuits generally coincides with the timing of neuronal addition in the surrounding brain regions. Thus, by hatching the majority of neurons comprising thalamic and archopallial regions including RA, lMAN, and DLM have been born (reviewed in (Thompson, 2011)). The striatum and nidopallium continue to increase in neuronal number after hatching, and so too do the song nuclei of these regions, specifically HVC and Area X (Alvarez-Buylla et al., 1992).

In zebra finches by as early as 12 days post-hatch, many of the song control nuclei begin to diverge between male and female songbirds, with male songbirds typically exhibiting greater neuronal number, degree of connectivity, and total volume (Bottjer et al., 1985). Interestingly, the emergence of sexual dimorphism is not only due to growth of the male circuitry, but also degeneration of the female circuitry. In male zebra finches, HVC and RA volumes increase 147% and 46%, respectively, between post-hatch days 12 and 25 (Bottjer et al., 1985). Between days 25 and 53 post-hatch, HVC and RA volumes of male zebra finches increase another 27% and 78%, respectively (Bottjer et al., 1985). HVC growth during this time results from prolonged developmental neurogenesis, with approximately 200,000 neurons being added between days 15 and 45 to an existing approximate 250,000 neurons (Kirn and DeVoogd, 1989). The increase in volume of RA in male zebra finches subsequent to day 25 is not due increased neuronal addition, as RA neuronal number stabilizes at between days 10 and 25 at nearly 15,000 neurons (Kirn and DeVoogd, 1989). Instead, male RA growth is due to increased soma size and spacing of neurons (Kirn and DeVoogd, 1989; Konishi and Akutagawa, 1990). Alternatively, in females HVC neuron number nearly halves from approximately 200,000 to 100,000 neurons likely via apoptotic mechanisms between post-hatch days 15 and 20 (Kirn and DeVoogd, 1989). Reduced RA volume in females result from apoptosis of nearly 40% of RA neurons after 20 days post-hatch in addition to lower neuronal densities and soma size when compared to males (Kirn and DeVoogd, 1989; Konishi and Akutagawa, 1985).

The nuclei of the anterior forebrain pathway also become sexual dimorphic. Area X in both male and female zebra finches is barely to non-detectable up to 12 days post-hatch using stereological techniques (Bottjer et al., 1985; Johnson and Bottjer, 1992). Between days 25 and 53 post-hatch, juvenile male zebra finches increase the volume of Area X by 47% (Bottjer et al., 1985; Johnson and Bottjer, 1992). Neuronal loss, specifically the appearance of pyknotic cells, in Area X of males peaks between post-hatch days 10 and 25 (Kirn and DeVoogd, 1989). Degeneration of Area X in females, however, has received little attention, as Area X in females fails to develop to a detectable degree. To this end, cell death in the dorsolateral zone in lobus parolfactorius (DL-LPO), the region that contains Area X of both males and females, exhibits a similar temporal pattern of cell death in females as in male (DL-LPO) and male Area X (Kirn and DeVoogd, 1989).

Also in the anterior forebrain pathway, lMAN has been reported to nearly halve in volume in both male and female zebra finches due to neuronal loss and refinement of projections from RA (Bottjer, 2004; Bottjer et al., 1985; Herrmann and Arnold, 1991). The degree of lMAN regression in males is debated, however. Using stereological techniques, studies in zebra finches report a 59% reduction in volume and neuron number in lMAN between 12 and 53 days post-hatch (Bottjer et al., 1985). Alternatively, using retrograde backfills of lMAN neurons projecting to RA studies report that lMAN to RA projection neuron number does not change after 25 days post-hatch in male zebra finches (Nordeen et al., 1992).

Lesions to lMAN of male zebra finches at 20 days post-hatch result in the loss of nearly 40% of RA neurons, indicating afferent input from lMAN is required for RA neuron survival (Johnson and Bottjer, 1994). This result further suggests that a reduction in input from afferent targets, not limited to the lMAN to RA projection, and mechanisms of Hebbian competition (Hebb et al., 1994) might influence differential neuronal survival in males and females during development of the song system. In males functional connectivity of developmentally-generated neurons generally coincides with usage of given nucleus of the song control system for learning and production of song. HVC neuronal projections to Area X along with many of the projections in the anterior forebrain pathway between Area X, DLM, lMAN and RA form their respective connections by 15–20 days post-hatch in the zebra finch, presumably to facilitate the sensory learning of the song template (Johnson and Bottjer, 1992). Seventy-six percent of HVC neurons projecting to Area X are born *in ovo* around day 9, whereas only 0.8% of HVC to RA projection neurons are born at this time (Alvarez-Buylla et al., 1988). By post-hatch day 20, the majority of HVC neurons projecting to Area X have formed their synapses (Mooney and Rao, 1994). Alternatively, HVC neurons synapsing in RA (i.e., the descending motor pathway necessary for crystalized song production; (Johnson and Bottjer, 1994)) do not establish connections in RA until 30–35 days post-hatch (Konishi and Akutagawa, 1985; Mooney and Rao, 1994).

In songbirds sexual differentiation of the song control system is not only determined by genetic sex and resulting gene expression, but also hormone levels, specifically sex steroids, during development. Sexual differentiation of the song control system depends on localization of sex steroid synthesis, levels of sex steroids, and critical windows for localization and levels (Arnold and Itoh, 2011; Maekawa et al., 2014). The first evidence that sex steroids influenced masculinization of the song circuits emerged from studies that manipulated levels of plasma estradiol, a metabolite of testosterone (Fig. 2 for sex steroid synthesis), with hormonal agonists and antagonists. Supplementation of female zebra finches with estradiol for 28 days beginning immediately after hatching masculinized the female brain and behavior: the volume of HVC and RA in females administered estradiol was significantly larger than females administered an androgen metabolite of testosterone, dihydrotestosterone (DHT), alone or no hormone (Gurney and Konishi, 1980). Moreover, unlike non-masculinized females, females treated with estradiol sang (Gurney and Konishi, 1980). Subsequently, the critical window for estradiol supplementation to impact

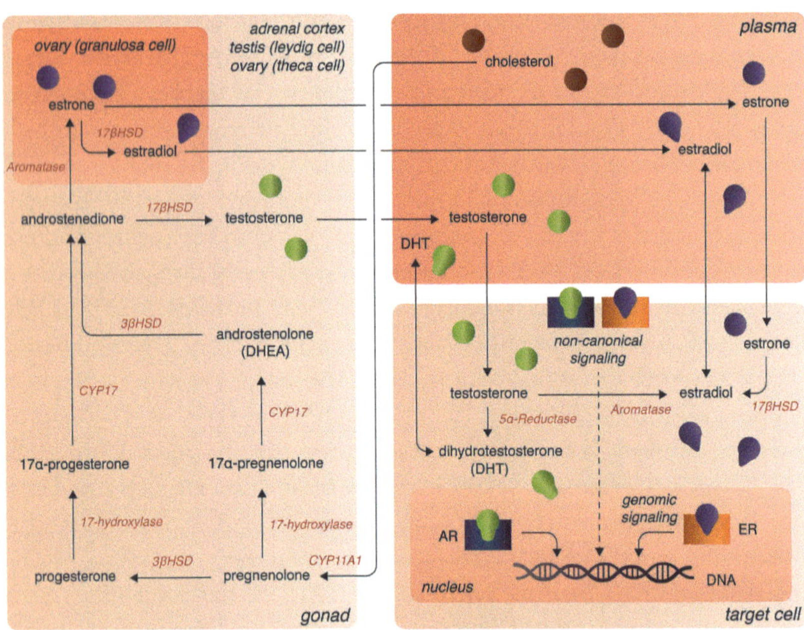

Fig. 2 The biosynthesis and localization of sex steroids and their synthetic enzymes (red italics). In the case of de novo sex steroid synthesis in the avian brain, the "target cell" would be a neuron or glial cell within the song control nuclei. *Modified from Larson, T.A., 2018. Sex steroids, adult neurogenesis, and inflammation in CNS homeostasis, degeneration, and repair. Front. Endocrinol. (Lausanne) 9, 205.*

sexual differentiation of the song control system was determined to be within 1 week of hatching (Adkins-Regan et al., 1994).

Reduction of estradiol synthesis with inhibition of aromatase, and estrogen synthase (Fig. 2), results in the development of testes in genetically female zebra finches, but does not induce a masculine song system even in the presence of testes (Springer and Wade, 1997). Likewise, castration of young male zebra finches at 9 days post-hatch does not alter song development (Arnold, 1975). The lack of sufficiency and necessity of testes development in genetic females and males, respectively, for the masculinization of the song circuits and singing behavior suggests that gonadal secretion of testosterone and its metabolites are not necessary for sexual differentiation of the brain.

In fact, peripheral levels of sex steroids, specifically those synthesized by the gonads, do not always correlate with sex steroid levels in the brain—sex steroids can be synthesized de novo in the brain (Fusani et al., 2000; Remage-Healey et al., 2008; Schlinger and Arnold, 1992). Steroidogenic enzymes expressed in the stem cell niche (i.e., the ventricular zone) and within the song control nuclei contribute to local synthesis of sex steroids and influence the differentiation of the song control system independent of gonadal sex steroids. More specifically, the genes necessary for de novo synthesis of androgens, progestins, and estrogens, including, but not limited to cytochrome P450 17α-hydroxylase 17,20-lyase (CYP17) and aromatase (Fig. 2), are present throughout the telencephalon, hypothalamus, thalamus, and midbrain between 1 and 5 days post-hatch of zebra finches (Jacobs et al., 1999; London and Schlinger, 2007; London et al., 2003; Perlman and Arnold, 2003). Likewise, the major receptors for DHT and E2, androgen receptor and estrogen receptors alpha and beta, respectively, are expressed as early as embryonic days 30–34 in the telencephalon of zebra finches (Perlman and Arnold, 2003; Perlman et al., 2003).

Conclusive evidence that sex steroid synthesis de novo in the brain impacts sexual differentiation comes from studies that supplement sex steroids locally within HVC of females and found localized sex steroids drive masculinization. In 10–13 days old female zebra finches, micro-implantation of estradiol near HVC increases the volumes and neuronal soma size in HVC and RA after 50 days (Grisham et al., 1994). Control female finches that have estradiol pellets implanted under the skin of the breast or in the peritoneal cavity retain feminized song control nuclei (Grisham et al., 1994), suggesting that locally synthesized estradiol rather than gonadal derived estradiol masculinizes the song control system. Local synthesis of sex steroids are likely curtailing cell death in the male brain directly through the repression of pro-apoptotic genes via genomic signaling and indirectly through the up regulation of neurotrophins and other pro-neurogenic factors, as discussed briefly below. For more in-depth review of the pro-neurogenic and survival mechanisms of sex steroids in the avian brain see (Brenowitz, 2013; Brenowitz and Larson, 2015; Duncan et al., 2013; Larson, 2018).

It is important to note that most of the work on the emergence of sexual dimorphism in song birds has been performed in species in which only the males sing. Female birds of some species, most numerous in the tropics, do, in fact, sing. As such, differences between males and females in the rate of cell death in the song system might not exist in all species, especially in those in which females sing. Limited evidence suggests, however, that even in species in which adult females sing complex songs, the development of the song system between males and females diverges. In blue-capped cordon bleus (*Uraeginthus cyanocephalus*), a species in which both sexes sing, the volume and neuron number of HVC and RA become higher in males when compared to females as early as 20 days post-hatch and remain elevated throughout adulthood (Lobato et al., 2015). Whether or not all singing female songbirds undergo substantial apoptosis of the song control system, and if so, how singing females compensate for this neuronal loss to learn and produce song, remains an open question.

5. Homeostasis and natural turnover in the song system

As with development, neurogenesis and neuronal death are critically involved in the maintenance of homeostasis within the song control system. In the adult avian brain new neurons are continuously generated from the proliferation of neural stem cells and neural progenitor cells residing primarily in ventricular zone along the ventrolateral wall of the lateral ventricle. The ventricular zone extends anterior to posterior throughout most of the avian telencephalon and directly overlies the song control nucleus HVC. Proliferative cells in the ventricular zone generate both radial glia that likely serve as neural progenitor cells in the avian brain and new neurons (Alvarez-Buylla et al., 2002; Goldman et al., 1996). In the canary brain, the neuroblast progeny depart the ventricular zone 1–4 days after birth and migrate across long distances incorporate as mature neurons over most of the telencephalon (Alvarez-Buylla and Nottebohm, 1988; Vellema et al., 2010). Specifically within HVC, neuroblasts migrate primarily via undirected, tortuous wandering over 1–2 weeks before incorporating into

HVC (Alvarez-Buylla and Nottebohm, 1988; Scott et al., 2012). As early as 8 days, most, if not all, neuroblasts within HVC differentiate into RA projecting neurons (Scott and Lois, 2007; Scotto-Lomassese et al., 2007). By 8 months, at least 75% of newborn neurons project to RA, with the remaining 25% presumably assuming an interneuron fate (Kirn et al., 1999; Paton et al., 1985; Scott and Lois, 2007). However, co-labeling of adult-born neurons with several interneuronal markers including gamma-aminobutyric acid (GABA), parvalbumin, and calcium binding proteins have failed to label new HVC neurons (Scotto-Lomassese et al., 2007). Thus, the question of whether all new HVC neurons project to RA remains unresolved. Notably, no adult-born neurons migrating into HVC send projections to Area X (Scharff et al., 2000). Outside of HVC, new neurons are also added to NCM, Field L, Area X, and lMAN, among other regions (Vellema et al., 2010). However, the majority of work exploring the phenomenon and mechanisms of adult neurogenesis and neuronal turnover in the avian brain has focused on HVC homeostasis and plasticity. As such, subsequent discussion regarding neuronal turnover in the song system will focus on characterization and hypothesis testing in HVC.

Not all adult-born neurons persist to functional incorporation into the existing circuitry. In fact, an estimated 50% of progeny die during migration and nearly another half of all adult-born post-migratory neurons die within 2–3 weeks of their arrival in HVC (Fig. 3; (Brenowitz and Larson, 2015; Kirn et al., 1999)). The new neurons that persist through this initial culling can survive months (Kirn et al., 1991; Nottebohm et al., 1994) to years (Walton et al., 2012), depending on the time of their birth and the physiological state of the bird (discussed below).

As in humans, primates, and rodents (Riddle and Lichtenwalner, 2007), levels of neurogenesis decrease in the aging brain of songbirds. In adult songbirds, the replacement of mature neurons by new neurons begins as

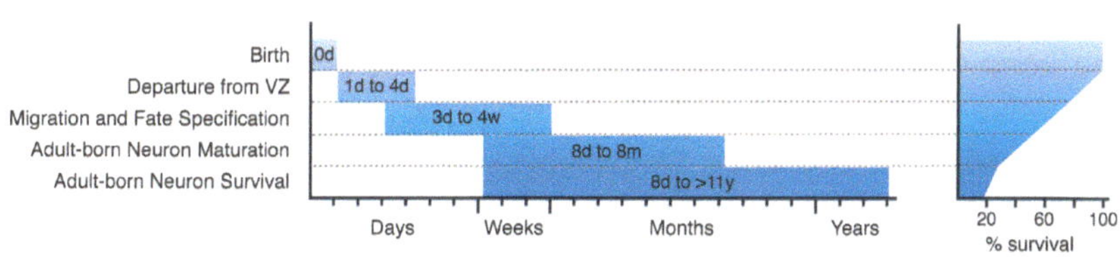

Fig. 3 A timeline of the production of adult-born neurons and their likelihood to survive through the given phase of functional neurogenesis. *Modified from Brenowitz, E.A., Larson, T.A., 2015. Neurogenesis in the adult avian song-control system. Cold Spring Harb. Perspect. Biol. 7(6).*

early as 4 months after hatching (Alvarez-Buylla et al., 1992). After 4 months the rates of neuronal cell death, neural progenitor cell proliferation, neuronal recruitment, and adult-born neuronal survival remain fairly constant to at least post-hatch day 240 (Alvarez-Buylla et al., 1988), but gradually and steadily decline with age (Adar et al., 2008; Larson et al., 2014; Wang et al., 2002; Wilbrecht and Nottebohm, 2004). While neurogenesis occurs at comparatively low levels in aged birds, cell proliferation and neuronal addition have not yet been observed to stop entirely: adult-born neurons have been observed to incorporate into HVC at 11 years of age in zebra finches (Walton et al., 2012).

Neuronal death and replacement by adult-born neurons has been proposed to serve many different functions. In addition to models that posit that new neurons allow for plasticity in size of the song control nuclei to enable seasonal learning and performance of singing behavior (discussed below), neuronal turnover in the adult avian brain has been proposed to enable adult birds to replace neurons damaged by over excitation or weakened by activity-related DNA damage and to maintain song quality (reviewed in (Brenowitz and Larson, 2015)). Although no studies in birds have directly manipulated levels of neurogenesis to determine the impact of adult neurogenesis loss, many of the hypothetical functions of adult-born neurons are supported by correlational observations. For instance, male birds sing at extremely high rates for several months during the breeding season. Underlying this high rate of song production is a breeding-season-dependent elevation of oxidative metabolism via cytochrome oxidase activity in HVC, Area X, and RA neurons (Wennstrom et al., 2001). This higher rate of metabolism likely corresponds to increased neuronal excitatory activity and high gene transcription rate during singing. Elevation in excitatory activity would increase the activation of AMPA/kainate and NMDA-type glutamate receptors present in HVC and RA (Dutar et al., 2000) and could result in glutamatergic-dependent excitotoxic cell death (for an in-depth review of the molecular mechanisms and consequences of excitotoxicity see (Dong et al., 2009)). Likewise, increased metabolic rate and production of reactive oxygen species from elevated rates of transcription could also facilitate the accumulation of double stranded DNA breaks in the post-mitotic cells of HVC, Area X, and RA, as is the case in rat neurons and human lymphocytes (Dickey et al., 2012). Several observations, however, confound the hypotheses that neuronal loss due to excitotocity or double strand DNA breaks results in neuronal addition. Specifically, the neurons of HVC, a highly neurogenic nucleus, do not increase spontaneous firing during the breeding season. Alternatively, the neurons of RA, a nucleus that does not add adult-born neurons, do increase both evoked and spontaneous firing (Fee et al., 2004; Meitzen et al., 2007). Moreover, Area X spiny neurons exhibit song-related firing (Woolley et al., 2014), but neuronal addition in Area X does not differ seasonally (Thompson and Brenowitz, 2005). This evidence suggests that if cell death and replacement occurs in the song system as a result of protection from over excitation or DNA breaks, then neuronal turnover likely serves another function as well.

One such function might be to provide a plastic substrate for maintaining learned song. Maintenance of learned song in adults is a continuous process that requires auditory feedback and error correction to match motor output to the previously learned song. Disruption to auditory feedback, in the form of deafening (Lombardino and Nottebohm, 2000; Nordeen and Nordeen, 1992; Woolley and Rubel, 1997), masking of frequency (Tumer and Brainard, 2007), or delayed auditory feedback (Leonardo and Konishi, 1999) all result in changes to crystalized song. The addition of adult-born neurons to HVC and area X have been proposed to provide the substrate for the plasticity required to fine tune song motor output (Brenowitz and Larson, 2015; Wilbrecht and Kirn, 2004). In this model the death of existing HVC neurons previously programmed to produce aspects of learned song and the addition of naive neurons might allow for the error correction necessary for song maintenance.

6. Natural plasticity of neuronal birth and death in the song system

The overarching hypothesis that neuronal birth in the avian brain is functionally linked to cell death arose primarily from work examining the process and mechanisms of seasonal growth and regression of the descending motor pathway of the song system. Thus, I will primarily discuss the processes and impacts of cell death on neuronal recruitment and singing behavior during seasonal growth and regression of HVC and RA. For an extensive review of the plasticity of the song control system in adults, including the processes, mechanisms and proposed function of adult neurogenesis both within and outside of HVC, refer to reviews by (Ball et al., 2004; Brenowitz, 2008; Brenowitz and Larson, 2015; Wilbrecht and Kirn, 2004).

In nearly all seasonally breeding songbirds examined, the seasonal emergence of breeding singing behavior is accompanied by changes in physiology, morphology and gene expression at both the whole organismal and neural circuit levels (Figs. 4 and 5; (Ball et al., 2004; Brenowitz and Larson, 2015; Tramontin and Brenowitz, 2000)). These dramatic seasonal changes in morphology and behavior are driven primarily by changes in day length and gonadal sex steroid levels. As day length increases (generally to greater than 12h of light each day), gonadal production of testosterone in males increases (Wingfield and Farner, 1978). Growth of the song system occurs rapidly thereafter: in both wild and laboratory-reared white-crowned sparrows, the growth of HVC begins as soon as 1 day following elevation of plasma testosterone levels (Fig. 4; (Smith et al., 1995, 1997)). Other song nuclei including RA, Area X, and nXIIts all increase in volume as well, but with rates seemingly dependent on timing of usage and to varying degrees (Thompson and Brenowitz, 2005; Tramontin et al., 2000). Unlike the majority of other song nuclei, the growth in volume within HVC is due primarily to the addition of over 50,000 adult-born neurons, though the exact number

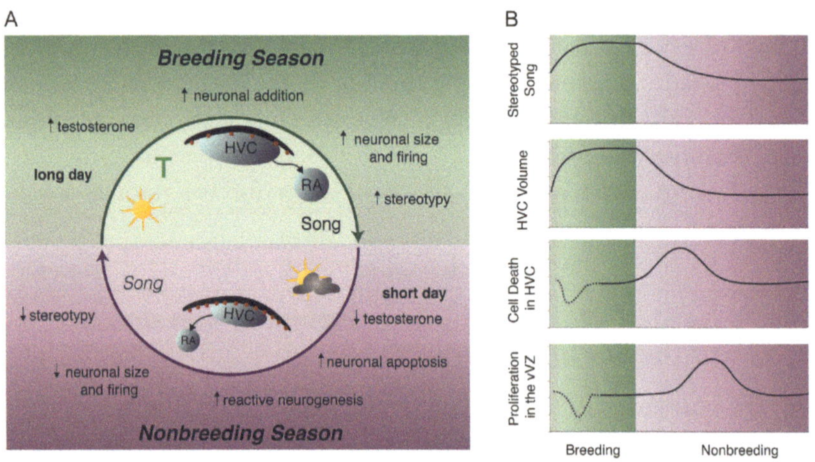

Fig. 4 The seasonal changes in physiology, morphology, and behavior of songbirds. (A) Cyclical representation of the seasonal changes in whole organism physiology and the morphology of the descending motor pathway. At the onset of the breeding season, plasma testosterone (T) levels increase, driving an increase in neuronal number in HVC and increased spontaneous and evoked firing of RA neurons. The changes in morphology and physiology of the descending motor pathway result in the production of stereotyped, breeding condition song. At the onset of the non-breeding season, T levels drop, HVC neurons die, reactive neurogenesis occurs in the nearby ventricular zone (indicated by the red dots), and song degrades in quality. (B) A graphic representation of the changes in singing behavior, HVC morphology and cell death, and proliferation in the nearby ventricular zone (VZ). Solid line represents the percent maximum for the given attribute as determined from existing data, whereas the dotted line represents predicted but un-verified values.

varies depending on the species. In wild-caught song sparrows (*Melospiza melodia*) HVC neuron number increases to approximately 250,000 from 150,000 in the fall breeding season (Smith et al., 1997), whereas HVC increases from around 90,000 to 150,000 neurons in breeding-like conditions in captive white-crowned sparrows (Tramontin et al., 2000; Tramontin and Brenowitz, 1999).

Several key factors promote an increase in survival of an initial cohort of adult-born neurons to populate HVC at breeding season neuronal numbers. Briefly, gonadal and locally synthesized sex steroids promote the survival of an initial wave of new neurons (Larson et al., 2019) through the expression of neurotrophic factors, specifically neurotrophic factor (VEGF) and brain derived neurotrophic factor (BDNF; (Louissaint et al. 2002)). Following aromatization from testosterone, estradiol up regulates endothelial cell expression of VEGF and its receptor, Quek1/VEGF-R2 tyrosine kinase (Figs. 2 and 5; (Hidalgo et al., 1995; Louissaint et al. 2002)). Increase in local VEGF not only facilitates the expansion of the microvasculature of HVC, but also stimulates BDNF production by preexisting and new endothelial cells (Louissaint et al. 2002). In turn, BDNF promotes the survival of mature and nascent neurons within HVC. This pro-survival effect of BDNF on new neurons is precisely timed: endothelial cells increase expression and secretion of BDNF within 1 week following exposure to testosterone (Louissaint et al. 2002; Thompson et al., 2012; Wissman and Brenowitz, 2009), which coincides with the critical window of 14–20 days after the birth of a new neuron for BDNF exposure to promote survival (Rasika et al., 1999). For a more comprehensive review of the interactions between sex steroids and neurotrophins see (Brenowitz, 2013).

As fall approaches, seasonally breeding male songbirds discontinue production of gonadal testosterone (Wingfield and Farner, 1978), aromatase expression in the neostriatum decreases (Fusani et al., 2000), the song system regresses in morphology (Nottebohm, 1981), and birds discontinue singing breeding quality song and sing less often (Larson et al., 2014; Meitzen et al., 2009; Nottebohm et al., 1986). Within 1 day of transition into non-breeding condition (i.e., short photoperiods and low plasma testosterone levels), nearly 30% of neurons in HVC of white-crowned sparrows die via activated-caspase-3 mediated mechanisms (Larson et al., 2014). Dying cells in HVC also express cleaved Poly ADP-ribose polymerase (PARP) as early as 3 days following transition into non-breeding conditions (Thompson and Brenowitz, 2008). This dramatic decrease in HVC neuron number is driven by a 50% loss of all HVC neurons projecting to RA

(Cohen et al., 2016). The seasonal death of HVC neurons can be prevented for at least 3 days following the decrease in plasma testosterone with intracerebral infusions of testosterone (Thompson and Brenowitz, 2010) or a cocktail of caspase inhibitors (Larson et al., 2014; Thompson and Brenowitz, 2008). The sufficiency of testosterone and caspase inhibitors to prevent neuronal loss even in the absence of systemic testosterone suggests the following scenario for regression (Figs. 4 and 5). As gonadal testosterone decreases, local synthesis of estradiol also decreases within HVC. The drop in local sex steroids result in a decrease in local neurotrophin synthesis (Wissman and Brenowitz, 2009) and relaxation of sex steroid genomic and non-canonical signaling (Larson et al., 2015; Thompson et al., 2012). The decrease in sex steroid dependent transcription of anti-apoptotic genes

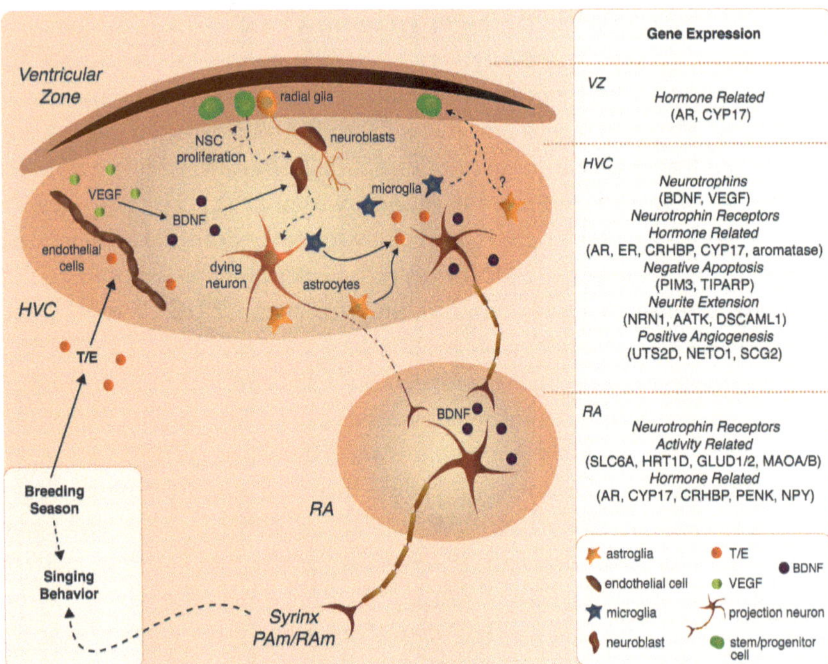

Fig. 5 A schematic of various factors that influence the birth and death of neurons in HVC. Known direct interactions are represented by solid black arrows, whereas unknown or indirect interactions of factors are represented by dashed arrows. The break-out panel highlights gene expression related to hormones, neurotrophins, and apoptosis that influence the development, sexual differentiation, homeostasis, and plasticity of HVC and RA. *Modified from Brenowitz, E.A., Larson, T.A., 2015. Neurogenesis in the adult avian song-control system. Cold Spring Harb. Perspect. Biol. 7(6); Larson, T.A., 2018. Sex steroids, adult neurogenesis, and inflammation in CNS homeostasis, degeneration, and repair. Front. Endocrinol. (Lausanne) 9, 205.*

among others primes HVC neurons for caspase-mediated cell death. The specific signals that promote neuronal apoptosis in HVC and the cues that determine which specific neurons will die remain unknown.

The rate of neural stem and progenitor cell proliferation in the nearby, but distinct, ventricular zone increases following the death of mature neurons in HVC. As mature HVC neurons die at the end of the breeding season, new differentiated neurons can be observed within HVC as soon as 4 weeks thereafter (Alvarez-Buylla et al., 1990; Kirn et al., 1994). The presence of newly born neurons in HVC following seasonal neuronal death and HVC volume regression is supported by findings of increased proliferation in the ventricular zone as early as 2 days following the peak in HVC cell death (Larson et al., 2014). The increase in stem cell and neural progenitor cell proliferation and the presence of differentiated nascent neurons in HVC does not, however, result in an immediate regrowth of HVC (Larson et al., 2014). These nascent neurons will not receive neurotrophic support during the critical window for survival and so will not survive to form functional projections to RA. Infusion of a cocktail of caspase inhibitors, including inhibitors to caspases-3, −9, and −1, into HVC as systemic testosterone levels decrease not only protects HVC from neuronal loss, but also prevents the increase in ventricular zone proliferation following HVC cell death and the addition of new adult-born neurons (Larson et al., 2014; Thompson and Brenowitz, 2008).

The observed increases in new neurons in HVC following neuronal loss led to the hypothesis that neuronal death in HVC promotes recruitment of new neurons by creating "vacancies" in HVC (Alvarez-Buylla and Kirn, 1997). Not necessarily contradicting this hypothesis, reactive proliferation, or the birth of new neurons following cell death, likely occurs as an active and direct response to the death of cells in HVC, rather than as a passive response to the creation of vacancies. As in the mammalian brain, cleanup of apoptotic neurons in HVC likely occurs via phagocytosis by brain resident macrophages called microglia and other neuroglia including astrocytes. Microglia and astroglia have recently emerged as having profound beneficial roles beyond classic inflammatory responses in the central nervous system. During early neural development, microglia and astroglia remove debris, phagocytose excess neurons, prune synapses, and provide trophic support to mature neurons (reviewed in (Larson, 2018)). In the mammalian brain, microglia also guide neuronal differentiation of neuroblasts (Walton et al., 2006) and phagocytose the excess neuroblasts that do not differentiate or functionally incorporate (Sierra et al., 2010).

Within HVC of white-crowned sparrows, inhibition of microglia activation with the oral anti-inflammatory minocycline prevents reactive neurogenesis following seasonal neuronal loss in HVC (unpublished observation, Larson, Tokareva, Meritt Cole, & Brenowitz. Inflammation regulates ventricular zone proliferation in a plastic neural circuit of the healthy adult brain.). The necessity for microglia to activate following natural neuronal loss in HVC suggests that, as in the mammalian brain, microglia confer a still unknown signal of HVC cell death directly to the neural stem and progenitor cells of the nearby, but distinct ventricular zone to increase their proliferation and replace recently lost neurons.

7. Neural insults and damage in the song system

The majority of work examining the processes and mechanisms of cell death has been performed exploiting the "natural" regression of the descending motor pathway. Beyond natural regression, lesion to both the telencephalon in general and targeted to specific cell populations within HVC have also contributed to our understanding of the mechanisms of cell death and replacement in the song control system.

One such study tested the hypothesis that cell death is inextricably linked to compensatory neurogenesis by selectively killing HVC projection neurons (Scharff et al., 2000). As discussed above, two populations of HVC projection neurons exist: (1) HVC to Area X projection neurons that are primarily generated in ovo and (2) HVC to RA projection neurons that are generated post-hatch and continuously throughout the lifespan of songbirds. In male zebra finches selective death of HVC to RA neurons increases the number of newly incorporated HVC to RA projections neurons, supporting the hypothesis that neuronal death leads to replacement (Scharff et al., 2000). Interestingly, selective death of the HVC to Area X projection neurons does not promote the incorporation of new HVC to Area X neurons. Rather, as new neurons incorporate into HVC to compensate for the loss of HVC to Area X neurons, the majority of nascent neurons send projections to RA while the rest presumably differentiate into interneurons (Scharff et al., 2000). These results indicated that not all neuronal death in HVC results in reactive neurogenesis to replace the same neuronal subtype that was lost. Moreover, the projection of new neurons to RA rather than Area X suggests that a guidance cue selective to RA, and perhaps even emanating from RA itself, likely exists. Whether this guidance cue is structural support from existing HVC to RA neurons, neural activity in RA (Larson et al., 2013), trophic support from RA (Thompson et al., 2012; Wissman and Brenowitz, 2009), some other factor generated within HVC or RA (Larson et al., 2015; Thompson et al., 2012; Wade et al., 2004), or some combination of two or more of these factors, remains to be tested.

A few studies provide further insight into mechanisms that link cell death in the avian telencephalon to compensatory neurogenesis. Stab injury to the telencephalon of zebra finch increases local aromatase mRNA expression within 24h (Peterson et al., 2001). Elevated aromatase expression persists for up to 1 week thereafter (Peterson et al., 2001) and co-localizes with vimentin positive glial cells—presumably reactive astrocytes (Duncan and Saldanha, 2011; Saldanha et al., 2005). Simultaneous stab injury and injection of fadrazole, an inhibitor of aromatase, into the telencephalon of zebra finch induces greater amounts of TUNEL positive cell death and a larger subsequent lesion within 72h when compared to saline control stab injections (Wynne and Saldanha, 2004). This up regulation of cell death upon local aromatase inhibition suggests that aromatase expression might limit the extent of cell death in the avian brain (Wynne and Saldanha, 2004). Further supporting this role of aromatase in ameliorating cell death, injection of estradiol, the downstream metabolite of aromatase, with fadrazole into the telencephalon rescues the effect of fadrazole alone on apoptosis (Saldanha et al., 2005). Although the connection between increased aromatase expression following neuronal death and subsequent compensatory neurogenesis has not been formally tested in the context of injury, in mammals up regulation of aromatase in the context of neural injury and inflammation is generally pro-neurogenic (Larson, 2018). Likewise, given that in songbirds, the activation of microglia, a non-glial-derived macrophage, is necessary for reactive neurogenesis following the natural, seasonal regression of HVC and that in mammals microglia are responsive to estradiol (Lenz et al., 2013), a functional connection between aromatase expression and compensatory neurogenesis following injury seems probable.

8. Future directions and conclusions

In the field of cell death biology, traditional model systems such as *Drosophila melanogaster* (fruit fly), *Caenorhabditis elegans* (roundworm), and *Mus musculus* (mouse) offer the advantages of well-characterized tissue and cell level organization, genetic and genomic tools, targeted mutagenesis, and in vivo imaging of cellular interactions. These model systems, however, often struggle to link the processes and mechanisms of cell death to a biologically relevant behavior and to place the spatiotemporal patterns of cell death in an evolutionary context. The song control system provides a unique opportunity to study both the homeostatic and neurodegenerative mechanisms of cell death in a sensorimotor circuit that underlies a natural and robust behavior, song learning and production. Until recently, the major limitation for testing questions regarding the mechanisms of cell death in the song system has been the lack of available genomic and genetic tools. Within the last 10 years whole genome alignments have been performed for 50 different bird species (as of November 2019 in NCBI Genomes), including the zebra finch (Warren et al., 2010), Tibetan ground tit (*Pseudopodoces humilis*; (Qu et al., 2013)), Darwin's finch (*Geospiza fortis*; (Parker et al., 2012)), and domesticated canary (Frankl-Vilches et al., 2015). Moreover, the annotation quality of the zebra finch and canary genomes are currently being improved using HiC and pac-bio sequencing to generate more accurate and longer contigs (https://www.ncbi.nlm.nih.gov/assembly/GCF_003957565.1/, (Rhie et al., n.d., in preparation); and https://www.ncbi.nlm.nih.gov/assembly/GCF_007115625.1/, respectively). Tens to hundreds of additional songbirds species, including the great tit (*Parus major*; (Laine et al., 2019)) and white-throated sparrow (*Zonotrichia albicollis*; (Romanov et al., 2011)), have varying degrees of sequence information available in publicly accessible repositories with NCBI and Ensembl.

Efforts are currently underway to establish a more feasible procedure for generating transgenic lineage marker lines and CRISPR mutagenesis in zebra finch (Jung et al., 2019; Lee et al., 2017), with the goal to broaden the approaches to other songbird species. To date, transgenic line establishment has relied on viral transfection of primordial germline cells (Agate et al., 2009), whereas gene knockdown has relied on lentivirus-mediated RNA interference (Haesler et al., 2007) and cell lineage marking has relied on mosaic transfection with micro-injections of viral constructs directly into the neural region of interest (Roberts et al., 2010). Most viral techniques used so far have proved challenging and more inefficient in birds when compared to mammals. The establishment of more feasible and efficient methods for generating lineage marker transgenic lines will not only allow for targeted mutagenesis and lineage analyses, but will also transform the power of established in vivo imaging techniques using a head-mounted fluorescent microscopes (Markowitz et al., 2015) or two-photon imaging (Roberts et al., 2010) and allow for more detailed imaging of multifaceted cell–cell interactions.

With the development of these resources in songbirds, several outstanding questions in regards to mechanisms and function of cell death can be addressed. These include, but are certainly not limited to:

- Is cell death during developmental sexual differentiation driven by differential expression of genes on the sex chromosomes? Do female birds that exhibit "male-equivalent" abilities to learn and produce song have neural circuits similar in content and function to male songbirds? If the answer to both questions are "yes," do female songbirds that sing have sex chromosome dose compensation (Ellegren, 2009) of factors necessary for the masculinization of the song control system?
- Is cell death that occurs for maintenance of homeostasis and during regression of HVC limited to caspase-3 mediated apoptosis or are other types of cell death like autophagy or caspase-independent programmed cell death involved as well?
- What are the specific signals that promote neuronal apoptosis in HVC? What are the cues that determine which specific neurons will die versus survive during HVC seasonal regression?
- What other cell types play a role in the death of neurons in the song system? We currently have evidence that microglia and astroglia are involved in mediating reactive neurogenesis and increasing pro-survival signals via increase aromatase expression. The extent to which each type of neuroglia contributes to homeostasis, neuronal and synaptic pruning, cell debris removal, and stimulation of neural stem or progenitor cell proliferation remains unknown. Likewise, many other cell types predicted to play a role in neural degeneration and regeneration including, but not limited to, infiltrating leucocytes, oligodendroctyes, endothelial cells and ependymal cells have received little attention in the context of cell death in the avian song control system.
- Are the molecular mechanisms involved in natural turnover and plasticity of the song system shared with those that mediate cell death and reactive neurogenesis following damage or insult to the cellular architecture of the song circuits?
- How have ultimate mechanisms shaped the spatiotemporal patterns of cell death within the avian brain? The large number of birds species as a whole exhibit a vast repertoire of learned (e.g., tool use in New Caledonian crows [*Corvus moneduloides*; (Bayern et al., 2018)], etc.) and innate behaviors (e.g., lekking and dancing behaviors in golden-collared manakins [(*Manacus vitellinus*); (Day et al., 2007)], etc.). Underlying this diversity in behavior is likely diversity in neural

cytoarchitecture and synaptic connectivity across behavior types and species. Comparative neuroanatomical and behavioral genomic analyses can be integrated with established and emerging cellular and molecular techniques to provide a unique perspective into the evolutionary origins and function of diverse spatiotemporal patterns of cell birth, maintenance, and death.

Exploring these and related questions using the avian model will undoubtably provide novel insights into the roles of cell death in the developmental refinement of tissue level organization, the proper turnover of cells mature tissues, and the return to homeostasis after substantial cell death.

References

Adar, E., Nottebohm, F., Barnea, A., 2008. The relationship between nature of social change, age, and position of new neurons and their survival in adult zebra finch brain. J. Neurosci. 28 (20), 5394–5400.

Adkins-Regan, E., Mansukhani, V., Seiwert, C., Thompson, R., 1994. Sexual differentiation of brain and behavior in the zebra finch: critical periods for effects of early estrogen treatment. J. Neurobiol. 25 (7), 865–877.

Agate, R.J., Scott, B.B., Haripal, B., Lois, C., Nottebohm, F., 2009. Transgenic songbirds offer an opportunity to develop a genetic model for vocal learning. Proc. Natl. Acad. Sci. U.S.A. 106 (42), 17963–17967.

Altman, J., Das, G.D., 1965. Autoradiographic and histological evidence of postnatal hippocampal neurogenesis in rats. J. Comp. Neurol. 124 (3), 319.

Alvarez-Buylla, A., Kirn, J.R., 1997. Birth, migration, incorporation, and death of vocal control neurons in adult songbirds. J. Neurobiol. 33 (5), 585–601.

Alvarez-Buylla, A., Nottebohm, F., 1988. Migration of young neurons in adult avian brain. Nature 335 (6188), 353–354.

Alvarez-Buylla, A., Theelen, M., Nottebohm, F., 1988. Birth of projection neurons in the higher vocal center of the canary forebrain before, during, and after song learning. Proc. Natl. Acad. Sci. U.S.A. 85 (22), 8722–8726.

Alvarez-Buylla, A., Kirn, J.R., Nottebohm, F., 1990. Birth of projection neurons in adult avian brain may be related to perceptual or motor learning. Science 249 (4975), 1444–1446.

Alvarez-Buylla, A., Ling, C.Y., Nottebohm, F., 1992. High vocal center growth and its relation to neurogenesis, neuronal replacement and song acquisition in juvenile canaries. J. Neurobiol. 23 (4), 396–406.

Alvarez-Buylla, A., Seri, B., Doetsch, F., 2002. Identification of neural stem cells in the adult vertebrate brain. Brain Res. Bull. 57 (6), 751–758.

Arnold, A.P., 1975. The effects of castration and androgen replacement on song, courtship, and aggression in zebra finches (Poephila guttata). J. Exp. Zool. 191 (3), 309–326.

Arnold, A.P., Itoh, Y., 2011. Factors causing sex differences in birds. Avian Biol. Res. 4 (2), 44–51.

Aronov, D., Andalman, A.S., Fee, M.S., 2008. A specialized forebrain circuit for vocal babbling in the juvenile songbird. Science 320 (5876), 630–634.

Ball, G.F., Auger, C.J., Bernard, D.J., Charlier, T.D., Sartor, J.J., Riters, L.V., Balthazart, J., 2004. Seasonal plasticity in the song control system: multiple brain sites of steroid hormone action and the importance of variation in Song behavior. Ann. N. Y. Acad. Sci. 1016, 586–610.

Bauer, E.E., Coleman, M.J., Roberts, T.F., Roy, A., Prather, J.F., Mooney, R., 2008. A synaptic basis for auditory-vocal integration in the songbird. J. Neurosci. 28 (6), 1509–1522.

Bayern, A., Danel, S., Auersperg, A.M.I., Mioduszewska, B., Kacelnik, A., 2018. Compound tool construction by new Caledonian crows. Sci. Rep. 8 (1), 15676.

Bottjer, S.W., 2004. Developmental regulation of basal ganglia circuitry during the sensitive period for vocal learning in songbirds. Ann. N. Y. Acad. Sci. 1016, 395–415.

Bottjer, S.W., Miesner, E.A., Arnold, A.P., 1984. Forebrain lesions disrupt development but not maintenance of song in passerine birds. Science 224 (4651), 901–903.

Bottjer, S.W., Glaessner, S.L., Arnold, A.P., 1985. Ontogeny of brain nuclei controlling song learning and behavior in zebra finches. J. Neurosci. 5 (6), 1556–1562.

Brenowitz, E.A., 2008. Plasticity of the song control system in adult birds. In: Zeigler, H.P., Marler, P. (Eds.), Neuroscience of Birdsong. Cambridge University Press, Cambridge, pp. 332–349.

Brenowitz, E.A., 2013. Testosterone and brain-derived neurotrophic factor interactions in the avian song control system. Neuroscience 239, 115–123.

Brenowitz, E.A., Larson, T.A., 2015. Neurogenesis in the adult avian song-control system. Cold Spring Harb. Perspect. Biol. 7 (6).

Catchpole, C.K., Slater, P.J.B., 1995. Bird Song: Biological Themes and Variations. Cambridge University Press, Cambridge, UK.

Cohen, R.E., Macedo-Lima, M., Miller, K.E., Brenowitz, E.A., 2016. Adult neurogenesis leads to the functional reconstruction of a telencephalic neural circuit. J. Neurosci. 36 (34), 8947–8956.

Day, L.B., Fusani, L., Hernandez, E., Billo, T.J., Sheldon, K.S., Wise, P.M., Schlinger, B.A., 2007. Testosterone and its effects on courtship in golden-collared manakins (Manacus vitellinus): seasonal, sex, and age differences. Horm. Behav. 51 (1), 69–76.

Dickey, J.S., Baird, B.J., Redon, C.E., Avdoshina, V., Palchik, G., Wu, J.F., Kondratyev, A., Bonner, W.M., Martin, O.A., 2012. Susceptibility to bystander DNA damage is influenced by replication and transcriptional activity. Nucleic Acids Res. 40 (20), 10274–10286.

Dong, X.-x., Wang, Y., Qin, Z.-h., 2009. Molecular mechanisms of excitotoxicity and their relevance to pathogenesis of neurodegenerative diseases. Acta Pharmacol. Sin. 30 (4), 379–387.

Duncan, K.A., Saldanha, C.J., 2011. Neuroinflammation induces glial aromatase expression in the uninjured songbird brain. J. Neuroinflammation 8, 81.

Duncan, K.A., Walters, B.J., Saldanha, C.J., 2013. Traumatized and inflamed—but resilient: glial aromatization and the avian brain. Horm. Behav. 63 (2), 208–215.

Dutar, P., Petrozzino, J.J., Vu, H.M., Schmidt, M.F., Perkel, D.J., 2000. Slow synaptic inhibition mediated by metabotropic glutamate receptor activation of GIRK channels. J. Neurophysiol. 84 (5), 2284–2290.

Ellegren, H., 2009. Genomic evidence for a large-Z effect. Proc. Biol. Sci. 276 (1655), 361–366.

Fee, M.S., Kozhevnikov, A.A., Hahnloser, R.H., 2004. Neural mechanisms of vocal sequence generation in the songbird. Ann. N. Y. Acad. Sci. 1016, 153–170.

Fortune, E.S., Margoliash, D., 1995. Parallel pathways and convergence onto HVc and adjacent neostriatum of adult zebra finches (*Taeniopygia guttata*). J. Comp. Neurol. 360 (3), 413–441.

Frankl-Vilches, C., Kuhl, H., Werber, M., Klages, S., Kerick, M., Bakker, A., de Oliveira, E.H.C., Reusch, C., Capuano, F., Vowinckel, J., Leitner, S., Ralser, M., Timmermann, B., Gahr, M., 2015. Using the canary genome to decipher the evolution of hormone-sensitive gene regulation in seasonal singing birds. Genome Biol. 16, 19.

Fusani, L., Van't Hof, T., Hutchison, J.B., Gahr, M., 2000. Seasonal expression of androgen receptors, estrogen receptors, and aromatase in the canary brain in relation to circulating androgens and estrogens. J. Neurobiol. 43 (3), 254–268.
Gahr, M., Leitner, S., Fusani, L., Rybak, F., 2002. What is the adaptive role of neurogenesis in adult birds? Prog. Brain Res. 138, 233–254.
Goldman, S.A., Williams, S., Barami, K., Lemmon, V., Nedergaard, M., 1996. Transient coupling of Ng-CAM expression to NgCAM-dependent calcium signaling during migration of new neurons in the adult songbird brain. Mol. Cell. Neurosci. 7 (1), 29–45.
Grisham, W., Mathews, G.A., Arnold, A.P., 1994. Local intracerebral implants of estrogen masculinize some aspects of the zebra finch song system. J. Neurobiol. 25 (2), 185–196.
Gurney, M., Konishi, M., 1980. Hormone-induced sexual differentiation of brain and behavior in zebra finches. Science 208, 1380–1383.
Haesler, S., Rochefort, C., Georgi, B., Licznerski, P., Osten, P., Scharff, C., 2007. Incomplete and inaccurate vocal imitation after knockdown of FoxP2 in songbird basal ganglia nucleus area X. PLoS Biol. 5 (12), e321.
Hebb, D.O., Martinez, J.L., Glickman, S.E., 1994. The organization of behavior—a neuropsychological theory—Hebb,Do. Contemp. Psychol. 39 (11), 1018–1020.
Herrmann, K., Arnold, A.P., 1991. The development of afferent projections to the robust archistriatal nucleus in male zebra finches: a quantitative electron microscopic study. J. Neurosci. 11 (7), 2063–2074.
Hidalgo, A., Barami, K., Iversen, K., Goldman, S.A., 1995. Estrogens and non-estrogenic ovarian influences combine to promote the recruitment and decrease the turnover of new neurons in the adult female canary brain. J. Neurobiol. 27 (4), 470–487.
Jabes, A., Lavenex, P.B., Amaral, D.G., Lavenex, P., 2010. Quantitative analysis of postnatal neurogenesis and neuron number in the macaque monkey dentate gyrus. Eur. J. Neurosci. 31 (2), 273–285.
Jacobs, E.C., Arnold, A.P., Campagnoni, A.T., 1999. Developmental regulation of the distribution of aromatase- and estrogen-receptor- mRNA-expressing cells in the zebra finch brain. Dev. Neurosci. 21 (6), 453–472.
Johnson, F., Bottjer, S.W., 1992. Growth and regression of thalamic efferents in the song-control system of male zebra finches. J. Comp. Neurol. 326 (3), 442–450.
Johnson, F., Bottjer, S.W., 1994. Afferent influences on cell death and birth during development of a cortical nucleus necessary for learned vocal behavior in zebra finches. Development 120 (1), 13–24.
Jung, K.M., Kim, Y.M., Keyte, A.L., Biegler, M.T., Rengaraj, D., Lee, H.J., Mello, C.V., Velho, T.A.F., Fedrigo, O., Haase, B., Jarvis, E.D., Han, J.Y., 2019. Identification and characterization of primordial germ cells in a vocal learning Neoaves species, the zebra finch. FASEB J. 33, 13825–13836.
Kaplan, M.S., Hinds, J.W., 1977. Neurogenesis in the adult rat: electron microscopic analysis of light radioautographs. Science 197 (4308), 1092–1094.
Kirn, J.R., DeVoogd, T.J., 1989. Genesis and death of vocal control neurons during sexual differentiation in the zebra finch. J. Neurosci. 9 (9), 3176–3187.
Kirn, J.R., Nottebohm, F., 1993. Direct evidence for loss and replacement of projection neurons in adult canary brain. J. Neurosci. 13 (4), 1654–1663.
Kirn, J.R., Alvarez-Buylla, A., Nottebohm, F., 1991. Production and survival of projection neurons in a forebrain vocal center of adult male canaries. J. Neurosci. 11 (6), 1756–1762.
Kirn, J., O'Loughlin, B., Kasparian, S., Nottebohm, F., 1994. Cell death and neuronal recruitment in the high vocal center of adult male canaries are temporally related to changes in song. Proc. Natl. Acad. Sci. U.S.A. 91 (17), 7844–7848.
Kirn, J.R., Fishman, Y., Sasportas, K., Alvarez-Buylla, A., Nottebohm, F., 1999. Fate of new neurons in adult canary high vocal center during the first 30 days after their formation. J. Comp. Neurol. 411 (3), 487–494.

Konishi, M., Akutagawa, E., 1985. Neuronal growth, atrophy and death in a sexually dimorphic song nucleus in the zebra finch brain. Nature 315 (6015), 145–147.
Konishi, M., Akutagawa, E., 1990. Growth and atrophy of neurons labeled at their birth in a song nucleus of the zebra finch. Proc. Natl. Acad. Sci. U.S.A. 87 (9), 3538–3541.
Laine, V.N., Verhagen, I., Mateman, A.C., Pijl, A., Williams, T.D., Gienapp, P., van Oers, K., Visser, M.E., 2019. Exploration of tissue-specific gene expression patterns underlying timing of breeding in contrasting temperature environments in a song bird. BMC Genomics 20 (1), 693.
Larson, T.A., 2018. Sex steroids, adult neurogenesis, and inflammation in CNS homeostasis, degeneration, and repair. Front. Endocrinol. (Lausanne) 9, 205.
Larson, T.A., Wang, T.W., Gale, S.D., Miller, K.E., Thatra, N.M., Caras, M.L., Perkel, D.J., Brenowitz, E.A., 2013. Postsynaptic neural activity regulates neuronal addition in the adult avian song control system. Proc. Natl. Acad. Sci. U.S.A. 110, 16640–16644.
Larson, T.A., Thatra, N.M., Lee, B.H., Brenowitz, E.A., 2014. Reactive neurogenesis in response to naturally occurring apoptosis in an adult brain. J. Neurosci. 34 (39), 13066–13076.
Larson, T.A., Lent, K.L., Bammler, T.K., MacDonald, J.W., Wood, W.E., Caras, M.L., Thatra, N.M., Budzillo, A., Perkel, D.J., Brenowitz, E.A., 2015. Network analysis of microRNA and mRNA seasonal dynamics in a highly plastic sensorimotor neural circuit. BMC Genomics 16, 905.
Larson, T.A., Thatra, N.M., Hou, D., Hu, R.A., Brenowitz, E.A., 2019. Seasonal changes in neuronal turnover in a forebrain nucleus in adult songbirds. J. Comp. Neurol. 527 (4), 767–779.
Lee, H.J., Kim, Y.M., Ono, T., Han, J.Y., 2017. Genome modification technologies and their applications in avian species. Int. J. Mol. Sci. 18 (11), 2245–2257.
Lenz, K.M., Nugent, B.M., Haliyur, R., McCarthy, M.M., 2013. Microglia are essential to masculinization of brain and behavior. J. Neurosci. 33 (7), 2761–2772.
Leonardo, A., Konishi, M., 1999. Decrystallization of adult birdsong by perturbation of auditory feedback. Nature 399 (6735), 466–470.
Lobato, M., Vellema, M., Gahr, C., Leitao, A., de Lima, S.M.A., Geberzahn, N., Gahr, M., 2015. Mismatch in sexual dimorphism of developing song and song control system in blue-capped cordon-bleus, a songbird species with singing females and males. Front. Ecol. Evol. 3, 117.
Lombardino, A.J., Nottebohm, F., 2000. Age at deafening affects the stability of learned song in adult male zebra finches. J. Neurosci. 20 (13), 5054–5064.
London, S.E., Schlinger, B.A., 2007. Steroidogenic enzymes along the ventricular proliferative zone in the developing songbird brain. J. Comp. Neurol. 502 (4), 507–521.
London, S.E., Boulter, J., Schlinger, B.A., 2003. Cloning of the zebra finch androgen synthetic enzyme CYP17: a study of its neural expression throughout posthatch development. J. Comp. Neurol. 467 (4), 496–508.
Louissaint Jr., A., Rao, S., Leventhal, C., Goldman, S.A., 2002. Coordinated interaction of neurogenesis and angiogenesis in the adult songbird brain. Neuron 34 (6), 945–960.
Maekawa, F., Tsukahara, S., Kawashima, T., Nohara, K., Ohki-Hamazaki, H., 2014. The mechanisms underlying sexual differentiation of behavior and physiology in mammals and birds: relative contributions of sex steroids and sex chromosomes. Front. Neurosci. 8, 242.
Markowitz, J.E., Liberti 3rd, W.A., Guitchounts, G., Velho, T., Lois, C., Gardner, T.J., 2015. Mesoscopic patterns of neural activity support songbird cortical sequences. PLoS Biol. 13 (6), e1002158.
Meitzen, J., Perkel, D.J., Brenowitz, E.A., 2007. Seasonal changes in intrinsic electrophysiological activity of song control neurons in wild song sparrows. J. Comp. Physiol. A Neuroethol. Sens. Neural Behav. Physiol. 193 (6), 677–683.

Meitzen, J., Thompson, C.K., Choi, H., Perkel, D.J., Brenowitz, E.A., 2009. Time course of changes in Gambel's white-crowned sparrow song behavior following transitions in breeding condition. Horm. Behav. 55 (1), 217–227.
Mooney, R., 2009. Neural mechanisms for learned birdsong. Learn. Mem. 16 (11), 655–669.
Mooney, R., Rao, M., 1994. Waiting periods versus early innervation: the development of axonal connections in the zebra finch song system. J. Neurosci. 14 (11 Pt 1), 6532–6543.
Mooney, R., Hoese, W., Nowicki, S., 2001. Auditory representation of the vocal repertoire in a songbird with multiple song types. Proc. Natl. Acad. Sci. U.S.A. 98 (22), 12778–12783.
Nordeen, K.W., Nordeen, E.J., 1992. Auditory feedback is necessary for the maintenance of stereotyped song in adult zebra finches. Behav. Neural Biol. 57 (1), 58–66.
Nordeen, E.J., Grace, A., Burek, M.J., Nordeen, K.W., 1992. Sex-dependent loss of projection neurons involved in avian song learning. J. Neurobiol. 23 (6), 671–679.
Nottebohm, F., 1981. A brain for all seasons: cyclical anatomical changes in song control nuclei of the canary brain. Science 214 (4527), 1368–1370.
Nottebohm, F., Stokes, T.M., Leonard, C.M., 1976. Central control of song in the canary, Serinus canarius. J. Comp. Neurol. 165 (4), 457–486.
Nottebohm, F., Nottebohm, M.E., Crane, L., 1986. Developmental and seasonal changes in canary song and their relation to changes in the anatomy of song-control nuclei. Behav. Neural Biol. 46 (3), 445–471.
Nottebohm, F., O'Loughlin, B., Gould, K., Yohay, K., Alvarez-Buylla, A., 1994. The life span of new neurons in a song control nucleus of the adult canary brain depends on time of year when these cells are born. Proc. Natl. Acad. Sci. U.S.A. 91 (17), 7849–7853.
Parker, P., Li, B., Li, H., Wang, J., 2012. The genome of Darwin's finch (Geospiza fortis). GigaScience.
Paton, J.A., Nottebohm, F.N., 1984. Neurons generated in the adult brain are recruited into functional circuits. Science 225 (4666), 1046–1048.
Paton, J.A., O'Loughlin, B.E., Nottebohm, F., 1985. Cells born in adult canary forebrain are local interneurons. J. Neurosci. 5 (11), 3088–3093.
Perlman, W.R., Arnold, A.P., 2003. Expression of estrogen receptor and aromatase mRNAs in embryonic and posthatch zebra finch brain. J. Neurobiol. 55 (2), 204–219.
Perlman, W.R., Ramachandran, B., Arnold, A.P., 2003. Expression of androgen receptor mRNA in the late embryonic and early posthatch zebra finch brain. J. Comp. Neurol. 455 (4), 513–530.
Peterson, R.S., Saldanha, C.J., Schlinger, B.A., 2001. Rapid upregulation of aromatase mRNA and protein following neural injury in the zebra finch (Taeniopygia guttata). J. Neuroendocrinol. 13 (4), 317–323.
Qu, Y.H., Zhao, H.W., Han, N.J., Zhou, G.Y., Song, G., Gao, B., Tian, S.L., Zhang, J.B., Zhang, R.Y., Meng, X.H., Zhang, Y., Zhang, Y., Zhu, X.J., Wang, W.J., Lambert, D., Ericson, P.G.P., Subramanian, S., Yeung, C., Zhu, H.M., Jiang, Z., Li, R.Q., Lei, F.M., 2013. Ground tit genome reveals avian adaptation to living at high altitudes in the Tibetan plateau. Nat. Commun. 4, 2071.
Rasika, S., Alvarez-Buylla, A., Nottebohm, F., 1999. BDNF mediates the effects of testosterone on the survival of new neurons in an adult brain. Neuron 22 (1), 53–62.
Reiner, A., Perkel, D.J., Mello, C.V., Jarvis, E.D., 2004. Songbirds and the revised avian brain nomenclature. Ann. N. Y. Acad. Sci. 1016, 77–108.
Remage-Healey, L., Maidment, N.T., Schlinger, B.A., 2008. Forebrain steroid levels fluctuate rapidly during social interactions. Nat. Neurosci. 11 (11), 1327–1334.
Rhie, A., S. McCarthy, E. W. Myers, R. Durbin, A. M. Phillippy and E. D. Jarvis n.d. "The vertebrate genomes project: developing standards and a path towards complete and error free genome assemblies (in preparation)."

Riddle, D.R., Lichtenwalner, R.J., 2007. Neurogenesis in the adult and aging brain. In: Riddle, D.R. (Ed.), Brain Aging: Models, Methods, and Mechanisms. NCBI Bookshelf.

Roberts, T.F., Tschida, K.A., Klein, M.E., Mooney, R., 2010. Rapid spine stabilization and synaptic enhancement at the onset of behavioural learning. Nature 463 (7283), 948–952.

Romanov, M.N., Dodgson, J.B., Gonser, R.A., Tuttle, E.M., 2011. Comparative BAC-based mapping in the white-throated sparrow, a novel behavioral genomics model, using interspecies overgo hybridization. BMC. Res. Notes 4, 211.

Saldanha, C.J., Rohmann, K.N., Coomaralingam, L., Wynne, R.D., 2005. Estrogen provision by reactive glia decreases apoptosis in the zebra finch (Taeniopygia guttata). J. Neurobiol. 64 (2), 192–201.

Scharff, C., Kirn, J.R., Grossman, M., Macklis, J.D., Nottebohm, F., 2000. Targeted neuronal death affects neuronal replacement and vocal behavior in adult songbirds. Neuron 25 (2), 481–492.

Schlinger, B.A., Arnold, A.P., 1992. Plasma sex steroids and tissue aromatization in hatchling zebra finches: implications for the sexual differentiation of singing behavior. Endocrinology 130 (1), 289–299.

Scott, B.B., Lois, C., 2007. Developmental origin and identity of song system neurons born during vocal learning in songbirds. J. Comp. Neurol. 502 (2), 202–214.

Scott, B.B., Gardner, T., Ji, N., Fee, M.S., Lois, C., 2012. Wandering neuronal migration in the postnatal vertebrate forebrain. J. Neurosci. 32 (4), 1436–1446.

Scotto-Lomassese, S., Rochefort, C., Nshdejan, A., Scharff, C., 2007. HVC interneurons are not renewed in adult male zebra finches. Eur. J. Neurosci. 25 (6), 1663–1668.

Sierra, A., Encinas, J.M., Deudero, J.J.P., Chancey, J.H., Enikolopov, G., Overstreet-Wadiche, L.S., Tsirka, S.E., Maletic-Savatic, M., 2010. Microglia shape adult hippocampal neurogenesis through apoptosis-coupled phagocytosis. Cell Stem Cell 7 (4), 483–495.

Smith, G.T., Brenowitz, E.A., Wingfield, J.C., Baptista, L.F., 1995. Seasonal changes in song nuclei and song behavior in Gambel's white-crowned sparrows. J. Neurobiol. 28 (1), 114–125.

Smith, G.T., Brenowitz, E.A., Wingfield, J.C., 1997. Roles of photoperiod and testosterone in seasonal plasticity of the avian song control system. J. Neurobiol. 32 (4), 426–442.

Springer, M.L., Wade, J., 1997. The effects of testicular tissue and prehatching inhibition of estrogen synthesis on the development of courtship and copulatory behavior in zebra finches. Horm. Behav. 32 (1), 46–59.

Thompson, C.K., 2011. Cell death and the song control system: a model for how sex steroid hormones regulate naturally-occurring neurodegeneration. Dev. Growth Differ. 53 (2), 213–224.

Thompson, C.K., Brenowitz, E.A., 2005. Seasonal change in neuron size and spacing but not neuronal recruitment in a basal ganglia nucleus in the avian song control system. J. Comp. Neurol. 481 (3), 276–283.

Thompson, C.K., Brenowitz, E.A., 2008. Caspase inhibitor infusion protects an avian song control circuit from seasonal-like neurodegeneration. J. Neurosci. 28 (28), 7130–7136.

Thompson, C.K., Brenowitz, E.A., 2010. Neuroprotective effects of testosterone in a naturally occurring model of neurodegeneration in the adult avian song control system. J. Comp. Neurol. 518 (23), 4760–4770.

Thompson, C.K., Bentley, G.E., Brenowitz, E.A., 2007. Rapid seasonal-like regression of the adult avian song control system. Proc. Natl. Acad. Sci. U.S.A. 104 (39), 15520–15525.

Thompson, C.K., Meitzen, J., Replogle, K., Drnevich, J., Lent, K.L., Wissman, A.M., Farin, F.M., Bammler, T.K., Beyer, R.P., Clayton, D.F., Perkel, D.J., Brenowitz, E.A., 2012. Seasonal changes in patterns of gene expression in avian song control brain regions. PLoS ONE 7 (4), e35119.
Tramontin, A.D., Brenowitz, E.A., 1999. A field study of seasonal neuronal incorporation into the song control system of a songbird that lacks adult song learning. J. Neurobiol. 40 (3), 316–326.
Tramontin, A.D., Brenowitz, E.A., 2000. Seasonal plasticity in the adult brain. Trends Neurosci. 23 (6), 251–258.
Tramontin, A.D., Hartman, V.N., Brenowitz, E.A., 2000. Breeding conditions induce rapid and sequential growth in adult avian song control circuits: a model of seasonal plasticity in the brain. J. Neurosci. 20 (2), 854–861.
Tumer, E.C., Brainard, M.S., 2007. Performance variability enables adaptive plasticity of 'crystallized' adult birdsong. Nature 450 (7173), 1240–U1211.
Van Hout, A.J., Eens, M., Balthazart, J., Pinxten, R., 2009. Complex modulation of singing behavior by testosterone in an open-ended learner, the European Starling. Horm. Behav. 56 (5), 564–573.
Vates, G.E., Nottebohm, F., 1995. Feedback circuitry within a song-learning pathway. Proc. Natl. Acad. Sci. U.S.A. 92 (11), 5139–5143.
Vellema, M., van der Linden, A., Gahr, M., 2010. Area-specific migration and recruitment of new neurons in the adult songbird brain. J. Comp. Neurol. 518 (9), 1442–1459.
Wade, J., Peabody, C., Coussens, P., Tempelman, R.J., Clayton, D.F., Liu, L., Arnold, A.P., Agate, R., 2004. A cDNA microarray from the telencephalon of juvenile male and female zebra finches. J. Neurosci. Methods 138 (1–2), 199–206.
Walton, N.M., Sutter, B.M., Laywell, E.D., Levkoff, L.H., Kearns, S.M., Marshall 2nd, G.P., Scheffler, B., Steindler, D.A., 2006. Microglia instruct subventricular zone neurogenesis. Glia 54 (8), 815–825.
Walton, C., Pariser, E., Nottebohm, F., 2012. The zebra finch paradox: song is little changed, but number of neurons doubles. J. Neurosci. 32 (3), 761–774.
Wang, N., Hurley, P., Pytte, C., Kirn, J.R., 2002. Vocal control neuron incorporation decreases with age in the adult zebra finch. J. Neurosci. 22 (24), 10864–10870.
Warren, W.C., Clayton, D.F., Ellegren, H., Arnold, A.P., Hillier, L.W., Kunstner, A., Searle, S., White, S., Vilella, A.J., Fairley, S., Heger, A., Kong, L., Ponting, C.P., Jarvis, E.D., Mello, C.V., Minx, P., Lovell, P., Velho, T.A., Ferris, M., Balakrishnan, C.N., Sinha, S., Blatti, C., London, S.E., Li, Y., Lin, Y.C., George, J., Sweedler, J., Southey, B., Gunaratne, P., Watson, M., Nam, K., Backstrom, N., Smeds, L., Nabholz, B., Itoh, Y., Whitney, O., Pfenning, A.R., Howard, J., Volker, M., Skinner, B.M., Griffin, D.K., Ye, L., McLaren, W.M., Flicek, P., Quesada, V., Velasco, G., Lopez-Otin, C., Puente, X.S., Olender, T., Lancet, D., Smit, A.F., Hubley, R., Konkel, M.K., Walker, J.A., Batzer, M.A., Gu, W., Pollock, D.D., Chen, L., Cheng, Z., Eichler, E.E., Stapley, J., Slate, J., Ekblom, R., Birkhead, T., Burke, T., Burt, D., Scharff, C., Adam, I., Richard, H., Sultan, M., Soldatov, A., Lehrach, H., Edwards, S.V., Yang, S.P., Li, X., Graves, T., Fulton, L., Nelson, J., Chinwalla, A., Hou, S., Mardis, E.R., Wilson, R.K., 2010. The genome of a songbird. Nature 464 (7289), 757–762.
Wennstrom, K.L., Reeves, B.J., Brenowitz, E.A., 2001. Testosterone treatment increases the metabolic capacity of adult avian song control nuclei. J. Neurobiol. 48 (4), 256–264.
Wilbrecht, L., Kirn, J.R., 2004. Neuron addition and loss in the Song system: regulation and function. Ann. N. Y. Acad. Sci. 1016, 659–683.
Wilbrecht, L., Nottebohm, F., 2003. Vocal learning in birds and humans. Ment. Retard. Dev. Disabil. Res. Rev. 9 (3), 135–148.

Wilbrecht, L., Nottebohm, F., 2004. Age and experience affect the recruitment of new neurons to the song system of zebra finches during the sensitive period for song learning: ditto for vocal learning in humans? Ann. N. Y. Acad. Sci. 1021, 404–409.
Wild, J.M., 1993. Descending projections of the songbird nucleus robustus archistriatalis. J. Comp. Neurol. 338 (2), 225–241.
Wingfield, J.C., Farner, D.S., 1978. The annual cycle of plasma irLH and steroid hormones in feral populations of the white-crowned sparrow, Zonotrichia leucophrys gambelii. Biol. Reprod. 19 (5), 1046–1056.
Wissman, A.M., Brenowitz, E.A., 2009. The role of neurotrophins in the seasonal-like growth of the avian song control system. J. Neurosci. 29 (20), 6461–6471.
Woolley, S.M., 2012. Early experience shapes vocal neural coding and perception in songbirds. Dev. Psychobiol. 54 (6), 612–631.
Woolley, S.M., Rubel, E.W., 1997. Bengalese finches *Lonchura striata* domestica depend upon auditory feedback for the maintenance of adult song. J. Neurosci. 17 (16), 6380–6390.
Woolley, S.C., Rajan, R., Joshua, M., Doupe, A.J., 2014. Emergence of context-dependent variability across a basal ganglia network. Neuron 82 (1), 208–223.
Wynne, R.D., Saldanha, C.J., 2004. Glial aromatization decreases neural injury in the zebra finch (*Taeniopygia guttata*): influence on apoptosis. J. Neuroendocrinol. 16 (8), 676–683.

CHAPTER FOUR

Crosstalk between apoptosis and autophagy signaling pathways

W. Douglas Fairlie[a,b,c], Sharon Tran[a,b], Erinna F. Lee[a,b,c,*]
[a]Olivia Newton-John Cancer Research Institute, Heidelberg, VIC, Australia
[b]School of Cancer Medicine, La Trobe University, Melbourne, VIC, Australia
[c]La Trobe Institute for Molecular Science, La Trobe University, Melbourne, VIC, Australia
*Corresponding author: e-mail address: erinna.lee@latrobe.edu.au

Contents

International Review of Cell and Molecular Biology, Volume 352
ISSN 1937-6448
https://doi.org/10.1016/bs.ircmb.2020.01.003

Abstract

The fate of a cell is determined by multiple signaling pathways in response to a range of stimuli. Probably the most prominent cell death mechanism is apoptosis which can be triggered by both internal stresses, as well as extracellular stimuli, and is executed by two well-characterized pathways, the intrinsic and the extrinsic apoptosis pathways. Although autophagy can also lead to cell death under certain circumstances, its major function is as a cell survival process. Given that the life of a cell is at stake, it is not surprising that there is significant molecular crosstalk between these pathways. The nature of these interconnections is diverse and ranges from protein-protein interactions and post-translational modifications through to the degradation of molecular components by different proteins and organelles. In this chapter we review these mechanisms in detail.

1. Introduction

The fate of a cell is dictated by multiple molecular pathways. The number of ways cells can be driven to die in response to various stimuli seems to be ever-increasing (reviewed in Galluzzi et al., 2018), though perhaps the best known and most widely studied is apoptosis. Given the finality of apoptosis induction and the potential for pathological consequences if it is induced inappropriately, it is not surprising that there is a strong interconnection between the apoptotic pathways and other pathways that can promote cell survival. Most prominent of these is autophagy, with the "crosstalk" between the pathways being regulated by a multitude of mechanisms that can result in quite different outcomes (i.e., life or death) depending on the cell type, as well as the nature of the stimulus. Although autophagy itself can also elicit cell death under certain circumstances (see Bialik et al., 2018; Denton et al., 2015; Galluzzi et al., 2018; Maiuri et al., 2007c for reviews), this chapter will predominantly focus on its survival function and the myriad molecular events that directly connect autophagy with both the intrinsic (Fig. 1) and extrinsic (Fig. 2) pathways of apoptosis. Below we will first introduce the various pathways and their key players involved, then expand on the molecular connections between them.

2. Autophagy

2.1 General features of autophagy

Autophagy is defined as the lysosomal degradation of intracellular components. It is a catabolic process that enables the recycling of virtually most

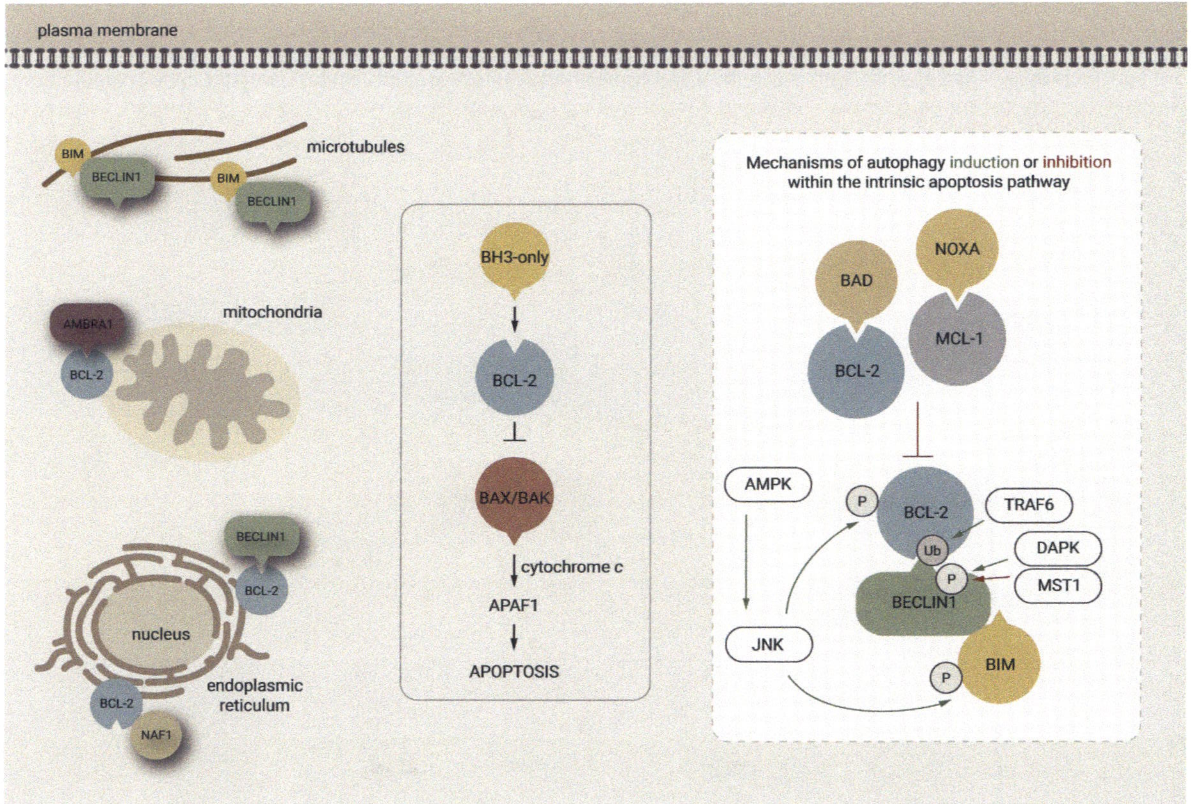

Fig. 1 The intrinsic pathway of apoptosis and its crosstalk with the autophagy pathway. Crosstalk between components of the intrinsic apoptosis and autophagy pathways occurs at multiple points and is mediated by mechanisms including protein-protein interactions and post-translational modifications. This review largely focuses on the crosstalk between BECLIN1 and regulators of the intrinsic apoptosis pathway. Left panel: Interactions between BECLIN1 with various proteins on different organelles; Center panel: Schematic of intrinsic apoptosis pathway; Right panel: Some post-translational modifications to BECLIN1 that influence autophagy outcomes.

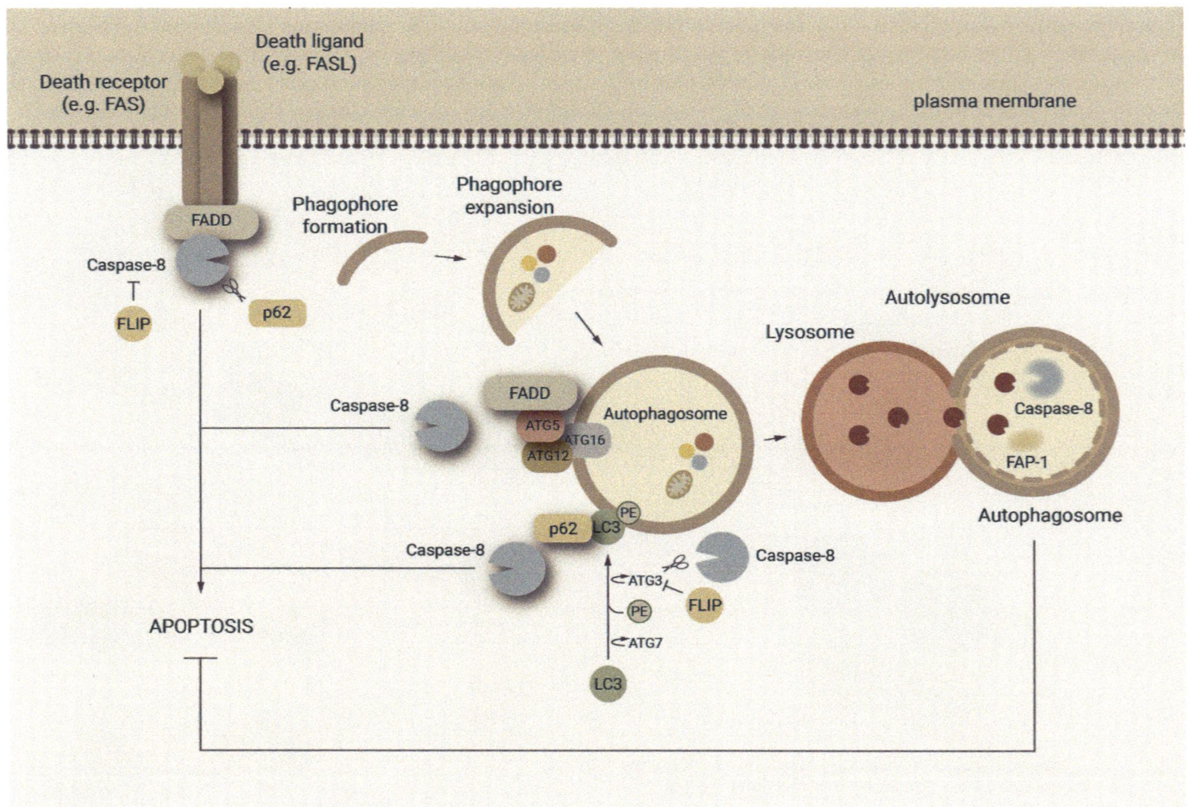

Fig. 2 The extrinsic pathway of apoptosis and its crosstalk with the autophagy pathway. As with the intrinsic apoptosis pathway, multiple points of convergence occur between regulators of the extrinsic apoptosis pathway and the autophagy pathway that include protein-protein interactions, post-translational modifications, protein cleavage and degradation. In addition, components within each pathway are often co-opted to a signaling axis distinct from their métier.

materials found throughout the cytoplasm, including proteins, whole or portions of organelles such as the mitochondria and nucleus, as well as invading pathogens (Galluzzi et al., 2017). These cargoes can be specifically or non-specifically targeted to the lysosome and broken-down into their constituent macromolecules such as amino acids and nucleosides, in a bioenergetic reaction (Levine and Kroemer, 2019). Resembling the ubiquitin proteasome degradation system, basal levels of autophagy are important for maintaining physiological cellular homeostasis by ensuring the clearance of damaged cellular entities. In addition to this role, adaptive autophagy is inducible with environmental stimuli, most commonly nutrient deprivation, infection and hypoxia, amongst others. It is thus no surprise that autophagy is highly conserved in eukaryotes, and dysregulation of this process contributes to the pathology of a diverse range of disorders, including cancer and neurodegeneration (Dikic and Elazar, 2018; Mercer et al., 2018; Wesselborg and Stork, 2015).

Autophagy is divisible into three broad categories depending on the route of cargo delivery to the lysosome; (i) Chaperone-mediated autophagy is the direct translocation of proteins bearing a KFERQ-like motif through a translocation complex consisting of HSPA-8 and LAMP2A, (ii) Microautophagy delivers cargo via the invagination of the lysosomal membrane, (iii) Macroautophagy, which is the focus of this review (and will be referred to throughout just as autophagy), involves the generation of a double membrane vesicle, termed the autophagosome, to encapsulate target cargo before fusion of this vesicle with the endo-lysosomal compartment. Of these three categories of autophagy, macroautophagy requires the cell to undergo the most obvious morphological changes and has been extensively studied for well over 20 years. Research in this field has converged to detail a classical morphological and molecular pathway for this process, which continues to undergo revision and refinement.

2.2 Overview of the stages of autophagy

Morphologically, autophagy begins with the nucleation of the nascent autophagosomal membrane, termed the phagophore or isolation membrane (IM). These can form from Ω-shaped entities on the endoplasmic reticulum (ER) enriched with both phosphatidylinositol-3-phosphate (PI3P) and double-FYVE domain-containing protein 1 (DFCP1) (Axe et al., 2008; Hayashi-Nishino et al., 2009). In addition to the ER, various other organelles are also thought to contribute membranes for autophagosome biogenesis. These include mitochondria (Hailey et al., 2010; Hamasaki et al., 2013), the plasma membrane (Nascimbeni et al., 2017; Ravikumar et al., 2010), and the ER-Golgi intermediate compartment (Ge et al., 2013, 2014). Following nucleation, the IM elongates and closes to form a sealed autophagosome. Autophagosomes undergo a maturation process where surface proteins are modified, allowing cellular recognition of the vesicle as ready for docking and fusion with endolysosomes. Cargo degradation is carried out by lysosomal hydrolytic enzymes, and the redistribution of the degradation products back into the cytosolic milieu is *via* lysosomal permeases. Finally, autophagolysosomes can bud to re-generate and replenish the lysosomal pool in a process termed "autophagic lysosome reformation" (Chen and Yu, 2017).

2.3 Protein complexes mediating the stages of autophagy

The core molecular machinery governing the stages of autophagy described above are clustered into six distinct groups based on functionality, and each act at various stages in the pathway. These groups include: (1) Unc-51 like kinase 1/2 (ULK1/2) complex, (2) class III phosphatidylinositol 3-kinase (PI3KCIII) complexes, (3) Autophagy-related (ATG)-9 (ATG9) vesicles, (4) WD-repeat phosphoinositide interacting (WIPI) and ATG2 proteins, (5) ATG5-ATG12/ATG16L1 conjugation machinery, and (6) the ATG8 protein family and associated conjugation machinery. As not all of these are involved in the crosstalk with the apoptosis pathways, we will only discuss those that are most relevant to the focus of this chapter.

Autophagy-stimulating conditions prompt nutrient-sensing machineries such as the mechanistic target of rapamycin (mTOR) complex or AMP-activated kinase (AMPK) to activate the ULK complex at the earliest stages of autophagosome formation, i.e., the nucleation of the nascent phagophore (Ganley et al., 2009; Gwinn et al., 2008; Inoki et al., 2003; Jung et al., 2009; Kim et al., 2011; Mack et al., 2012; Puente et al., 2016; Sanchez et al., 2012; Shang et al., 2011). The activated ULK complex, consisting of ULK1/2, FIP200, ATG13 and ATG101, triggers autophagy (Chan et al., 2009; Ganley et al., 2009; Loffler et al., 2011; Petherick et al., 2015) when the serine-kinase activity of ULK phosphorylates components of the ULK complex itself, or other members of the core autophagy machinery. Such substrates include ATG9 and ATG4B, as well as subunits of the PI3KCIII complexes (described in more detail below) such as BECLIN1, UVRAG and ATG14 (Park et al., 2016; Pengo et al., 2017; Russell et al., 2013; Zhou et al., 2017). The consequence of these phosphorylation events is modulation of their catalytic activities or subcellular distribution to initiate autophagy, and in the case of ATG4B phosphorylation, to inhibit processing of the ubiquitin-like protein LC3 (described below). It should be mentioned that aside from supporting kinase activity, members of the ULK complex may also be recruitment factors for targeting of the autophagy machinery to the nascent autophagosome. For example, FIP200 interacts with ATG16L and p62, bringing downstream conjugation machinery into the vicinity of the autophagic membrane (Nishimura et al., 2013; Turco et al., 2019).

There are two PI3KCIII complexes, which consist of the subunits VPS34, BECLIN1, VPS15/p150 and either ATG14 (PI3KCIII Complex 1, PI3KCIII-1) or UVRAG (PI3KCIII Complex 2, PI3KCIII-2) (Itakura et al., 2008; Liang et al., 2006; Matsunaga et al., 2009). The phosphorylation of VPS34 by ULK triggers its lipid kinase activity, enabling VPS34 to unleash its critical role of phosphatidylinositol 3-phosphate (PI3P) production by phosphorylation of phosphatidylinositol (PI). It should be noted that the ULK complex is not the only kinase to trigger VPS34 activity as both the mTOR complex and AMPK have also been shown to directly phosphorylate components of the PI3KCIII complexes (Kim et al., 2013; Yuan et al., 2013). Localized production of this phospholipid on the nascent phagophore leads to the recruitment of PI3P effector proteins, which include the WIPI family and DFCP1, all of which possess PI3P-binding domains (Axe et al., 2008; Bakula et al., 2017; Baskaran et al., 2012; Dooley et al., 2014; Gaugel et al., 2012; Jeffries et al., 2004; Krick et al., 2006, 2012; Obara et al., 2008; Polson et al., 2010; Proikas-Cezanne et al., 2004). Cumulatively the combination of the ULK complex, PI3KCIII, PI3P and PI3P-effectors drives the recruitment of the downstream machineries to the nucleation site to facilitate the membrane rearrangements required for autophagosome elongation and biogenesis.

Two ubiquitin-like conjugation systems are essential to phagophore elongation. The first is the ATG5-ATG12/ATG16 complex. The first step is the conjugation of ATG12 (a ubiquitin-like molecule) to ATG5 catalyzed by ATG7 (an E1-ubiquitin-activating-like enzyme) and ATG10 (an E2 ubiquitin-conjugating-like enzyme ATG10) (Ohsumi, 2001; Tanida et al., 2002). This ATG5-ATG12 conjugate subsequently associates non-covalently with ATG16 *via* ATG5 (Kuma et al., 2002). This complex is then recruited to the phagophore *via* WIPI2 (Dooley et al., 2014).

Once autophagosome formation is complete, the ternary complex dissociates (Mizushima et al., 2001, 2003). The second ubiquitin-like conjugation system involves the ATG8 proteins, which in mammals comprise two sub-families, LC3 and GABARAP (Tanida et al., 2004). Of these, LC3 is perhaps the best-characterized member of the family. Following cleavage by the cysteine protease ATG4 to expose its C-terminal glycine residue (Kabeya et al., 2000; Kirisako et al., 2000), LC3-I, a cytosolic ubiquitin-like protein, is conjugated to phosphatidylethanolamine (PE). This is performed by ATG7 (E1-like) and ATG3 (E2-like), together with the ATG5-ATG12/ATG16 complex, resulting in the lipidated form, LC3-II, which decorates the inner and outer membrane of the growing autophagosome (Kabeya et al., 2000; Tanida et al., 2002). Lipidation of ATG8 is a reversible process as ATG4 is also capable of cleaving the LC3-PE bond (Tanida et al., 2004). Lipidated ATG8 has been shown to possess inherent membrane tethering abilities that facilitates membrane fusion *in vitro*, which potentially explains how autophagosomal membrane growth occurs by addition of membranes from other vesicles (Nakatogawa et al., 2007).

3. Apoptosis

Apoptosis is probably the most extensively characterized form of programmed or regulated cell death. Two separate apoptotic pathways, called intrinsic and extrinsic, converge on different cysteinyl-aspartate proteases (caspases), that upon activation, can cleave hundreds of intracellular substrates resulting in the demise of the cell. As their name suggests, each pathway is invoked in response to different types of stimuli—the extrinsic pathway in response to the binding of extracellular ligands to cell surface receptors, and the intrinsic pathway to intracellular "stresses" that range from DNA damage, reactive oxygen species through to viral infection and growth factor deprivation. Both pathways have direct molecular connections to autophagy signaling, though perhaps the best studied of these is through the intrinsic apoptosis pathway.

4. The intrinsic apoptosis pathway

The intrinsic apoptosis pathway (also known as the mitochondrial pathway) is regulated by the BCL-2 family of proteins comprising more than a dozen members (Czabotar et al., 2014) (Fig. 1). These are all related to each other by the presence of up to four BCL-2 Homology (BH) domains. Each family member is generally categorized depending on its function—either pro-survival or pro-death. In mammals, the pro-survival members include BCL-2 itself, BCL-X_L, BCL-W, MCL-1 and BFL-1/A1 (though BCL-B is also sometimes included in this group). The pro-apoptotic members are divided into two further sub-groups: the "effector" proteins, BAX, BAK and BOK and the "BH3-only proteins" which includes BIM, PUMA, NOXA, BAD, BID, BIK, BMF and HRK, that initiate the pathway.

Structurally, the pro-survival proteins and the BAX/BAK/BOK molecules are very similar, despite their opposing functions. Both groups possess all four BH domains (BH1–4), and fold into similar globular three-dimensional structures consisting of eight helices (Czabotar et al., 2007; Lee and Fairlie, 2019; Moldoveanu et al., 2006; Petros et al., 2000; Sattler et al., 1997; Suzuki et al., 2000). In contrast, the BH3-only proteins only have the BH3 domain, and are mostly intrinsically unstructured (Hinds et al., 2007).

A notable feature on the pro-survival proteins is a large hydrophobic groove. This is the critical site for regulation of the pathway as it can engage both classes of the pro-apoptotic proteins *via* their BH3 domains which bind into the groove as an extended amphipathic helix (Czabotar et al., 2007; Lee and Fairlie, 2019; Petros et al., 2000; Sattler et al., 1997) (Fig. 3). The effector proteins BAX and BAK can also undergo a major structural change to an activated conformation following contact with membrane lipids that exposes their BH3-domains (Brouwer et al., 2014; Czabotar et al., 2013). This enables their homo-oligomerization through the exposed BH3 domain binding into a hydrophobic groove on another BAX/BAK molecule, in a manner directly analogous to how pro-survival proteins bind BH3 domains.

The sequence of events that lead to the demise of the cell *via* this pathway is reasonably well understood (Chen et al., 2015; Letai et al., 2002; Llambi et al., 2011; Willis et al., 2005, 2007). Cellular stresses lead to transcriptional upregulation of the BH3-only proteins or their "activation" through post-translational modifications. The BH3-only proteins then bind to the pro-survival proteins, displacing pre-bound, activated BAX/BAK. Certain BH3-only proteins known as "activators" (e.g., BIM and BID) can also engage the hydrophobic grooves on BAX and BAK to induce the same major conformational change. Once unrestrained, activated BAX and BAK oligomerize forming pore-like structures in the outer mitochondrial membrane resulting in the release of cytochrome *c*. Cytochrome *c* then binds to APAF-1 which assembles into a large platform (the "apoptosome") that enables conversion of inactive caspase-3/7/9 zymogens to active proteases, leading ultimately to the death of the cell due to cleavage of vital cellular substrates.

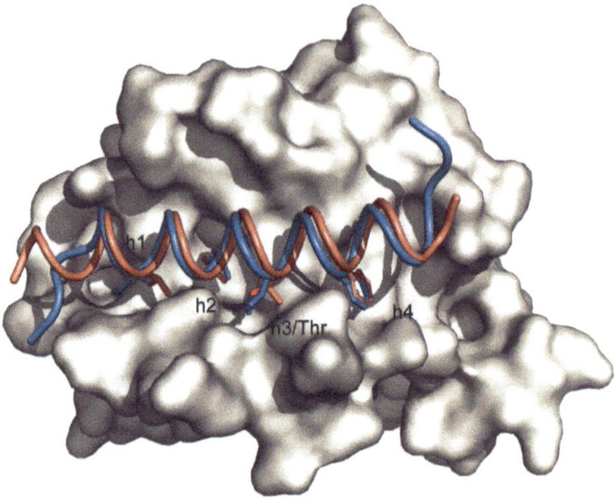

Fig. 3 Pro-apoptotic and pro-autophagic BH3 domains engage pro-survival proteins similarly. The BH3 domain from the pro-apoptotic BH3-only protein BIM (blue) engages the conserved hydrophobic groove of BCL-X_L (white). This high affinity interaction is mediated by the projection of four conserved hydrophobic residues (h1–h4) into corresponding pockets along the groove. The BH3 domain from the pro-autophagic protein BECLIN1 (pink) also engages pro-survival proteins in a similar manner. The main point of difference is the polar residue (threonine) in the h3 position, which is a DAPK phosphorylation site that leads to the disengagement of BECLIN1 from its pro-survival protein inhibitor. BIM BH3: BCL-X_L PDB ID: 3FDL. BECLIN1 BH3: BCL-X_L PDB ID: 2P1L.

Whether, the principal role of the pro-survival proteins is to restrain BAX and BAK or alternatively to prevent the activator BH3-only proteins from binding and activating BAX and BAK has been the subject of many studies. Most recently, some quite definitive studies using engineered cells and mice have pointed to the key mechanism of apoptotic inhibition being by the pro-survival proteins primarily interacting with BAX and BAK (Huang et al., 2019; Lee et al., 2016a; O'Neill et al., 2016). However, it is likely both modes of restraint could play a role depending on the context. Importantly, the fundamental molecular basis of apoptotic regulation (i.e., the "BH3-in-groove" interaction) is also central to how the intrinsic apoptosis and autophagy pathways talk to each other.

5. Crosstalk between intrinsic apoptosis and autophagy pathways

5.1 BECLIN1 is a BCL-2-interacting protein

Arguably, the best example of a direct molecular connection between the autophagy and apoptosis signaling pathways involves the autophagy regulator, BECLIN1. BECLIN1 is key component in the class III phosphotidylinositol-3-kinase complex that enables activation of VPS34 required for generation of phosphatidylinositol-3-phosphate and expansion of the phagophore and initiation of autophagy. BECLIN1 was actually discovered in a yeast two hybrid screen for BCL-2 interacting proteins in an attempt to understand why BCL-2 expression could protect cells against viral infection (hence, BECLIN1 was so-named because it is a BCL-2 (Becl) interacting protein) (Liang et al., 1998). BECLIN1 protein is homologous (24%) to the yeast autophagy gene, *apg6/vps30*, and can restore autophagic activity in *apg6/vps30*-disrupted yeast cells (Liang et al., 1999).

The human *BECLIN1* gene encodes a 451 amino acid protein and is located on chromosome 17q21 (Aita et al., 1999), a breast cancer susceptibility locus, and its monoallelic deletion is observed in a number of human malignancies (Liang et al., 1999). Initial studies using BECLIN1 truncation constructs mapped the BCL-2 (and BCL-X_L) interaction site to amino acids 88–150 which precedes a central coiled-coil domain (Liang et al., 1998). Subsequent studies identified F123A (the "h4" position in Fig. 3) as a critical residue for interaction with BCL-2 (Pattingre et al., 2005). Further investigation showed that BCL-2 expression could inhibit BECLIN1-dependent autophagy in yeast, and starvation-induced autophagy in mammalian cells, though not when the BECLIN1 F123A mutant was expressed, or with the BCL-2 G145A mutant that fails to bind BECLIN1 (Pattingre et al., 2005). Similarly, expression of BECLIN1 F123A induced excessive basal autophagy in cells. Strikingly, mice engineered to carry the equivalent murine sequence mutation (F121A) have increased healthspan and lifespan compared to littermates (Fernandez et al., 2018), which is associated with higher levels of basal autophagic flux. Similarly, these mice crossed onto an Alzheimer's disease (AD) mouse model show reduced amyloid accumulation that prevents cognitive decline and increased survival compared to the mice without the BECLIN1 mutation (Rocchi et al., 2017). All these data strongly point to a direct physical and functional link between apoptotic and autophagic regulators.

5.2 BECLIN1 is a novel BH3-only protein

It was not until 2007, however, that a flurry of studies recognized the BCL-2 binding domain in BECLIN1 (encompassing F123) as being a BH3 domain (Erlich et al., 2007; Feng et al., 2007; Maiuri et al., 2007b; Oberstein et al., 2007). Structural studies (both NMR and X-ray crystallographic) confirmed that the proposed BH3 sequence engaged BCL-X_L in a very similar manner to that shown for BH3 domains from pro-apoptotic BCL-2 proteins (Fig. 3). Similar structures have more recently been obtained for the interaction with BCL-2 (Lee et al., 2019). Notably, biophysical studies have consistently shown that the affinity of the BECLIN1 BH3 domain for BCL-X_L and BCL-2 was considerably weaker than for many corresponding pro-apoptotic BH3 sequences, with the K_D being in the low μM range (Feng et al., 2007; Lee et al., 2016b, 2019; Maiuri et al., 2007b; Noble et al., 2008; Oberstein et al., 2007; Reljic et al., 2016) *versus* low nM range for some of the pro-apoptotic BH3 domains (Chen et al., 2005). This weaker affinity is likely due to the presence of at least one unusual amino acid residue in BECLIN1 BH3, a threonine at position 119 (in the human sequence) that is normally occupied by a bulky hydrophobic residue in pro-apoptotic BH3 domains (Feng et al., 2007; Lee et al., 2019; Oberstein et al., 2007) (the "h3" position in Fig. 3). The weaker affinity of the BECLIN1 BH3 sequence likely accounts for why BECLIN1 is itself not pro-apoptotic, even when over-expressed in cells (Ciechomska et al., 2009a,b). This is despite an early study showing micro-injection of a BECLIN1 BH3 peptide into cells induced apoptosis, though this result mostly likely reflects the peptide being introduced at very high concentrations, rather than a normal physiological function of BECLIN1 (Maiuri et al., 2007b). Recognition of the BECLIN1 BH3 domain led to additional studies showing that mutations that weaken the interaction with BCL-2/BCL-X_L enhance autophagic responses, providing further evidence for the direct physical connection between the pathways (Fernandez et al., 2018; Maiuri et al., 2007b).

5.3 Regulation of PI3KCIII assembly *via* the BECLIN1:BCL-2 interaction

At the most fundamental level, binding of BECLIN1 to BCL-2 inhibits autophagy by preventing formation of the PI3K complex (Pattingre et al., 2005). However, there are multiple further levels of regulation that will be discussed in the following sections (summarized in Fig. 1 and Table 1).

As described above, BECLIN1 itself is a constituent of two class III PI3K complexes that contain the core subunits BECLIN1, VPS34 and VPS15 (Itakura et al., 2008; Liang et al., 2008; Matsunaga et al., 2009; Sun et al., 2008; Zhong et al., 2009). However, while there are now structures of BECLIN1 in the context of both these complexes (Fig. 4) (Baskaran et al., 2014; Rostislavleva et al., 2015), it is still unclear how its interaction with BCL-2 interferes with their formation, though inferences about the mechanism(s) can be drawn from various lines of evidence.

Table 1 Some key regulatory mechanisms involving BECLIN1 and intrinsic apoptosis proteins.

Regulator	Mechanism	Outcome
BCL-2/BCL-XL/ MCL-1/Viral BCL-2 proteins	Interaction with BECLIN1 BH3 domain	Inhibits autophagy
EGFR tyrosine kinase	BECLIN1 phosphorylation at Y229, Y233, Y352. Promotes interaction with BCL-2	Inhibits autophagy
JNK1	BCL-2 phosphorylation at S69, T70, S87. Reduces BCL-2 affinity for BECLIN1	Promotes autophagy
DAPK	BECLIN1 phosphorylation at T119 in its BH3 domain. Promotes dissociation of BCL-2	Promotes autophagy
MAPKAK2/MAPAK3	BECLIN1 phosphorylation at S90. Promotes dissociation of BCL-2	Promotes autophagy
MST1	BECLIN1 phosphorylation at T108 in BH3 domain. Promotes association of BCL-2	Inhibits autophagy
TRAF6	BECLIN1 ubiquitination at K117 in BH3 domain. Promotes dissociation of BCL-2	Promotes autophagy
NAF-1	Increases association of BECLIN1 with BCL-2 at ER	Inhibits autophagy
AMBRA1	Interacts with BECLIN1 preventing association with BCL-2	Promotes autophagy
HMGB1	Oxidized form interacts with BECLIN1 and displaces BCL-2. Also promotes ERK1/2 phosphorylation of BCL-2	Promotes autophagy
NOXA/BAD	BH3-only proteins displace BCL-2 pro-survival proteins from BECLIN1	Promotes autophagy
BIM-EL/BIM-L	BH3-only protein isoforms sequester BECLIN1 to dynein light chain LC8	Inhibits autophagy
BIF-1	Promotes association with UVRAG	Promotes autophagy
Caspase-3	Cleavage of BECLIN1 at D149	Inhibits autophagy

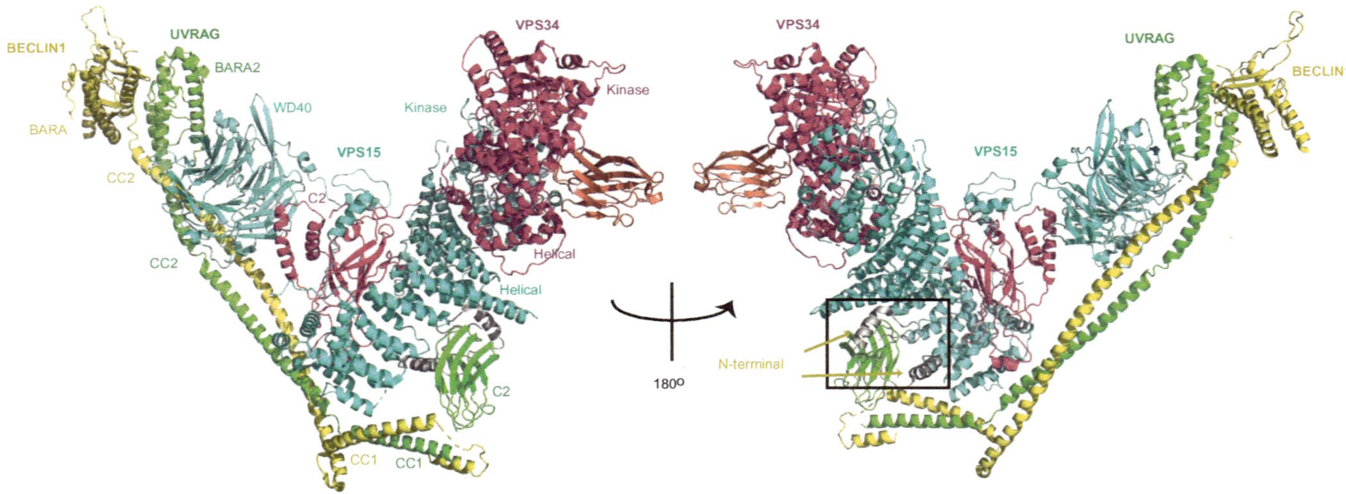

Fig. 4 Structure of the phosphatidylinositol 3-kinase (PI3KCIII) Complex 2. Both PI3KCIII complexes adopt a V-shaped architecture. Central to the complex is VPS15 (cyan), with one arm of the "V" comprising the BECLIN1 (yellow) and ATG14/UVRAG (green) heterodimer and the opposite arm made up of the catalytically active component of the complex, VPS34 (pink). The N-terminal of BECLIN1, which comprises the BH3 domain engaged by BCL-2 pro-survival proteins to inhibit autophagy and other sites of post-translational modification, is thought to be located at the apex of the "V" and potentially includes the gray helices (boxed). PDB ID: 5DFZ. BARA: β-α repeated, autophagy-specific; CC: coiled-coil.

BECLIN1 has an extended structure consisting of an unstructured N-terminal region that contains the BH3 domain (Lee et al., 2016b; Yao et al., 2016), a central coiled-coiled domain and a C-terminal β-α repeated autophagic (BARA) domain (also referred to as the evolutionary conserved [ECD] domain) that docks the complex to the membrane (Baskaran et al., 2014; Huang et al., 2012; Li et al., 2012; Noda et al., 2012; Rostislavleva et al., 2015). Both class III PI3K complexes have a similar V-shaped architecture with the left arm formed by the long coiled-coils of BECLIN1 that interact with the coiled-coils either of ATG14 or UVRAG (Baskaran et al., 2014; Rostislavleva et al., 2015). These coiled-coiled domains make multiple contacts with VPS15 and VPS34 that comprise most of the right arm of the "V." The BECLIN1 BARA domain is at tip of the left arm while the N-terminal region is at the apex of the "V." In the first structures of the PI3K complexes, the N-terminal domain of BECLIN1 was poorly resolved though in the yeast PI3KCIII-2 crystal structure, density corresponding to three helices that contact the coiled-coils domains were tentatively assigned to the BECLIN1 N-terminal region (Fig. 4) (Rostislavleva et al., 2015; Young et al., 2016). More recently, a cryo-EM structure of the human PI3KCIII-1 in complex with a positive autophagy regulator nuclear receptor binding factor 2 (NRBF2) showed NRBF interacts with, a helix not seen in previous structures and was also tentatively assigned to the BECLIN1 BH3 domain (Young et al., 2019). This suggests the BH3 domain perhaps becomes ordered upon interaction with NRBF2 (similar to that observed upon binding BCL-XL or BCL-2 (Feng et al., 2007; Lee et al., 2016b, 2019; Oberstein et al., 2007)). Moreover, although the assignment of this stretch of electron density to the BECLIN1 N-terminal domain, and the BH3 domain in particular, is tentative, it seems reasonable to speculate that BCL-2 could inhibit assembly of the PI3KCIII complex by sterically hindering the interaction with VPS15 as the unassigned electron density makes contacts with it.

Beyond these structures, biophysical studies have also provided an alternative/complementary mechanism for how BCL-2 might be functioning to inhibit autophagy. In solution, BECLIN1 can form a homodimer *via* its coiled-coiled domains (Noble et al., 2008) (probably in an anti-parallel orientation (Li et al., 2012)) which can be destabilized by the UVRAG coiled-coil domain to form a heterodimer. Binding of BCL-X_L to the BH3 domain stabilizes the BECLIN1 homodimer and reduces the affinity of UVRAG for BECLIN1 (Noble et al., 2008) which in turn would be expected to inhibit assembly of the PI3KCIII complex, at least for Complex 2. Whether ATG14 functions similarly in Complex 1 is yet to be determined.

5.4 Regulation of the BECLIN1:BCL-2 complex by post-translational modifications

A number of post-translational modifications on both BCL-2 and BECLIN1 have been shown to regulate autophagy. Related to the role of the homodimer described above, phosphorylation of BECLIN1 by epidermal growth factor receptor tyrosine kinase at Y229, Y233 and Y352 increases BECLIN1 interactions with BCL-2 by promoting BECLIN1 homodimerization, thereby inhibiting autophagy (Wei et al., 2013). By contrast, in response to starvation and ceramide exposure, c-jun N-terminal protein kinase 1 (JNK1) is activated and can phosphorylate BCL-2 within an unstructured loop at S69, T70 and S87 (Pattingre et al., 2009a,b; Wei et al., 2008). This multisite phosphorylation instead results in dissociation of BCL-2, leading to autophagy induction. Here, phosphorylation promotes an interaction between the unstructured loop and a region adjacent to the BH3 binding groove which also impairs the interaction with pro-apoptotic BH3 domains (Follis et al., 2018). This provides a mechanism whereby short-term starvation can induce autophagy through low level BCL-2 phosphorylation and dissociation of BECLIN1, but longer-term nutrient deficiency leads to dissociation of pro-apoptotic molecules such as BAX from BCL-2, switching the cells from a survival mode to one where death is promoted. Notably, mice engineered to express a BCL-2 mutant that cannot be phosphorylated by JNK1 have altered glucose metabolism during exercise and impaired exercise-mediated protection against glucose intolerance due to a high fat diet (He et al., 2012). The corollary of this is that in situations of high glucose levels (hyperglycemia), AMPK activity is reduced, which in turn reduces AMPK-mediated activation of JNK1, thereby promoting the interaction of BECLIN1 with BCL-2 and suppressing autophagy (He et al., 2013). This mechanism has important implications in cardiac cells where it can switch cells to apoptosis, leading to diabetic cardiomyopathy.

Phosphorylation of BECLIN1 within the BH3 domain can both positively and negatively regulate autophagy by modulating interactions with BCL-2. Death-associated protein kinase (DAPK) is a calcium/calmodulin-dependent kinase that can induce autophagy. The underlying mechanism is that DAPK phosphorylates BECLIN1 at T119 (the residue that is unique to the BECLIN1 BH3 among other BH3 sequences) (Fig. 3). This phosphorylation event reduces the affinity of the interaction with BCL-X_L and thereby promotes autophagy (Zalckvar et al., 2009a,b).

Phosphorylation of BECLIN1 at S90, just adjacent to the BH3 domain, by MAP kinase associated kinases, MAPKAK2/MAPKAK3, in response to starvation also promotes autophagy (Wei et al., 2015). However, BCL-2 binding to BECLIN1 can sterically inhibit this phosphorylation event, thereby further inhibiting autophagy induction (Wei et al., 2015).

In contrast, mammalian Ste-20-like kinase 1 (MST1) also phosphorylates BECLIN1 within its BH3 domain at T108. However, unlike with T119 phosphorylation, the interaction between BECLIN1 and BCL-X_L or BCL-2 is stabilized by formation of an electrostatic interaction between the phosphate group and a conserved histidine residue on both pro-survival proteins (Maejima et al., 2013). In this context, BCL-2/BCL-X_L sequestration by BECLIN1 also promotes apoptosis by reducing the amount of free pro-survival proteins available to inhibit BAX. Our recent structural studies showed that this interaction can form, though likely requires some other factor (such as the presence of membranes) to reduce the mobility of the histidine (Lee et al., 2019).

Another post-translational modification within the BH3-domain that regulates autophagy is K63-linked ubiquitination. Toll-like receptor 4 (TLR4) signaling, IFN-γ and amino acid starvation all result in TRAF6-mediated ubiquitination of BECLIN1 at K117 within the BH3 domain, promoting autophagy (Shi and Kehrl, 2010a,b). Conversely, A20-mediated deubiquitination of BECLIN1 limits autophagy in response to TLR signaling. Although not directly shown, mechanistically it is likely this ubiquitination event results in BCL-2 dissociation from BECLIN1, which has been observed following TLR stimulation (Shi and Kehrl, 2008).

5.5 Regulation of BECLIN1:BCL-2 interactions by other proteins

The ability of BCL-2 to inhibit autophagy initiated by BECLIN1 has also been shown to be regulated by several other proteins in different contexts. Some of the initial studies on BECLIN1 showed BCL-2 only interacts with endoplasmic reticulum (ER)-associated, but not mitochondrial-localized BCL-2 (Pattingre et al., 2005). Studies looking for ER proteins that interact with BCL-2 identified nutrient-deprivation autophagy factor (NAF-1) as a protein that is required for association of BCL-2 with BECLIN1 (Chang et al., 2010). NAF-1 does not have a BH3 domain but instead binds BCL-2 on a surface involving the BH4 and BH3 domains, adjacent to the BH3-ligand binding groove (Tamir et al., 2014). The interaction of BCL-2 with NAF-1, however, can be disrupted by BH3-only proteins (e.g., BIK), probably due to conformational changes induced in BCL-2 upon engagement with BH3 sequences, resulting in autophagy in response to nutrient deprivation. BCL-2 binding of NAF-1, however, does not influence apoptotic responses, for example, due to BH3-only protein expression.

Another protein known as activating molecule in BECLIN1-regulated autophagy (AMBRA1) positively regulates autophagy through a direct interaction with BECLIN1 (Fimia et al., 2007), and has also been found to bind to BCL-2 (Strappazzon et al., 2011). In contrast to NAF-1, this interaction occurs at the mitochondria possibly by a putative BH3 domain within AMBRA1, thereby sequestering AMBRA1 from BECLIN1 (Di Rita and Strappazzon, 2017; Strappazzon et al., 2011, 2016). Autophagy induction results in AMBRA1 release from BCL-2 by an unknown mechanism, allowing it to then bind to BECLIN1 at a site close to the BH3 domain, thereby inhibiting the BCL-2:BECLIN1 interaction and, hence, promoting the autophagy response. Interestingly, AMBRA1 has also been shown to have a pro-apoptotic effect. This occurs downstream of caspase-3 activation which causes cleavage and release of the AMBRA1 C-terminal fragment containing the putative BH3 sequence, which in turn inhibits the pro-survival activity of BCL-2 (Di Rita and Strappazzon, 2017; Tamir et al., 2014).

Yet another protein, high mobility group box 1 (HMGB1), can also interact with BECLIN1 to regulate autophagy (Tang et al., 2010). HMGB1 normally resides in the nucleus, though can translocate to the cytosol under the oxidizing conditions that can occur, for example, following autophagy-inducing conditions such as starvation. This translocation event involves C106 on HMGB1, while oxidative conditions also promote formation of a disulfide bond between C23 and C45, which in turn, promotes interaction with BECLIN1 and competitive displacement of BCL-2. This mechanism is further complicated due to HMGB1 being able to regulate phosphorylation of ERK (ERK1/2) through an unknown mechanism and, as a consequence, of BCL-2, which like JNK1-mediated phosphorylation, reduces BCL-2 binding to BECLIN1. Hence, HMGB1 promotes autophagy though multiple mechanisms that converge on the dissociation of BECLIN1 from BCL-2.

5.6 Regulation of autophagy by other mammalian BCL-2 proteins

Beyond BCL-2 and BCL-X_L that have been studied exhaustively, other BCL-2 pro-survival proteins have also been shown to regulate BECLIN1-mediated autophagy. Probably, the best characterized of these is MCL-1, a prominent pro-survival protein in multiple contexts including embryonic development and tumorigenesis. Apoptosis induced by multiple stimuli is often associated with MCL-1 degradation (Cuconati et al., 2003; Czabotar et al., 2007). Interestingly, nutrient deprivation conditions also trigger MCL-1 degradation. In post-mitotic neurons, this is accompanied by a strong autophagic response and cell survival (though is converted to apoptosis if BECLIN1 is not present to initiate autophagy) (Germain et al., 2011). This is consistent with the observation of a direct physical interaction between MCL-1 and BECLIN1 (Erlich et al., 2007; Germain et al., 2011). The degree of crosstalk between the pathways is notably cell-type dependent as in neuronal progenitor cells, apoptosis is instead the major response following MCL-1 removal, with the differences between the progenitors and post-mitotic cells reflected in overall levels of pro-apoptotic and autophagic proteins (Germain et al., 2011).

MCL-1 is also involved in autophagic responses in human ovarian surface epithelial cell transformed to express oncogenic Ras (H-Ras^{V12}), though with a different outcome. Here, oncogene expression results in potent upregulation of the BH3-only protein NOXA which has preferential binding affinity for MCL-1 among the pro-survival proteins, resulting in displacement of BECLIN1 (Elgendy et al., 2011). However, in contrast to the survival response seen in neurons, BECLIN1 release from its pro-survival protein antagonist results in excessive autophagy and autophagic (not apoptotic) cell death.

While the two examples above demonstrate how MCL-1 can directly inhibit BECLIN1, another mechanism of modulation has also been observed. Studies in melanoma cells showed that MCL-1 and BECLIN1 can negatively co-regulate each other by competitively binding to the deubiquitinase USP9X (Elgendy et al., 2014). Displacement of either protein from the complex with USP9X results in its ubiquitination and proteasomal degradation. This regulation of the relative levels of the MCL-1 and BECLIN1 has been proposed to have a role in progression of melanoma. These data together further demonstrate how crosstalk between the pathways can lead to different cell fates depending on the cell type, relative protein levels, and the stimulus.

Studies have also shown that the lesser studied pro-survival proteins BFL-1/A1 and BCL-B can also both inhibit autophagy in response to diverse stimuli, primarily through interactions with BECLIN1 (Kathania et al., 2011; Robert et al., 2012). Hence, all pro-survival members of the intrinsic apoptotic pathway can impact autophagic responses, though which ones are involved is going to be influenced by a multitude of factors including tissue expression pattern, upregulation of different BH3-only proteins (as these can differentially bind different pro-survival family members (Chen et al., 2005)), and the stimulus triggering the response.

Finally, a somewhat less direct connection between BECLIN1 and BCL-2 proteins involves BIF-1 (also known as Endophilin B1). BIF-1 can regulate apoptosis through a direct interaction with pro-apoptotic BAX (Cuddeback et al., 2001). However, BIF-1 also interacts with BECLIN1 *via* UVRAG (Takahashi et al., 2007, 2008). Formation of this complex involving BECLIN1, UVRAG and BIF-1 acts to positively regulate PI3KCIII lipid kinase activity and promotes autophagy.

5.7 Regulation of autophagy by other viral BCL-2 proteins

As discussed, BECLIN1 was discovered in a screen for proteins that could interact with human BCL-2 to protect cells against viral infection (Liang et al., 1998). It subsequently became apparent that many viruses express their own BCL-2 orthologues that can also bind to BECLIN1 to inhibit or evade autophagy and promote infection (Banjara et al., 2019; Ku et al., 2008; Pattingre et al., 2005; Piya et al., 2011; Sinha et al., 2008; Su et al., 2014). Not surprisingly, these interactions are mediated *via* the BECLIN1 BH3 domain that binds the viral BCL-2 proteins analogously to the mammalian BCL-2 counterparts (i.e., as an α-helix within the large hydrophobic groove). The affinities of these interactions are typically in the μM range, similar to that for the mammalian BCL-2:BECLIN1 BH3 interaction, though a higher affinity (K_D 40 nM) was reported for γ-herpesvirus 68 M11 protein which correlated with its increased capacity to inhibit autophagy compared to human BCL-2 (Ku et al., 2008). Combined these studies provide further support for a conserved mechanism of autophagy regulation through interactions involving the BECLIN1 BH3 domain.

5.8 BH3-mimetics target the BECLIN1:BCL-2 interaction

In the last 10 years, a major advance in the field has been the development of a new class of small molecule drugs, the "BH3-mimetics," that can inhibit the function of the pro-survival BCL-2 proteins in a manner analogous to the BH3-only proteins (i.e., by engaging the same hydrophobic groove to which BH3 domains bind) (Merino et al., 2018). Some of these have now entered the clinic and showing considerable success for the treatment of certain cancers.

Once the presence of a BH3 domain in BECLIN1 was recognized, it naturally followed that BH3-mimetics could be used to test the importance of the BCL-2:BECLIN1 interaction, and moreover, whether they could induce autophagy through competitively displacing the BCL-2:BECLIN1 interaction (Maiuri et al., 2007a,b; Malik et al., 2011a,b). Indeed, initial reports showed ABT-737 which binds BCL-2, BCL-X_L and BCL-W could dissociate BCL-2:BECLIN1 complexes (Maiuri et al., 2007a,b) and induce classic markers of autophagy.

More recent studies by Lindqvist et al., however, have questioned whether the BH3-mimetics were acting in these experiments via the direct displacement mechanism proposed, and instead demonstrated that autophagic markers only become apparent following BAX/BAK activation and apoptosis induction (Lindqvist and Vaux, 2014; Lindqvist et al., 2014). Specifically, these data showed that concentrations of ABT-737 capable of inducing apoptosis (hence, capable of potently inhibiting BCL-2 proteins) had no effect on LC3B-I lipidation unless BAX and BAK were present (Lindqvist and Vaux, 2014). However, this study was later questioned by Pedro et al. using a different BAX/BAK-deficient cell line, and suggested differences in the experimental protocols including the time-frame of the treatments (i.e., they were too short) and concentrations of drugs (i.e., they were too low) could account for the discrepancies with previous observations (Pedro et al., 2015). Nevertheless, additional subsequent studies by Lindqvist et al. further showed that an enantiomer of ABT-737 incapable of engaging BCL-2 proteins or inhibiting their function in cells could still induce LC3B-I lipidation if used at the higher concentrations suggested by Pedro et al., hence providing evidence for an off-target effect of the drug, and questioning the direct role of BCL-2 proteins in inhibiting BECLIN1 (Reljic et al., 2016). Hence, whilst there are multiple lines of evidence supporting a critical role of BCL-2 proteins in inhibiting autophagy by engaging BECLIN1, these studies provide some reasonable grounds for further investigation of these interactions in understanding the direct interplay between the cell survival and cell death pathways.

5.9 Regulation of autophagy by caspase cleavage of BECLIN1

Interestingly, in contrast to the studies by Lindqvist et al. that show markers of autophagy are induced by apoptosis, other studies have shown autophagy is inhibited downstream of apoptosis induction due to BECLIN1 cleavage by caspases, most prominently caspase-3 (Cho et al., 2009;

Djavaheri-Mergny et al., 2010; Luo and Rubinsztein, 2010; Wirawan et al., 2010; Zhu et al., 2010). This cleavage has been shown to occur at BECLIN1 D149 in response to a range of stimuli including growth factor withdrawal, treatment with TRAIL as well as nutrient deprivation and oxidative stress. One study showed the N- and C-terminal BECLIN1 fragments mislocalize to the nucleus and mitochondria, respectively, potentially explaining the loss of autophagy activity (Wirawan et al., 2010). Surprisingly, the C-terminal fragment, and not the N-terminal region which contains the BH3-domain, was found to possess pro-apoptotic activity, further amplifying the cell death response due to growth factor withdrawal. Hence, these data support a model whereby the pro-survival function of autophagy is shut-down to promote the cell death.

5.10 BH3-only proteins regulate, and are regulated by, autophagy

An obvious mechanism by which autophagy is regulated is through the upregulation of BH3-only proteins such as BAD that can bind to BCL-2 and BCL-X_L to displace BECLIN1 (Maiuri et al., 2007b). A sub-group of BH3-only proteins called the BNIPs have also been shown to act in the same way as a survival mechanism in response to hypoxia (Bellot et al., 2009). A distinct mechanism of autophagy regulation by proteins of the intrinsic pathway involves the BH3-only protein BIM, though expression of the BIM-EL and BIM-L isoforms do not have the same effect as, e.g., BAD (Luo and Rubinsztein, 2013). This is unexpected given the high affinity of the BIM BH3 domain for all BCL-2 pro-survival proteins (Chen et al., 2005). Instead, BIM is a negative regulator of autophagy due to its ability to mislocalize BECLIN1 to the dynein light chain LC8 through a direct interaction that does not involve either the BIM or BECLIN1 BH3 domains (Luo and Rubinsztein, 2013). Starvation conditions leads to JNK-mediated phosphorylation of BIM and its release, together with BECLIN1 from LC8. Therefore, depending on the conditions, BIM can switch from being autophagy-inhibitory to pro-apoptotic.

Finally, there is one particularly interesting example whereby autophagy can regulate apoptotic responses. Here, levels of the BH3-only protein PUMA are regulated by autophagy in a p62-dependent mechanism (Thorburn et al., 2014). This has significant effect on both the kinetics and extent of the cell death response to the extrinsic apoptotic pathway ligand, TRAIL. Hence, this study provides an example where molecular components of both intrinsic and extrinsic apoptosis converge with autophagy.

In the next section we will describe multiple further examples of direct interactions between the extrinsic apoptotic pathway and autophagy.

6. The extrinsic apoptosis pathway

In contrast to the intrinsic pathway of apoptosis whereby intracellular death signals converge to activate the upstream BH3-only proteins, the extrinsic pathway is initiated by the binding of cognate ligands to transmembrane receptors at the cell surface (Green and Llambi, 2015). These so-called death receptors are members of the tumor necrosis factor (TNF) receptor superfamily and include FAS, TNFR1, TRAIL-R1/R2 and their corresponding ligands FASL, TNF-α and TRAIL.

In the case of FAS (Kischkel et al., 1995) and TRAIL-R1/R2 (Sprick et al., 2000), ligand binding results in formation of a caspase activation platform known as the death-inducing signaling complex (DISC), which is mediated by a series of homotypic protein-protein interactions. Here, the adaptor protein FADD binds the cytoplasmic tail of the death receptor *via* the "death" domain (DD). This adaptor protein in turn recruits inactive pro-caspase-8 *via* the "death effector" domain (DED) enabling the two-step proximity-induced autocatalytic cleavage of caspase-8. The first step is stabilization of the active dimer (Boatright et al., 2003) which is followed by its full activation and autocatalytic processing that requires its further aggregation. This is followed by its subsequent release from the DISC to exert its function. Once caspase-8 is released from its activating platform, apoptosis ensues as the active enzyme proceeds to activate either the executioner caspases (caspase-3 and caspase-7) that dismantle critical intracellular targets critical (Crawford and Wells, 2011) or cleave the BH3-only protein BID which leads to BAX and/or BAK activation (Jost et al., 2009).

With TNFR1, the outcome of ligand binding is significantly more complex and instead of a direct path to cell death via the activation of caspases, different protein complexes dictate whether a cell lives, or the mechanism by which it dies (e.g., necroptosis or apoptosis). As the focus of the review is on the factors mediating the crosstalk between autophagy and apoptosis, we will not go into further detail on this as there are many excellent reviews on the TNFR1-signaling pathway and how it determines cell survival or death (Brenner et al., 2015; Green and Llambi, 2015; Peltzer et al., 2016).

There have been multiple postulated points of intersection between the extrinsic apoptosis pathway and the autophagic machinery (summarized in Fig. 2 and specific regulators summarized in Table 2). In many instances "traditional" autophagy regulators undertake functions not necessarily

Table 2 Crosstalk between extrinsic apoptosis and autophagy regulators.

Autophagy regulator(s)	Mechanism	Outcome
ATG5	Protein-protein interaction	Increased apoptosis Activation of Caspase-8
p62, LC3	Protein-protein interaction	Increased apoptosis Activation of Caspase-8
ATG3	Protein-protein interaction	Decreased autophagy Impairment of autophagic vacuole formation and LC3 lipidation
ATG3	Cleavage of ATG3 by Caspase-8	Decreased autophagy Inhibition of LC3 lipidation
p62	Cleavage of p62 by Caspase-8	Decreased autophagy Removal of autophagy-related domains on p62
Lysosome	Degradation of Caspase-8	Decreased apoptosis Lysosome-mediated degradation of Caspase-8
p62	Degradation of FAP-1	Increased apoptosis Autophagic degradation of FAP-1

related to their canonical roles within what is traditionally a survival process. Instead these autophagy regulators go rogue exerting cell death by interacting with regulators of the extrinsic pathway.

7. Crosstalk between extrinsic apoptosis and autophagy pathways

7.1 Autophagosomes remove cellular garbage, but also enable cell death

Autophagosomes are the epicenter of autophagy due to their ability to encapsulate cytoplasmic contents targeted for lysosomal degradation. In addition to serving as the waste and recycling facility of cells to enable survival, quite unexpectedly, they can also serve as a platform for the formation of complexes that include components of the extrinsic apoptosis pathway to deliver the coup de grâce once a cell is beyond salvation.

The first insights that autophagosomes can lead to cell death following autophagy induction came when disruption of their formation with chemical treatment using 3-methyladenine (3MA) which inhibits PI3KCIII, or by the genetic removal of genes essential to autophagosome formation (e.g., ATG5 or ATG7 (Geng and Klionsky, 2008)) inhibited apoptosis (Bell et al., 2008; Jiang et al., 2011; Laussmann et al., 2011; Pyo et al., 2005). In contrast, treatment with chemicals (such as bafilomycin A_1 or chloroquine) that inhibit autolysosomal acidification or autophagosome-lysosome fusion (Jiang et al., 2011; Pan et al., 2011), or deletion of genes such as ATG2 which causes the accumulation of unsealed immature autophagosomal membranes (Kishi-Itakura et al., 2014; Tang et al., 2017; Velikkakath et al., 2012), results in the promotion of cell death.

The obvious question from these observations is how does the extended presence of autophagosomes, which are a major component of a cellular survival program, lead to cell death? It was soon realized that the autophagosome can serve as a site on which an intracellular death-inducing signaling complex (iDISC) forms. This is akin to the DISC that forms at the plasma membrane following engagement of the death receptors by their ligands. Like the cell surface DISC, the end result of iDISC formation on the autophagosome is the recruitment, activation and subsequent release of activated caspase-8 to execute cell death. In addition to apoptosis, autophagy-activated caspase-8 is also thought to play non-apoptotic roles such as the regulation of mitogen-induced T cell proliferation (Bell et al., 2008).

7.2 ATG5:FADD:Caspase-8 recruitment to autophagosomes

There are now two well-accepted mechanisms by which caspase-8 is recruited to the autophagosome, both of which involve interactions with autophagy regulators either directly or indirectly. The first requires ATG5, which as described earlier is a key component of the ATG12 conjugation system (Bell et al., 2008; Deegan et al., 2014; Geng and Klionsky, 2008; Jiang et al., 2011; Laussmann et al., 2011; Pyo et al., 2005; Tang et al., 2017; Young et al., 2012). Using techniques that include co-immunoprecipitation, fluorescence microscopy and yeast two hybrid screens, ATG5 was shown to interact *via* its central and C-terminal region with the death domain of FADD (Pyo et al., 2005). The ability of ATG5 to bind FADD is dependent on the conjugation of the former to ATG12 and their association with ATG16 (Pyo et al., 2005). The binding of FADD to the ATG5-ATG12/ATG16 complex in turn enables the recruitment of caspase-8 to the autophagosomal membrane.

Notably, disabling ATG5-mediated recruitment of FADD to the autophagosome or inhibition of caspase-8 with a caspase inhibitor prevents cell death but not vacuole formation. This clearly delineates the pathway to autophagy-dependent cell death as two distinct steps: (1) autophagosome formation and (2) cell death mediated by caspase-8 (Pyo et al., 2005).

7.3 Caspase-8:p62:LC3 recruitment to autophagosomes

The second reported mechanism by which caspase-8 is recruited to autophagosomes is *via* its interaction with the ubiquitin-binding protein p62 and LC3, which is involved in phagophore elongation and autophagosome-lysosome fusion (Huang et al., 2013; Jin et al., 2009; Moscat and Diaz-Meco, 2009; Nguyen et al., 2016; Pan et al., 2011; Yan et al., 2019). Formation of this complex has been seen in cells where autophagy has been induced with chemotherapy or when treated with proteasomes inhibitors such as MG132. In support of the functional relevance of this complex in mediating autophagy-activated cell death, it was shown that p62 is the central mediator holding this tripartite complex together as its knock-down prevented caspase-8 binding to LC3 and attenuated caspase-8 activation, while knock-down of LC3 had little impact on the ability of caspase-8 to bind p62 (Huang et al., 2013; Pan et al., 2011). Further structure-function analysis demonstrated that an intact ubiquitin associated domain (UBA) and protein-protein interaction domain (PB1) on p62 is essential for effective caspase-8 activation to induce apoptosis (Huang et al., 2013). In turn, the LC3-interacting region (LIR) on p62 binds LC3 localizing the complex on autophagosomes. The recruitment of caspase-8 to autophagosomes by LC3 (and p62) appears to be essential for its aggregation and activation as an LC3 mutant defective in its ability to be lipidated, and be membrane localized, inhibited cell death (Huang et al., 2013; Pan et al., 2011).

The first report that caspase-8 recruitment to autophagosomes can be mediated by the p62:LC3 axis demonstrated that p62 in fact binds to ubiquitinated caspase-8 (Pan et al., 2011), which is not unexpected given the critical function of the UBA on p62. The E3 ligase cullin3 has been shown to induce the polyubiquitination of caspase-8 following death receptor ligation (Jin et al., 2009). However, it is still unknown how, in the absence of extrinsic death receptor signaling, does caspase-8 get ubiquitinated to enable its capture by p62. Interestingly, Cullin3 has been shown to be an LC3-interacting protein so perhaps ubiquitination of caspase-8 could occur in an LC3-dependent manner? It is also a curious observation that p62 appears to preferentially engage caspase-8 over the initiator caspase of the intrinsic pathway, caspase-9 (Huang et al., 2013). What accounts for this differential ability for autophagosomal recruitment and activation remains to be determined.

The recruitment and activation of caspase-8 to autophagosomes by either the ATG5:FADD or p62:LC3 axes are functionally redundant as individual loss of either FADD or p62 only marginally attenuated caspase-8-mediated cell death, as opposed to their simultaneous deficiency which significantly suppressed caspase activity (Tang et al., 2017; Young et al., 2012). The ATG5-ATG12/ATG16 complex localizes to phagophores and pre-phagaphore structures but dissociates when autophagosomes are fully formed (Rubinsztein et al., 2012). In contrast, LC3-II stays associated until autophagosomes fuse with lysosomes. Perhaps therefore, the ability of caspase-8 to be recruited *via* the two axes ensures signal longevity under times of prolonged stress that induces autophagy to then trigger apoptosis.

7.4 What is the purpose of caspase-8 at autophagosomes?

The recruitment of caspase-8 to autophagosomes through the ATG5:FADD or p62:LC3 axes exemplify how alteration of the protein interaction network between regulators of extrinsic apoptosis and autophagy can act as a switch to dictate cell death or cell survival. This is in contrast to when the functions of these regulators are altered as a consequence of protein modification or cleavage. Notably, formation of the iDISC is independent of the canonical extrinsic pathway induced by death receptor signaling or the canonical intrinsic pathway leading to BH3-only protein activation (Huang et al., 2013; Laussmann et al., 2011).

Why then does the autophagy machinery that normally instructs a cell to survive have the capacity to recruit a non-canonical pathway to cell death? It is possible that the role of non-canonical caspase-8 mediated apoptosis is an alternative mechanism to cell death in the absence of extrinsic or intrinsic signals that normally signal canonical apoptosis pathways. Instead of the requirement of an extrinsic signal *via* ligand binding to death receptors, caspase-8 can also be activated in response to intracellular signals such as nutrient starvation acting as a key switch between cytoprotective autophagy and apoptosis. As high stress levels and autophagy induction are sustained and not removed, cells have the capacity to sense and respond to this directly using the autophagy machinery to trigger cell death. Consistently it appears that adaptor proteins such as FADD are upregulated following stress induction (Deegan et al., 2014).

7.5 FLIPing between cell death and cell survival

As discussed above, during autophagosome elongation, LC3 is cleaved by the cysteine protease ATG4 to expose its carboxy terminal glycine residue. This enables its activation by, and transfer to, ATG3 which in turn leads to the conjugation of processed LC3 to phosphatidylethanolamine. Another well-described mechanism whereby crosstalk between regulators of the extrinsic apoptosis and autophagy pathways occurs to impact cell survival is the binding of ATG3 to the anti-apoptotic protein FLICE (caspase-8)-inhibitory protein, FLIP. The FLIP proteins were originally identified through a genome search for novel DED-containing virus-encoded inhibitors of apoptosis (Thome et al., 1997) with the cellular homologue (cFLIP) subsequently identified (Irmler et al., 1997). The FLIP proteins protect against apoptosis triggered by several death receptor ligation events involving FAS, TNFR1, TRAIL receptor 1 (Budd et al., 2006). This anti-apoptotic activity is mediated by the presence of two DEDs, found in all FLIPS, which compete with caspase-8 for recruitment to the DISC via FADD, thereby preventing the processing and subsequent activation of the enzyme (Irmler et al., 1997; Krueger et al., 2001).

In the same way that the DEDs on FLIP mediate its anti-apoptotic function, they also mediate its interaction with ATG3 *via* two distinct regions within its C-terminus (Lee et al., 2009). Interestingly, mutational analysis revealed that the anti-autophagic role of FLIP can be dissociated from this anti-apoptotic activity despite the DEDs also being responsible for inhibiting caspase-8 recruitment to FADD. Not unexpectedly given the critical role of ATG3 in the generation of lipid conjugated LC3 essential to autophagosome elongation, the binding of FLIP to ATG3 impairs autophagic vacuole formation and LC3 lipidation (He and He, 2013; Lee et al., 2009). This inhibition of autophagy by FLIP can occur in response to various autophagy-inducing stimuli, including starvation or rapamycin treatment. Under normal conditions, and hence low levels of autophagy, FLIP efficiently competes with LC3 to bind ATG3. However, in response to stress conditions, FLIP dissociates from ATG3 enabling the latter to go on and activate LC3 to increase autophagy levels. Therefore, similar to the way that the BCL-2 proteins can inhibit autophagy by binding

BECLIN1, a constituent of the extrinsic pathway is also capable of inhibiting autophagy. Notably, these points of crosstalk inhibit distinct steps within the multistep process of autophagy, providing a multi-tier level of control of the two diametrically opposing pathways of cell survival and cell death.

7.6 Caspase-8 cleavage of autophagy regulators

There have been multiple reports of caspases and other proteases mediating the cleavage of autophagy-related proteins to modulate cell survival and death (Norman et al., 2010). For example, calpain cleaves ATG5 (Yousefi et al., 2006), caspase-3 has been shown to cleave ATG4D (Betin and Lane, 2009), and as discussed above, BECLIN1 has been reported to be a substrate for multiple caspases (Cho et al., 2009; Djavaheri-Mergny et al., 2010; Luo and Rubinsztein, 2010; Wirawan et al., 2010; Zhu et al., 2010). In a similar vein, caspase-8, which is a key initiator caspase within the extrinsic apoptosis pathway, has been shown cleave ATG3 (Norman et al., 2010; Oral et al., 2012). This ability of caspase-8 to cleave ATG3 was first proposed based on the presence of a canonical caspase-8 cleavage site that is mostly conserved throughout evolution, with the exception of in *Drosophila melanogaster* and lower organisms such as *Caenorhabditis elegans* and yeast. This cleavage site is thought to fall in the region required for its interaction with ATG7. Hence, cleavage here would in turn inhibit the lipidation of LC3 (Norman et al., 2010; Yamada et al., 2007). Furthermore, the ATG3 fragments generated following caspase-8 cleavage lead to their destabilization and degradation in a proteasome-dependent manner, further ensuring autophagy suppression.

In addition to ATG3, p62 cleavage by caspase-8 following TRAIL-mediated induction of apoptosis has also been reported. The consequence of this cleavage is postulated to be a loss of polyubiquitin binding (Norman et al., 2010). As described above, p62 recruits polyubiquitinated caspase-8 to autophagosomes resulting in activation of the enzyme. One could therefore hypothesize that the ability of activated caspase-8 to cleave p62 may serve as a negative feedback mechanism to regulate the extent to which caspase-8-mediated apoptosis occurs. Interestingly, a recent report has described that p62 can also be cleaved by caspase-8 activated downstream of Toll-like receptor 3 (TLR3) signaling (Sanchez-Garrido et al., 2018). This TLR3/RIPK1-dependent activation of caspase-8 generates a stable fragment of p62, $p62^{TRM}$. This cleavage event abrogates the autophagic function of p62 due to the removal of conserved domains critical to the autophagic capacity of p62 (including the UBA and LIR). Instead of participating in autophagy, here p62^{TRM} instead promotes leucine sensing by mTORC1. Unlike other cleavage events of autophagy regulators mediated by caspase-8 to modulate levels of cell survival or cell death, this results in an autophagy regulator to be co-opted to an alternate signaling axis.

7.7 Autophagic degradation of extrinsic apoptosis regulators

An alternate way in which regulators of the extrinsic apoptosis pathway communicate with the autophagic pathway is not *via* direct interaction or protein modification as has been described above but instead, occurs *via* their autophagic degradation. In apoptosis-deficient cells, cytoprotective autophagy is induced following receipt of stress stimuli such as cytotoxic drugs (Carew et al., 2007; Katayama et al., 2007) or death receptor signaling (Corcelle et al., 2009; Herrero-Martin et al., 2009). However, when autophagy is inhibited, the balance shifts back in favor of cell death. It has been shown that under such circumstances, caspase-8, in particular the cleaved form and not the full-length protein, mediates this switch (Hou et al., 2010). In the absence of a fully functional apoptosis pathway (e.g., in the absence of BAX), activated caspase-8 is sequestered by autophagosomes and subsequently degraded by lysosomes. However, when autophagosome nucleation is diminished due to the knock-down of BECLIN1 or VPS34, cleaved caspase-8 is no longer eliminated by autophagy. It should be noted though that the autophagic repression of caspase-8 appears to be cell-type specific and furthermore, it remains to be determined what dictates the preferential sequestration of cleaved caspase-8 over the full-length protein by autophagosomes.

Another component within the extrinsic apoptosis pathway whose levels are regulated by autophagic degradation is FAP-1 (Gump et al., 2014). This tyrosine phosphatase specifically modulates FAS (but not TRAIL) signaling through its ability to dephosphorylate FAS to reduce its cell surface expression and activity, and hence serves to negatively regulate FAS-induced apoptosis (Ivanov et al., 2003, 2006; Li et al., 2000; Sato et al., 1995). The p62 protein has been shown to directly bind FAP-1 and direct it to autophagosomes for autophagic degradation. As a consequence, cells become more sensitive to FAS-mediated apoptosis. Significant stochastic variability in basal autophagy flux has been observed between cells within a population (though it should be noted that this variability is only transient). Therefore, this autophagy variation is thought to contribute to the cell-to-cell variability in apoptotic response seen within a population of cells. It should be noted that the impact of autophagic degradation of FAP-1 on cell survival is stimulus- and cell type-specific and the reasons for this are as yet unclear. However, one can postulate that the multiple levels of crosstalk between autophagy and members of the extrinsic apoptosis pathway that have been described all ultimately contribute to the impact of autophagy on cellular outcome.

Finally, there is a somewhat different example of how autophagic proteins can influence levels of extrinsic apoptotic regulators that does not directly involve autophagic degradation. It was shown that BECLIN1 can bind to Survivin, an inhibitor of apoptosis protein (Niu et al., 2010). This interaction stabilizes Survivin which is normally a short-lived protein. Moreover, in glioma cells, depletion of BECLIN1 results in more rapid turnover of Survivin, which results in increased sensitivity to TRAIL-induced apoptosis. Hence, these data further demonstrate how BECLIN1 can act to influence both intrinsic and extrinsic apoptosis.

8. Concluding remarks

Components of the intrinsic and extrinsic apoptotic pathways engage with autophagy regulators to dictate the fate of a cell in multiple different ways. For example, crosstalk between autophagy and intrinsic apoptosis can be directly mediated by BECLIN1:BCL-2 protein interaction and the post-translational modifications that enhance or disrupt this interaction. The extrinsic pathway is linked to autophagy by recruitment of caspase-8 and FLIP to the autophagosomes. Autophagy and apoptosis can also regulate each other by autophagic degradation of apoptotic proteins, or alternatively, cleavage of autophagic proteins by caspases downstream of apoptosis induction. However, as has been highlighted numerous times throughout this chapter, the mechanisms employed by cells to promote their survival or execute their own death varies enormously depending on the stimulus and cell type, and the relative levels of each of the molecular players at the time of the triggering event. Accordingly, deregulated expression of any of the proteins involved in either autophagy or apoptosis can have significant pathological impact. A simple but an important example of this is in tumor cells that often over-express BCL-2 pro-survival proteins. While this can obviously enhance their ability to evade cell death and promote tumorigenesis by inhibition of BAX/BAK, their oncogenic potential may also be further promoted due to the ability of BCL-2 proteins to also inhibit autophagy

(Pattingre and Levine, 2006). Hence, while our current knowledge of how apoptotic and autophagy pathways intersect is extensive, it is critical we continue to consider all of the potential interconnections between these pathways rather than viewing each in isolation if we are to fully understand why cells live or die under different conditions.

References

Aita, V.M., Liang, X.H., Murty, V.V., Pincus, D.L., Yu, W., Cayanis, E., Kalachikov, S., Gilliam, T.C., Levine, B., 1999. Cloning and genomic organization of beclin 1, a candidate tumor suppressor gene on chromosome 17q21. Genomics 59, 59–65.

Axe, E.L., Walker, S.A., Manifava, M., Chandra, P., Roderick, H.L., Habermann, A., Griffiths, G., Ktistakis, N.T., 2008. Autophagosome formation from membrane compartments enriched in phosphatidylinositol 3-phosphate and dynamically connected to the endoplasmic reticulum. J. Cell Biol. 182, 685–701.

Bakula, D., Muller, A.J., Zuleger, T., Takacs, Z., Franz-Wachtel, M., Thost, A.K., Brigger, D., Tschan, M.P., Frickey, T., Robenek, H., Macek, B., Proikas-Cezanne, T., 2017. WIPI3 and WIPI4 beta-propellers are scaffolds for LKB1-AMPK-TSC signalling circuits in the control of autophagy. Nat. Commun. 8, 15637.

Banjara, S., Shimmon, G.L., Dixon, L.K., Netherton, C.L., Hinds, M.G., Kvansakul, M., 2019. Crystal structure of African swine fever virus A179L with the autophagy regulator beclin. Viruses 11, E789.

Baskaran, S., Ragusa, M.J., Boura, E., Hurley, J.H., 2012. Two-site recognition of phosphatidylinositol 3-phosphate by PROPPINs in autophagy. Mol. Cell 47, 339–348.

Baskaran, S., Carlson, L.A., Stjepanovic, G., Young, L.N., Kim, D.J., Grob, P., Stanley, R.E., Nogales, E., Hurley, J.H., 2014. Architecture and dynamics of the autophagic phosphatidylinositol 3-kinase complex. Elife 3, e05115.

Bell, B.D., Leverrier, S., Weist, B.M., Newton, R.H., Arechiga, A.F., Luhrs, K.A., Morrissette, N.S., Walsh, C.M., 2008. FADD and caspase-8 control the outcome of autophagic signaling in proliferating T cells. Proc. Natl. Acad. Sci. U. S. A. 105, 16677–16682.

Bellot, G., Garcia-Medina, R., Gounon, P., Chiche, J., Roux, D., Pouyssegur, J., Mazure, N.M., 2009. Hypoxia-induced autophagy is mediated through hypoxia-inducible factor induction of BNIP3 and BNIP3L via their BH3 domains. Mol. Cell. Biol. 29, 2570–2581.

Betin, V.M., Lane, J.D., 2009. Caspase cleavage of Atg4D stimulates GABARAP-L1 processing and triggers mitochondrial targeting and apoptosis. J. Cell Sci. 122, 2554–2566.

Bialik, S., Dasari, S.K., Kimchi, A., 2018. Autophagy-dependent cell death—where, how and why a cell eats itself to death. J. Cell Sci. 131, jcs215152.

Boatright, K.M., Renatus, M., Scott, F.L., Sperandio, S., Shin, H., Pedersen, I.M., Ricci, J.E., Edris, W.A., Sutherlin, D.P., Green, D.R., Salvesen, G.S., 2003. A unified model for apical caspase activation. Mol. Cell 11, 529–541.

Brenner, D., Blaser, H., Mak, T.W., 2015. Regulation of tumour necrosis factor signalling: live or let die. Nat. Rev. Immunol. 15, 362–374.

Brouwer, J.M., Westphal, D., Dewson, G., Robin, A.Y., Uren, R.T., Bartolo, R., Thompson, G.V., Colman, P.M., Kluck, R.M., Czabotar, P.E., 2014. Bak core and latch domains separate during activation, and freed core domains form symmetric homodimers. Mol. Cell 55, 938–946.

Budd, R.C., Yeh, W.C., Tschopp, J., 2006. cFLIP regulation of lymphocyte activation and development. Nat. Rev. Immunol. 6, 196–204.

Carew, J.S., Nawrocki, S.T., Kahue, C.N., Zhang, H., Yang, C., Chung, L., Houghton, J.A., Huang, P., Giles, F.J., Cleveland, J.L., 2007. Targeting autophagy augments the anticancer activity of the histone deacetylase inhibitor SAHA to overcome Bcr-Abl-mediated drug resistance. Blood 110, 313–322.
Chan, E.Y., Longatti, A., Mcknight, N.C., Tooze, S.A., 2009. Kinase-inactivated ULK proteins inhibit autophagy via their conserved C-terminal domains using an Atg13-independent mechanism. Mol. Cell. Biol. 29, 157–171.
Chang, N.C., Nguyen, M., Germain, M., Shore, G.C., 2010. Antagonism of Beclin 1-dependent autophagy by BCL-2 at the endoplasmic reticulum requires NAF-1. EMBO J. 29, 606–618.
Chen, Y., Yu, L., 2017. Recent progress in autophagic lysosome reformation. Traffic (Copenhagen, Denmark) 18, 358–361.
Chen, L., Willis, S.N., Wei, A., Smith, B.J., Fletcher, J.I., Hinds, M.G., Colman, P.M., Day, C.L., Adams, J.M., Huang, D.C., 2005. Differential targeting of prosurvival Bcl-2 proteins by their BH3-only ligands allows complementary apoptotic function. Mol. Cell 17, 393–403.
Chen, H.C., Kanai, M., Inoue-Yamauchi, A., Tu, H.C., Huang, Y., Ren, D., Kim, H., Takeda, S., Reyna, D.E., Chan, P.M., Ganesan, Y.T., Liao, C.P., Gavathiotis, E., Hsieh, J.J., Cheng, E.H., 2015. An interconnected hierarchical model of cell death regulation by the BCL-2 family. Nat. Cell Biol. 17, 1270–1281.
Cho, D.H., Jo, Y.K., Hwang, J.J., Lee, Y.M., Roh, S.A., Kim, J.C., 2009. Caspase-mediated cleavage of ATG6/Beclin-1 links apoptosis to autophagy in HeLa cells. Cancer Lett. 274, 95–100.
Ciechomska, I.A., Goemans, C.G., Tolkovsky, A.M., 2009a. Why doesn't Beclin 1, a BH3-only protein, suppress the anti-apoptotic function of Bcl-2? Autophagy 5, 880–881.
Ciechomska, I.A., Goemans, G.C., Skepper, J.N., Tolkovsky, A.M., 2009b. Bcl-2 complexed with Beclin-1 maintains full anti-apoptotic function. Oncogene 28, 2128–2141.
Corcelle, E.A., Puustinen, P., Jaattela, M., 2009. Apoptosis and autophagy: targeting autophagy signalling in cancer cells -'trick or treats'? FEBS J. 276, 6084–6096.
Crawford, E.D., Wells, J.A., 2011. Caspase substrates and cellular remodeling. Annu. Rev. Biochem. 80, 1055–1087.
Cuconati, A., Mukherjee, C., Perez, D., White, E., 2003. DNA damage response and MCL-1 destruction initiate apoptosis in adenovirus-infected cells. Genes Dev. 17, 2922–2932.
Cuddeback, S.M., Yamaguchi, H., Komatsu, K., Miyashita, T., Yamada, M., Wu, C., Singh, S., Wang, H.G., 2001. Molecular cloning and characterization of Bif-1. A novel Src homology 3 domain-containing protein that associates with Bax. J. Biol. Chem. 276, 20559–20565.
Czabotar, P.E., Lee, E.F., Van Delft, M.F., Day, C.L., Smith, B.J., Huang, D.C., Fairlie, W.D., Hinds, M.G., Colman, P.M., 2007. Structural insights into the degradation of Mcl-1 induced by BH3 domains. Proc. Natl. Acad. Sci. U. S. A. 104, 6217–6222.
Czabotar, P.E., Westphal, D., Dewson, G., Ma, S., Hockings, C., Fairlie, W.D., Lee, E.F., Yao, S., Robin, A.Y., Smith, B.J., Huang, D.C., Kluck, R.M., Adams, J.M., Colman, P.M., 2013. Bax crystal structures reveal how BH3 domains activate Bax and nucleate its oligomerization to induce apoptosis. Cell 152, 519–531.
Czabotar, P.E., Lessene, G., Strasser, A., Adams, J.M., 2014. Control of apoptosis by the BCL-2 protein family: implications for physiology and therapy. Nat. Rev. Mol. Cell Biol. 15, 49–63.
Deegan, S., Saveljeva, S., Logue, S.E., Pakos-Zebrucka, K., Gupta, S., Vandenabeele, P., Bertrand, M.J., Samali, A., 2014. Deficiency in the mitochondrial apoptotic pathway reveals the toxic potential of autophagy under ER stress conditions. Autophagy 10, 1921–1936.

Denton, D., Xu, T., Kumar, S., 2015. Autophagy as a pro-death pathway. Immunol. Cell Biol. 93, 35–42.

Di Rita, A., Strappazzon, F., 2017. AMBRA1, a novel BH3-like protein: new insights into the AMBRA1-BCL2-family proteins relationship. Int. Rev. Cell Mol. Biol. 330, 85–113.

Dikic, I., Elazar, Z., 2018. Mechanism and medical implications of mammalian autophagy. Nat. Rev. Mol. Cell Biol. 19, 349–364.

Djavaheri-Mergny, M., Maiuri, M.C., Kroemer, G., 2010. Cross talk between apoptosis and autophagy by caspase-mediated cleavage of Beclin 1. Oncogene 29, 1717–1719.

Dooley, H.C., Razi, M., Polson, H.E., Girardin, S.E., Wilson, M.I., Tooze, S.A., 2014. WIPI2 links LC3 conjugation with PI3P, autophagosome formation, and pathogen clearance by recruiting Atg12-5-16L1. Mol. Cell 55, 238–252.

Elgendy, M., Sheridan, C., Brumatti, G., Martin, S.J., 2011. Oncogenic Ras-induced expression of Noxa and Beclin-1 promotes autophagic cell death and limits clonogenic survival. Mol. Cell 42, 23–35.

Elgendy, M., Ciro, M., Abdel-Aziz, A.K., Belmonte, G., Dal Zuffo, R., Mercurio, C., Miracco, C., Lanfrancone, L., Foiani, M., Minucci, S., 2014. Beclin 1 restrains tumorigenesis through Mcl-1 destabilization in an autophagy-independent reciprocal manner. Nat. Commun. 5, 5637.

Erlich, S., Mizrachy, L., Segev, O., Lindenboim, L., Zmira, O., Adi-Harel, S., Hirsch, J.A., Stein, R., Pinkas-Kramarski, R., 2007. Differential interactions between Beclin 1 and Bcl-2 family members. Autophagy 3, 561–568.

Feng, W., Huang, S., Wu, H., Zhang, M., 2007. Molecular basis of Bcl-xL's target recognition versatility revealed by the structure of Bcl-xL in complex with the BH3 domain of Beclin-1. J. Mol. Biol. 372, 223–235.

Fernandez, A.F., Sebti, S., Wei, Y., Zou, Z., Shi, M., Mcmillan, K.L., He, C., Ting, T., Liu, Y., Chiang, W.C., Marciano, D.K., Schiattarella, G.G., Bhagat, G., Moe, O.W., Hu, M.C., Levine, B., 2018. Disruption of the beclin 1-BCL2 autophagy regulatory complex promotes longevity in mice. Nature 558, 136–140.

Fimia, G.M., Stoykova, A., Romagnoli, A., Giunta, L., Di Bartolomeo, S., Nardacci, R., Corazzari, M., Fuoco, C., Ucar, A., Schwartz, P., Gruss, P., Piacentini, M., Chowdhury, K., Cecconi, F., 2007. Ambra1 regulates autophagy and development of the nervous system. Nature 447, 1121–1125.

Follis, A.V., Llambi, F., Kalkavan, H., Yao, Y., Phillips, A.H., Park, C.G., Marassi, F.M., Green, D.R., Kriwacki, R.W., 2018. Regulation of apoptosis by an intrinsically disordered region of Bcl-xL. Nat. Chem. Biol. 14, 458–465.

Galluzzi, L., Baehrecke, E.H., Ballabio, A., Boya, P., Bravo-San Pedro, J.M., Cecconi, F., Choi, A.M., Chu, C.T., Codogno, P., Colombo, M.I., Cuervo, A.M., Debnath, J., Deretic, V., Dikic, I., Eskelinen, E.L., Fimia, G.M., Fulda, S., Gewirtz, D.A., Green, D.R., Hansen, M., Harper, J.W., Jaattela, M., Johansen, T., Juhasz, G., Kimmelman, A.C., Kraft, C., Ktistakis, N.T., Kumar, S., Levine, B., Lopez-Otin, C., Madeo, F., Martens, S., Martinez, J., Melendez, A., Mizushima, N., Munz, C., Murphy, L.O., Penninger, J.M., Piacentini, M., Reggiori, F., Rubinsztein, D.C., Ryan, K.M., Santambrogio, L., Scorrano, L., Simon, A.K., Simon, H.U., Simonsen, A., Tavernarakis, N., Tooze, S.A., Yoshimori, T., Yuan, J., Yue, Z., Zhong, Q., Kroemer, G., 2017. Molecular definitions of autophagy and related processes. EMBO J. 36, 1811–1836.

Galluzzi, L., Vitale, I., Aaronson, S.A., Abrams, J.M., Adam, D., Agostinis, P., Alnemri, E.S., Altucci, L., Amelio, I., Andrews, D.W., Annicchiarico-Petruzzelli, M., Antonov, A.V., Arama, E., Baehrecke, E.H., Barlev, N.A., Bazan, N.G., Bernassola, F., Bertrand, M.J.M., Bianchi, K., Blagosklonny, M.V., Blomgren, K., Borner, C., Boya, P., Brenner, C., Campanella, M., Candi, E., Carmona-Gutierrez, D.,

Cecconi, F., Chan, F.K., Chandel, N.S., Cheng, E.H., Chipuk, J.E., Cidlowski, J.A., Ciechanover, A., Cohen, G.M., Conrad, M., Cubillos-Ruiz, J.R., Czabotar, P.E., D'angiolella, V., Dawson, T.M., Dawson, V.L., De Laurenzi, V., De Maria, R., Debatin, K.M., Deberardinis, R.J., Deshmukh, M., Di Daniele, N., Di Virgilio, F., Dixit, V.M., Dixon, S.J., Duckett, C.S., Dynlacht, B.D., El-Deiry, W.S., Elrod, J.W., Fimia, G.M., Fulda, S., Garcia-Saez, A.J., Garg, A.D., Garrido, C., Gavathiotis, E., Golstein, P., Gottlieb, E., Green, D.R., Greene, L.A., Gronemeyer, H., Gross, A., Hajnoczky, G., Hardwick, J.M., Harris, I.S., Hengartner, M.O., Hetz, C., Ichijo, H., Jaattela, M., Joseph, B., Jost, P.J., Juin, P.P., Kaiser, W.J., Karin, M., Kaufmann, T., Kepp, O., Kimchi, A., Kitsis, R.N., Klionsky, D.J., Knight, R.A., Kumar, S., Lee, S.W., Lemasters, J.J., Levine, B., Linkermann, A., Lipton, S.A., Lockshin, R.A., Lopez-Otin, C., Lowe, S.W., Luedde, T., Lugli, E., Macfarlane, M., Madeo, F., Malewicz, M., Malorni, W., Manic, G., et al., 2018. Molecular mechanisms of cell death: recommendations of the nomenclature committee on cell death 2018. Cell Death Differ. 25, 486–541.

Ganley, I.G., Lam Du, H., Wang, J., Ding, X., Chen, S., Jiang, X., 2009. ULK1.ATG13. FIP200 complex mediates mTOR signaling and is essential for autophagy. J. Biol. Chem. 284, 12297–12305.

Gaugel, A., Bakula, D., Hoffmann, A., Proikas-Cezanne, T., 2012. Defining regulatory and phosphoinositide-binding sites in the human WIPI-1 beta-propeller responsible for autophagosomal membrane localization downstream of mTORC1 inhibition. J. Mol. Signal. 7, 16.

Ge, L., Melville, D., Zhang, M., Schekman, R., 2013. The ER-Golgi intermediate compartment is a key membrane source for the LC3 lipidation step of autophagosome biogenesis. Elife 2, e00947.

Ge, L., Zhang, M., Schekman, R., 2014. Phosphatidylinositol 3-kinase and COPII generate LC3 lipidation vesicles from the ER-Golgi intermediate compartment. Elife 3, e04135.

Geng, J., Klionsky, D.J., 2008. The Atg8 and Atg12 ubiquitin-like conjugation systems in macroautophagy. 'Protein modifications: beyond the usual suspects' review series. EMBO Rep. 9, 859–864.

Germain, M., Nguyen, A.P., Le Grand, J.N., Arbour, N., Vanderluit, J.L., Park, D.S., Opferman, J.T., Slack, R.S., 2011. MCL-1 is a stress sensor that regulates autophagy in a developmentally regulated manner. EMBO J. 30, 395–407.

Green, D.R., Llambi, F., 2015. Cell death signaling. Cold Spring Harb. Perspect. Biol. 7, a006080.

Gump, J.M., Staskiewicz, L., Morgan, M.J., Bamberg, A., Riches, D.W., Thorburn, A., 2014. Autophagy variation within a cell population determines cell fate through selective degradation of Fap-1. Nat. Cell Biol. 16, 47–54.

Gwinn, D.M., Shackelford, D.B., Egan, D.F., Mihaylova, M.M., Mery, A., Vasquez, D.S., Turk, B.E., Shaw, R.J., 2008. AMPK phosphorylation of raptor mediates a metabolic checkpoint. Mol. Cell 30, 214–226.

Hailey, D.W., Rambold, A.S., Satpute-Krishnan, P., Mitra, K., Sougrat, R., Kim, P.K., Lippincott-Schwartz, J., 2010. Mitochondria supply membranes for autophagosome biogenesis during starvation. Cell 141, 656–667.

Hamasaki, M., Furuta, N., Matsuda, A., Nezu, A., Yamamoto, A., Fujita, N., Oomori, H., Noda, T., Haraguchi, T., Hiraoka, Y., Amano, A., Yoshimori, T., 2013. Autophagosomes form at ER-mitochondria contact sites. Nature 495, 389–393.

Hayashi-Nishino, M., Fujita, N., Noda, T., Yamaguchi, A., Yoshimori, T., Yamamoto, A., 2009. A subdomain of the endoplasmic reticulum forms a cradle for autophagosome formation. Nat. Cell Biol. 11, 1433–1437.

He, M.X., He, Y.W., 2013. A role for c-FLIP(L) in the regulation of apoptosis, autophagy, and necroptosis in T lymphocytes. Cell Death Differ. 20, 188–197.

He, C., Bassik, M.C., Moresi, V., Sun, K., Wei, Y., Zou, Z., An, Z., Loh, J., Fisher, J., Sun, Q., Korsmeyer, S., Packer, M., May, H.I., Hill, J.A., Virgin, H.W., Gilpin, C., Xiao, G., Bassel-Duby, R., Scherer, P.E., Levine, B., 2012. Exercise-induced BCL2-regulated autophagy is required for muscle glucose homeostasis. Nature 481, 511–515.

He, C., Zhu, H., Li, H., Zou, M.H., Xie, Z., 2013. Dissociation of Bcl-2-Beclin1 complex by activated AMPK enhances cardiac autophagy and protects against cardiomyocyte apoptosis in diabetes. Diabetes 62, 1270–1281.

Herrero-Martin, G., Hoyer-Hansen, M., Garcia-Garcia, C., Fumarola, C., Farkas, T., Lopez-Rivas, A., Jaattela, M., 2009. TAK1 activates AMPK-dependent cytoprotective autophagy in TRAIL-treated epithelial cells. EMBO J. 28, 677–685.

Hinds, M.G., Smits, C., Fredericks-Short, R., Risk, J.M., Bailey, M., Huang, D.C., Day, C.L., 2007. Bim, Bad and Bmf: intrinsically unstructured BH3-only proteins that undergo a localized conformational change upon binding to prosurvival Bcl-2 targets. Cell Death Differ. 14, 128–136.

Hou, W., Han, J., Lu, C., Goldstein, L.A., Rabinowich, H., 2010. Autophagic degradation of active caspase-8: a crosstalk mechanism between autophagy and apoptosis. Autophagy 6, 891–900.

Huang, W., Choi, W., Hu, W., Mi, N., Guo, Q., Ma, M., Liu, M., Tian, Y., Lu, P., Wang, F.L., Deng, H., Liu, L., Gao, N., Yu, L., Shi, Y., 2012. Crystal structure and biochemical analyses reveal Beclin 1 as a novel membrane binding protein. Cell Res. 22, 473–489.

Huang, S., Okamoto, K., Yu, C., Sinicrope, F.A., 2013. p62/sequestosome-1 up-regulation promotes ABT-263-induced caspase-8 aggregation/activation on the autophagosome. J. Biol. Chem. 288, 33654–33666.

Huang, K., O'neill, K.L., Li, J., Zhou, W., Han, N., Pang, X., Wu, W., Struble, L., Borgstahl, G., Liu, Z., Zhang, L., Luo, X., 2019. BH3-only proteins target BCL-xL/MCL-1, not BAX/BAK, to initiate apoptosis. Cell Res.

Inoki, K., Zhu, T., Guan, K.L., 2003. TSC2 mediates cellular energy response to control cell growth and survival. Cell 115, 577–590.

Irmler, M., Thome, M., Hahne, M., Schneider, P., Hofmann, K., Steiner, V., Bodmer, J.L., Schroter, M., Burns, K., Mattmann, C., Rimoldi, D., French, L.E., Tschopp, J., 1997. Inhibition of death receptor signals by cellular FLIP. Nature 388, 190–195.

Itakura, E., Kishi, C., Inoue, K., Mizushima, N., 2008. Beclin 1 forms two distinct phosphatidylinositol 3-kinase complexes with mammalian Atg14 and UVRAG. Mol. Biol. Cell 19, 5360–5372.

Ivanov, V.N., Lopez Bergami, P., Maulit, G., Sato, T.A., Sassoon, D., Ronai, Z., 2003. FAP-1 association with Fas (Apo-1) inhibits Fas expression on the cell surface. Mol. Cell. Biol. 23, 3623–3635.

Ivanov, V.N., Ronai, Z., Hei, T.K., 2006. Opposite roles of FAP-1 and dynamin in the regulation of Fas (CD95) translocation to the cell surface and susceptibility to Fas ligand-mediated apoptosis. J. Biol. Chem. 281, 1840–1852.

Jeffries, T.R., Dove, S.K., Michell, R.H., Parker, P.J., 2004. PtdIns-specific MPR pathway association of a novel WD40 repeat protein, WIPI49. Mol. Biol. Cell 15, 2652–2663.

Jiang, H., White, E.J., Rios-Vicil, C.I., Xu, J., Gomez-Manzano, C., Fueyo, J., 2011. Human adenovirus type 5 induces cell lysis through autophagy and autophagy-triggered caspase activity. J. Virol. 85, 4720–4729.

Jin, Z., Li, Y., Pitti, R., Lawrence, D., Pham, V.C., Lill, J.R., Ashkenazi, A., 2009. Cullin3-based polyubiquitination and p62-dependent aggregation of caspase-8 mediate extrinsic apoptosis signaling. Cell 137, 721–735.

Jost, P.J., Grabow, S., Gray, D., Mckenzie, M.D., Nachbur, U., Huang, D.C., Bouillet, P., Thomas, H.E., Borner, C., Silke, J., Strasser, A., Kaufmann, T., 2009. XIAP discriminates between type I and type II FAS-induced apoptosis. Nature 460, 1035–1039.

Jung, C.H., Jun, C.B., Ro, S.H., Kim, Y.M., Otto, N.M., Cao, J., Kundu, M., Kim, D.H., 2009. ULK-Atg13-FIP200 complexes mediate mTOR signaling to the autophagy machinery. Mol. Biol. Cell 20, 1992–2003.

Kabeya, Y., Mizushima, N., Ueno, T., Yamamoto, A., Kirisako, T., Noda, T., Kominami, E., Ohsumi, Y., Yoshimori, T., 2000. LC3, a mammalian homologue of yeast Apg8p, is localized in autophagosome membranes after processing. EMBO J. 19, 5720–5728.

Katayama, M., Kawaguchi, T., Berger, M.S., Pieper, R.O., 2007. DNA damaging agent-induced autophagy produces a cytoprotective adenosine triphosphate surge in malignant glioma cells. Cell Death Differ. 14, 548–558.

Kathania, M., Raje, C.I., Raje, M., Dutta, R.K., Majumdar, S., 2011. Bfl-1/A1 acts as a negative regulator of autophagy in mycobacteria infected macrophages. Int. J. Biochem. Cell Biol. 43, 573–585.

Kim, J., Kundu, M., Viollet, B., Guan, K.L., 2011. AMPK and mTOR regulate autophagy through direct phosphorylation of Ulk1. Nat. Cell Biol. 13, 132–141.

Kim, J., Kim, Y.C., Fang, C., Russell, R.C., Kim, J.H., Fan, W., Liu, R., Zhong, Q., Guan, K.L., 2013. Differential regulation of distinct Vps34 complexes by AMPK in nutrient stress and autophagy. Cell 152, 290–303.

Kirisako, T., Ichimura, Y., Okada, H., Kabeya, Y., Mizushima, N., Yoshimori, T., Ohsumi, M., Takao, T., Noda, T., Ohsumi, Y., 2000. The reversible modification regulates the membrane-binding state of Apg8/Aut7 essential for autophagy and the cytoplasm to vacuole targeting pathway. J. Cell Biol. 151, 263–276.

Kischkel, F.C., Hellbardt, S., Behrmann, I., Germer, M., Pawlita, M., Krammer, P.H., Peter, M.E., 1995. Cytotoxicity-dependent APO-1 (Fas/CD95)-associated proteins form a death-inducing signaling complex (DISC) with the receptor. EMBO J. 14, 5579–5588.

Kishi-Itakura, C., Koyama-Honda, I., Itakura, E., Mizushima, N., 2014. Ultrastructural analysis of autophagosome organization using mammalian autophagy-deficient cells. J. Cell Sci. 127, 4089–4102.

Krick, R., Tolstrup, J., Appelles, A., Henke, S., Thumm, M., 2006. The relevance of the phosphatidylinositolphosphat-binding motif FRRGT of Atg18 and Atg21 for the Cvt pathway and autophagy. FEBS Lett. 580, 4632–4638.

Krick, R., Busse, R.A., Scacioc, A., Stephan, M., Janshoff, A., Thumm, M., Kuhnel, K., 2012. Structural and functional characterization of the two phosphoinositide binding sites of PROPPINs, a beta-propeller protein family. Proc. Natl. Acad. Sci. U. S. A. 109, E2042–E2049.

Krueger, A., Schmitz, I., Baumann, S., Krammer, P.H., Kirchhoff, S., 2001. Cellular FLICE-inhibitory protein splice variants inhibit different steps of caspase-8 activation at the CD95 death-inducing signaling complex. J. Biol. Chem. 276, 20633–20640.

Ku, B., Woo, J.S., Liang, C., Lee, K.H., Hong, H.S., E, X., Kim, K.S., Jung, J.U., Oh, B.H., 2008. Structural and biochemical bases for the inhibition of autophagy and apoptosis by viral BCL-2 of murine gamma-herpesvirus 68. PLoS Pathog. 4, e25.

Kuma, A., Mizushima, N., Ishihara, N., Ohsumi, Y., 2002. Formation of the approximately 350-kDa Apg12-Apg5.Apg16 multimeric complex, mediated by Apg16 oligomerization, is essential for autophagy in yeast. J. Biol. Chem. 277, 18619–18625.

Laussmann, M.A., Passante, E., Dussmann, H., Rauen, J.A., Wurstle, M.L., Delgado, M.E., Devocelle, M., Prehn, J.H., Rehm, M., 2011. Proteasome inhibition can induce an autophagy-dependent apical activation of caspase-8. Cell Death Differ. 18, 1584–1597.

Lee, E.F., Fairlie, W.D., 2019. The structural biology of Bcl-xL. Int. J. Mol. Sci. 20, E2234.

Lee, J.S., Li, Q., Lee, J.Y., Lee, S.H., Jeong, J.H., Lee, H.R., Chang, H., Zhou, F.C., Gao, S.J., Liang, C., Jung, J.U., 2009. FLIP-mediated autophagy regulation in cell death control. Nat. Cell Biol. 11, 1355–1362.

Lee, E.F., Grabow, S., Chappaz, S., Dewson, G., Hockings, C., Kluck, R.M., Debrincat, M.A., Gray, D.H., Witkowski, M.T., Evangelista, M., Pettikiriarachchi, A., Bouillet, P., Lane, R.M., Czabotar, P.E., Colman, P.M., Smith, B.J., Kile, B.T., Fairlie, W.D., 2016a. Physiological restraint of Bak by Bcl-xL is essential for cell survival. Genes Dev. 30, 1240–1250.

Lee, E.F., Perugini, M.A., Pettikiriarachchi, A., Evangelista, M., Keizer, D.W., Yao, S., Fairlie, W.D., 2016b. The BECN1 N-terminal domain is intrinsically disordered. Autophagy 12, 460–471.

Lee, E.F., Smith, N.A., Soares Da Costa, T.P., Meftahi, N., Yao, S., Harris, T.J., Tran, S., Pettikiriarachchi, A., Perugini, M.A., Keizer, D.W., Evangelista, M., Smith, B.J., Fairlie, W.D., 2019. Structural insights into BCL2 pro-survival protein interactions with the key autophagy regulator BECN1 following phosphorylation by STK4/MST1. Autophagy 15, 785–795.

Letai, A., Bassik, M.C., Walensky, L.D., Sorcinelli, M.D., Weiler, S., Korsmeyer, S.J., 2002. Distinct BH3 domains either sensitize or activate mitochondrial apoptosis, serving as prototype cancer therapeutics. Cancer Cell 2, 183–192.

Levine, B., Kroemer, G., 2019. Biological functions of autophagy genes: a disease perspective. Cell 176, 11–42.

Li, Y., Kanki, H., Hachiya, T., Ohyama, T., Irie, S., Tang, G., Mukai, J., Sato, T., 2000. Negative regulation of Fas-mediated apoptosis by FAP-1 in human cancer cells. Int. J. Cancer 87, 473–479.

Li, X., He, L., Che, K.H., Funderburk, S.F., Pan, L., Pan, N., Zhang, M., Yue, Z., Zhao, Y., 2012. Imperfect interface of Beclin1 coiled-coil domain regulates homodimer and heterodimer formation with Atg14L and UVRAG. Nat. Commun. 3, 662.

Liang, X.H., Kleeman, L.K., Jiang, H.H., Gordon, G., Goldman, J.E., Berry, G., Herman, B., Levine, B., 1998. Protection against fatal Sindbis virus encephalitis by beclin, a novel Bcl-2-interacting protein. J. Virol. 72, 8586–8596.

Liang, X.H., Jackson, S., Seaman, M., Brown, K., Kempkes, B., Hibshoosh, H., Levine, B., 1999. Induction of autophagy and inhibition of tumorigenesis by beclin 1. Nature 402, 672–676.

Liang, C., Feng, P., Ku, B., Dotan, I., Canaani, D., Oh, B.H., Jung, J.U., 2006. Autophagic and tumour suppressor activity of a novel Beclin1-binding protein UVRAG. Nat. Cell Biol. 8, 688–699.

Liang, C., Lee, J.S., Inn, K.S., Gack, M.U., Li, Q., Roberts, E.A., Vergne, I., Deretic, V., Feng, P., Akazawa, C., Jung, J.U., 2008. Beclin1-binding UVRAG targets the class C Vps complex to coordinate autophagosome maturation and endocytic trafficking. Nat. Cell Biol. 10, 776–787.

Lindqvist, L.M., Vaux, D.L., 2014. BCL2 and related prosurvival proteins require BAK1 and BAX to affect autophagy. Autophagy 10, 1474–1475.

Lindqvist, L.M., Heinlein, M., Huang, D.C., Vaux, D.L., 2014. Prosurvival Bcl-2 family members affect autophagy only indirectly, by inhibiting Bax and Bak. Proc. Natl. Acad. Sci. U. S. A. 111, 8512–8517.

Llambi, F., Moldoveanu, T., Tait, S.W., Bouchier-Hayes, L., Temirov, J., Mccormick, L.L., Dillon, C.P., Green, D.R., 2011. A unified model of mammalian BCL-2 protein family interactions at the mitochondria. Mol. Cell 44, 517–531.

Loffler, A.S., Alers, S., Dieterle, A.M., Keppeler, H., Franz-Wachtel, M., Kundu, M., Campbell, D.G., Wesselborg, S., Alessi, D.R., Stork, B., 2011. Ulk1-mediated phosphorylation of AMPK constitutes a negative regulatory feedback loop. Autophagy 7, 696–706.

Luo, S., Rubinsztein, D.C., 2010. Apoptosis blocks Beclin 1-dependent autophagosome synthesis: an effect rescued by Bcl-xL. Cell Death Differ. 17, 268–277.

Luo, S., Rubinsztein, D.C., 2013. BCL2L11/BIM: a novel molecular link between autophagy and apoptosis. Autophagy 9, 104–105.

Mack, H.I., Zheng, B., Asara, J.M., Thomas, S.M., 2012. AMPK-dependent phosphorylation of ULK1 regulates ATG9 localization. Autophagy 8, 1197–1214.

Maejima, Y., Kyoi, S., Zhai, P., Liu, T., Li, H., Ivessa, A., Sciarretta, S., Del Re, D.P., Zablocki, D.K., Hsu, C.P., Lim, D.S., Isobe, M., Sadoshima, J., 2013. Mst1 inhibits autophagy by promoting the interaction between Beclin1 and Bcl-2. Nat. Med. 19, 1478–1488.

Maiuri, M.C., Criollo, A., Tasdemir, E., Vicencio, J.M., Tajeddine, N., Hickman, J.A., Geneste, O., Kroemer, G., 2007a. BH3-only proteins and BH3 mimetics induce autophagy by competitively disrupting the interaction between Beclin 1 and Bcl-2/Bcl-X(L). Autophagy 3, 374–376.

Maiuri, M.C., Le Toumelin, G., Criollo, A., Rain, J.C., Gautier, F., Juin, P., Tasdemir, E., Pierron, G., Troulinaki, K., Tavernarakis, N., Hickman, J.A., Geneste, O., Kroemer, G., 2007b. Functional and physical interaction between Bcl-X(L) and a BH3-like domain in Beclin-1. EMBO J. 26, 2527–2539.

Maiuri, M.C., Zalckvar, E., Kimchi, A., Kroemer, G., 2007c. Self-eating and self-killing: crosstalk between autophagy and apoptosis. Nat. Rev. Mol. Cell Biol. 8, 741–752.

Malik, S.A., Orhon, I., Morselli, E., Criollo, A., Shen, S., Marino, G., Benyounes, A., Benit, P., Rustin, P., Maiuri, M.C., Kroemer, G., 2011a. BH3 mimetics activate multiple pro-autophagic pathways. Oncogene 30, 3918–3929.

Malik, S.A., Shen, S., Marino, G., Benyounes, A., Maiuri, M.C., Kroemer, G., 2011b. BH3 mimetics reveal the network properties of autophagy-regulatory signaling cascades. Autophagy 7, 914–916.

Matsunaga, K., Saitoh, T., Tabata, K., Omori, H., Satoh, T., Kurotori, N., Maejima, I., Shirahama-Noda, K., Ichimura, T., Isobe, T., Akira, S., Noda, T., Yoshimori, T., 2009. Two Beclin 1-binding proteins, Atg14L and Rubicon, reciprocally regulate autophagy at different stages. Nat. Cell Biol. 11, 385–396.

Mercer, T.J., Gubas, A., Tooze, S.A., 2018. A molecular perspective of mammalian autophagosome biogenesis. J. Biol. Chem. 293, 5386–5395.

Merino, D., Kelly, G.L., Lessene, G., Wei, A.H., Roberts, A.W., Strasser, A., 2018. BH3-mimetic drugs: blazing the trail for new cancer medicines. Cancer Cell 34, 879–891.

Mizushima, N., Yamamoto, A., Hatano, M., Kobayashi, Y., Kabeya, Y., Suzuki, K., Tokuhisa, T., Ohsumi, Y., Yoshimori, T., 2001. Dissection of autophagosome formation using Apg5-deficient mouse embryonic stem cells. J. Cell Biol. 152, 657–668.

Mizushima, N., Kuma, A., Kobayashi, Y., Yamamoto, A., Matsubae, M., Takao, T., Natsume, T., Ohsumi, Y., Yoshimori, T., 2003. Mouse Apg16L, a novel WD-repeat protein, targets to the autophagic isolation membrane with the Apg12-Apg5 conjugate. J. Cell Sci. 116, 1679–1688.

Moldoveanu, T., Liu, Q., Tocilj, A., Watson, M., Shore, G., Gehring, K., 2006. The X-ray structure of a BAK homodimer reveals an inhibitory zinc binding site. Mol. Cell 24, 677–688.

Moscat, J., Diaz-Meco, M.T., 2009. p62 at the crossroads of autophagy, apoptosis, and cancer. Cell 137, 1001–1004.

Nakatogawa, H., Ichimura, Y., Ohsumi, Y., 2007. Atg8, a ubiquitin-like protein required for autophagosome formation, mediates membrane tethering and hemifusion. Cell 130, 165–178.

Nascimbeni, A.C., Giordano, F., Dupont, N., Grasso, D., Vaccaro, M.I., Codogno, P., Morel, E., 2017. ER-plasma membrane contact sites contribute to autophagosome biogenesis by regulation of local PI3P synthesis. EMBO J. 36, 2018–2033.

Nguyen, T.N., Padman, B.S., Usher, J., Oorschot, V., Ramm, G., Lazarou, M., 2016. Atg8 family LC3/GABARAP proteins are crucial for autophagosome-lysosome fusion but not autophagosome formation during PINK1/Parkin mitophagy and starvation. J. Cell Biol. 215, 857–874.
Nishimura, T., Kaizuka, T., Cadwell, K., Sahani, M.H., Saitoh, T., Akira, S., Virgin, H.W., Mizushima, N., 2013. FIP200 regulates targeting of Atg16L1 to the isolation membrane. EMBO Rep. 14, 284–291.
Niu, T.K., Cheng, Y., Ren, X., Yang, J.M., 2010. Interaction of Beclin 1 with survivin regulates sensitivity of human glioma cells to TRAIL-induced apoptosis. FEBS Lett. 584, 3519–3524.
Noble, C.G., Dong, J.M., Manser, E., Song, H., 2008. Bcl-xL and UVRAG cause a monomer-dimer switch in Beclin1. J. Biol. Chem. 283, 26274–26282.
Noda, N.N., Kobayashi, T., Adachi, W., Fujioka, Y., Ohsumi, Y., Inagaki, F., 2012. Structure of the novel C-terminal domain of vacuolar protein sorting 30/autophagy-related protein 6 and its specific role in autophagy. J. Biol. Chem. 287, 16256–16266.
Norman, J.M., Cohen, G.M., Bampton, E.T., 2010. The in vitro cleavage of the hAtg proteins by cell death proteases. Autophagy 6, 1042–1056.
Obara, K., Sekito, T., Niimi, K., Ohsumi, Y., 2008. The Atg18-Atg2 complex is recruited to autophagic membranes via phosphatidylinositol 3-phosphate and exerts an essential function. J. Biol. Chem. 283, 23972–23980.
Oberstein, A., Jeffrey, P.D., Shi, Y., 2007. Crystal structure of the Bcl-XL-Beclin 1 peptide complex: Beclin 1 is a novel BH3-only protein. J. Biol. Chem. 282, 13123–13132.
Ohsumi, Y., 2001. Molecular dissection of autophagy: two ubiquitin-like systems. Nat. Rev. Mol. Cell Biol. 2, 211–216.
O'Neill, K.L., Huang, K., Zhang, J., Chen, Y., Luo, X., 2016. Inactivation of prosurvival Bcl-2 proteins activates Bax/Bak through the outer mitochondrial membrane. Genes Dev. 30, 973–988.
Oral, O., Oz-Arslan, D., Itah, Z., Naghavi, A., Deveci, R., Karacali, S., Gozuacik, D., 2012. Cleavage of Atg3 protein by caspase-8 regulates autophagy during receptor-activated cell death. Apoptosis 17, 810–820.
Pan, J.A., Ullman, E., Dou, Z., Zong, W.X., 2011. Inhibition of protein degradation induces apoptosis through a microtubule-associated protein 1 light chain 3-mediated activation of caspase-8 at intracellular membranes. Mol. Cell. Biol. 31, 3158–3170.
Park, J.M., Jung, C.H., Seo, M., Otto, N.M., Grunwald, D., Kim, K.H., Moriarity, B., Kim, Y.M., Starker, C., Nho, R.S., Voytas, D., Kim, D.H., 2016. The ULK1 complex mediates MTORC1 signaling to the autophagy initiation machinery via binding and phosphorylating ATG14. Autophagy 12, 547–564.
Pattingre, S., Levine, B., 2006. Bcl-2 inhibition of autophagy: a new route to cancer? Cancer Res. 66, 2885–2888.
Pattingre, S., Tassa, A., Qu, X., Garuti, R., Liang, X.H., Mizushima, N., Packer, M., Schneider, M.D., Levine, B., 2005. Bcl-2 antiapoptotic proteins inhibit Beclin 1-dependent autophagy. Cell 122, 927–939.
Pattingre, S., Bauvy, C., Carpentier, S., Levade, T., Levine, B., Codogno, P., 2009a. Role of JNK1-dependent Bcl-2 phosphorylation in ceramide-induced macroautophagy. J. Biol. Chem. 284, 2719–2728.
Pattingre, S., Bauvy, C., Levade, T., Levine, B., Codogno, P., 2009b. Ceramide-induced autophagy: to junk or to protect cells? Autophagy 5, 558–560.
Pedro, J.M., Wei, Y., Sica, V., Maiuri, M.C., Zou, Z., Kroemer, G., Levine, B., 2015. BAX and BAK1 are dispensable for ABT-737-induced dissociation of the BCL2-BECN1 complex and autophagy. Autophagy 11, 452–459.
Peltzer, N., Darding, M., Walczak, H., 2016. Holding RIPK1 on the ubiquitin leash in TNFR1 signaling. Trends Cell Biol. 26, 445–461.

Pengo, N., Agrotis, A., Prak, K., Jones, J., Ketteler, R., 2017. A reversible phospho-switch mediated by ULK1 regulates the activity of autophagy protease ATG4B. Nat. Commun. 8, 294.
Petherick, K.J., Conway, O.J., Mpamhanga, C., Osborne, S.A., Kamal, A., Saxty, B., Ganley, I.G., 2015. Pharmacological inhibition of ULK1 kinase blocks mammalian target of rapamycin (mTOR)-dependent autophagy. J. Biol. Chem. 290, 28726.
Petros, A.M., Nettesheim, D.G., Wang, Y., Olejniczak, E.T., Meadows, R.P., Mack, J., Swift, K., Matayoshi, E.D., Zhang, H., Thompson, C.B., Fesik, S.W., 2000. Rationale for Bcl-xL/Bad peptide complex formation from structure, mutagenesis, and biophysical studies. Protein Sci. 9, 2528–2534.
Piya, S., White, E.J., Klein, S.R., Jiang, H., Mcdonnell, T.J., Gomez-Manzano, C., Fueyo, J., 2011. The E1B19K oncoprotein complexes with Beclin 1 to regulate autophagy in adenovirus-infected cells. PLoS One 6, e29467.
Polson, H.E., De Lartigue, J., Rigden, D.J., Reedijk, M., Urbe, S., Clague, M.J., Tooze, S.A., 2010. Mammalian Atg18 (WIPI2) localizes to omegasome-anchored phagophores and positively regulates LC3 lipidation. Autophagy 6, 506–522.
Proikas-Cezanne, T., Waddell, S., Gaugel, A., Frickey, T., Lupas, A., Nordheim, A., 2004. WIPI-1alpha (WIPI49), a member of the novel 7-bladed WIPI protein family, is aberrantly expressed in human cancer and is linked to starvation-induced autophagy. Oncogene 23, 9314–9325.
Puente, C., Hendrickson, R.C., Jiang, X., 2016. Nutrient-regulated phosphorylation of ATG13 inhibits starvation-induced autophagy. J. Biol. Chem. 291, 6026–6035.
Pyo, J.O., Jang, M.H., Kwon, Y.K., Lee, H.J., Jun, J.I., Woo, H.N., Cho, D.H., Choi, B., Lee, H., Kim, J.H., Mizushima, N., Oshumi, Y., Jung, Y.K., 2005. Essential roles of Atg5 and FADD in autophagic cell death: dissection of autophagic cell death into vacuole formation and cell death. J. Biol. Chem. 280, 20722–20729.
Ravikumar, B., Moreau, K., Jahreiss, L., Puri, C., Rubinsztein, D.C., 2010. Plasma membrane contributes to the formation of pre-autophagosomal structures. Nat. Cell Biol. 12, 747–757.
Reljic, B., Conos, S., Lee, E.F., Garnier, J.M., Dong, L., Lessene, G., Fairlie, W.D., Vaux, D.L., Lindqvist, L.M., 2016. BAX-BAK1-independent LC3B lipidation by BH3 mimetics is unrelated to BH3 mimetic activity and has only minimal effects on autophagic flux. Autophagy 12, 1083–1093.
Robert, G., Gastaldi, C., Puissant, A., Hamouda, A., Jacquel, A., Dufies, M., Belhacene, N., Colosetti, P., Reed, J.C., Auberger, P., Luciano, F., 2012. The anti-apoptotic Bcl-B protein inhibits BECN1-dependent autophagic cell death. Autophagy 8, 637–649.
Rocchi, A., Yamamoto, S., Ting, T., Fan, Y., Sadleir, K., Wang, Y., Zhang, W., Huang, S., Levine, B., Vassar, R., He, C., 2017. A Becn1 mutation mediates hyperactive autophagic sequestration of amyloid oligomers and improved cognition in Alzheimer's disease. PLoS Genet. 13, e1006962.
Rostislavleva, K., Soler, N., Ohashi, Y., Zhang, L., Pardon, E., Burke, J.E., Masson, G.R., Johnson, C., Steyaert, J., Ktistakis, N.T., Williams, R.L., 2015. Structure and flexibility of the endosomal Vps34 complex reveals the basis of its function on membranes. Science 350, aac7365.
Rubinsztein, D.C., Shpilka, T., Elazar, Z., 2012. Mechanisms of autophagosome biogenesis. Curr. Biol. 22, R29–R34.
Russell, R.C., Tian, Y., Yuan, H., Park, H.W., Chang, Y.Y., Kim, J., Kim, H., Neufeld, T.P., Dillin, A., Guan, K.L., 2013. ULK1 induces autophagy by phosphorylating Beclin-1 and activating VPS34 lipid kinase. Nat. Cell Biol. 15, 741–750.
Sanchez, A.M., Csibi, A., Raibon, A., Cornille, K., Gay, S., Bernardi, H., Candau, R., 2012. AMPK promotes skeletal muscle autophagy through activation of forkhead FoxO3a and interaction with Ulk1. J. Cell. Biochem. 113, 695–710.

Sanchez-Garrido, J., Sancho-Shimizu, V., Shenoy, A.R., 2018. Regulated proteolysis of p62/SQSTM1 enables differential control of autophagy and nutrient sensing. Sci. Signal. 11.
Sato, T., Irie, S., Kitada, S., Reed, J.C., 1995. FAP-1: a protein tyrosine phosphatase that associates with Fas. Science 268, 411–415.
Sattler, M., Liang, H., Nettesheim, D., Meadows, R.P., Harlan, J.E., Eberstadt, M., Yoon, H.S., Shuker, S.B., Chang, B.S., Minn, A.J., Thompson, C.B., Fesik, S.W., 1997. Structure of Bcl-xL-Bak peptide complex: recognition between regulators of apoptosis. Science 275, 983–986.
Shang, L., Chen, S., Du, F., Li, S., Zhao, L., Wang, X., 2011. Nutrient starvation elicits an acute autophagic response mediated by Ulk1 dephosphorylation and its subsequent dissociation from AMPK. Proc. Natl. Acad. Sci. U. S. A. 108, 4788–4793.
Shi, C.S., Kehrl, J.H., 2008. MyD88 and Trif target Beclin 1 to trigger autophagy in macrophages. J. Biol. Chem. 283, 33175–33182.
Shi, C.S., Kehrl, J.H., 2010a. Traf6 and A20 differentially regulate TLR4-induced autophagy by affecting the ubiquitination of Beclin 1. Autophagy 6, 986–987.
Shi, C.S., Kehrl, J.H., 2010b. TRAF6 and A20 regulate lysine 63-linked ubiquitination of Beclin-1 to control TLR4-induced autophagy. Sci. Signal. 3, ra42.
Sinha, S., Colbert, C.L., Becker, N., Wei, Y., Levine, B., 2008. Molecular basis of the regulation of Beclin 1-dependent autophagy by the gamma-herpesvirus 68 Bcl-2 homolog M11. Autophagy 4, 989–997.
Sprick, M.R., Weigand, M.A., Rieser, E., Rauch, C.T., Juo, P., Blenis, J., Krammer, P.H., Walczak, H., 2000. FADD/MORT1 and caspase-8 are recruited to TRAIL receptors 1 and 2 and are essential for apoptosis mediated by TRAIL receptor 2. Immunity 12, 599–609.
Strappazzon, F., Vietri-Rudan, M., Campello, S., Nazio, F., Florenzano, F., Fimia, G.M., Piacentini, M., Levine, B., Cecconi, F., 2011. Mitochondrial BCL-2 inhibits AMBRA1-induced autophagy. EMBO J. 30, 1195–1208.
Strappazzon, F., Di Rita, A., Cianfanelli, V., D'orazio, M., Nazio, F., Fimia, G.M., Cecconi, F., 2016. Prosurvival AMBRA1 turns into a proapoptotic BH3-like protein during mitochondrial apoptosis. Autophagy 12, 963–975.
Su, M., Mei, Y., Sanishvili, R., Levine, B., Colbert, C.L., Sinha, S., 2014. Targeting gamma-herpesvirus 68 Bcl-2-mediated down-regulation of autophagy. J. Biol. Chem. 289, 8029–8040.
Sun, Q., Fan, W., Chen, K., Ding, X., Chen, S., Zhong, Q., 2008. Identification of Barkor as a mammalian autophagy-specific factor for Beclin 1 and class III phosphatidylinositol 3-kinase. Proc. Natl. Acad. Sci. U. S. A. 105, 19211–19216.
Suzuki, M., Youle, R.J., Tjandra, N., 2000. Structure of Bax: coregulation of dimer formation and intracellular localization. Cell 103, 645–654.
Takahashi, Y., Coppola, D., Matsushita, N., Cualing, H.D., Sun, M., Sato, Y., Liang, C., Jung, J.U., Cheng, J.Q., Mule, J.J., Pledger, W.J., Wang, H.G., 2007. Bif-1 interacts with Beclin 1 through UVRAG and regulates autophagy and tumorigenesis. Nat. Cell Biol. 9, 1142–1151.
Takahashi, Y., Meyerkord, C.L., Wang, H.G., 2008. BARgaining membranes for autophagosome formation: regulation of autophagy and tumorigenesis by Bif-1/Endophilin B1. Autophagy 4, 121–124.
Tamir, S., Rotem-Bamberger, S., Katz, C., Morcos, F., Hailey, K.L., Zuris, J.A., Wang, C., Conlan, A.R., Lipper, C.H., Paddock, M.L., Mittler, R., Onuchic, J.N., Jennings, P.A., Friedler, A., Nechushtai, R., 2014. Integrated strategy reveals the protein interface between cancer targets Bcl-2 and NAF-1. Proc. Natl. Acad. Sci. U. S. A. 111, 5177–5182.

Tang, D., Kang, R., Livesey, K.M., Cheh, C.W., Farkas, A., Loughran, P., Hoppe, G., Bianchi, M.E., Tracey, K.J., Zeh 3rd, H.J., Lotze, M.T., 2010. Endogenous HMGB1 regulates autophagy. J. Cell Biol. 190, 881–892.
Tang, Z., Takahashi, Y., Chen, C., Liu, Y., He, H., Tsotakos, N., Serfass, J.M., Gebru, M.T., Chen, H., Young, M.M., Wang, H.G., 2017. Atg2A/B deficiency switches cytoprotective autophagy to non-canonical caspase-8 activation and apoptosis. Cell Death Differ. 24, 2127–2138.
Tanida, I., Tanida-Miyake, E., Komatsu, M., Ueno, T., Kominami, E., 2002. Human Apg3p/Aut1p homologue is an authentic E2 enzyme for multiple substrates, GATE-16, GABARAP, and MAP-LC3, and facilitates the conjugation of hApg12p to hApg5p. J. Biol. Chem. 277, 13739–13744.
Tanida, I., Ueno, T., Kominami, E., 2004. LC3 conjugation system in mammalian autophagy. Int. J. Biochem. Cell Biol. 36, 2503–2518.
Thome, M., Schneider, P., Hofmann, K., Fickenscher, H., Meinl, E., Neipel, F., Mattmann, C., Burns, K., Bodmer, J.L., Schroter, M., Scaffidi, C., Krammer, P.H., Peter, M.E., Tschopp, J., 1997. Viral FLICE-inhibitory proteins (FLIPs) prevent apoptosis induced by death receptors. Nature 386, 517–521.
Thorburn, J., Andrysik, Z., Staskiewicz, L., Gump, J., Maycotte, P., Oberst, A., Green, D.R., Espinosa, J.M., Thorburn, A., 2014. Autophagy controls the kinetics and extent of mitochondrial apoptosis by regulating PUMA levels. Cell Rep. 7, 45–52.
Turco, E., Witt, M., Abert, C., Bock-Bierbaum, T., Su, M.Y., Trapannone, R., Sztacho, M., Danieli, A., Shi, X., Zaffagnini, G., Gamper, A., Schuschnig, M., Fracchiolla, D., Bernklau, D., Romanov, J., Hartl, M., Hurley, J.H., Daumke, O., Martens, S., 2019. FIP200 claw domain binding to p62 promotes autophagosome formation at ubiquitin condensates. Mol. Cell 74, 330–346.e11.
Velikkakath, A.K., Nishimura, T., Oita, E., Ishihara, N., Mizushima, N., 2012. Mammalian Atg2 proteins are essential for autophagosome formation and important for regulation of size and distribution of lipid droplets. Mol. Biol. Cell 23, 896–909.
Wei, Y., Pattingre, S., Sinha, S., Bassik, M., Levine, B., 2008. JNK1-mediated phosphorylation of Bcl-2 regulates starvation-induced autophagy. Mol. Cell 30, 678–688.
Wei, Y., Zou, Z., Becker, N., Anderson, M., Sumpter, R., Xiao, G., Kinch, L., Koduru, P., Christudass, C.S., Veltri, R.W., Grishin, N.V., Peyton, M., Minna, J., Bhagat, G., Levine, B., 2013. EGFR-mediated Beclin 1 phosphorylation in autophagy suppression, tumor progression, and tumor chemoresistance. Cell 154, 1269–1284.
Wei, Y., An, Z., Zou, Z., Sumpter, R., Su, M., Zang, X., Sinha, S., Gaestel, M., Levine, B., 2015. The stress-responsive kinases MAPKAPK2/MAPKAPK3 activate starvation-induced autophagy through Beclin 1 phosphorylation. Elife 4, e50289.
Wesselborg, S., Stork, B., 2015. Autophagy signal transduction by ATG proteins: from hierarchies to networks. Cell. Mol. Life Sci. 72, 4721–4757.
Willis, S.N., Chen, L., Dewson, G., Wei, A., Naik, E., Fletcher, J.I., Adams, J.M., Huang, D.C., 2005. Proapoptotic Bak is sequestered by Mcl-1 and Bcl-xL, but not Bcl-2, until displaced by BH3-only proteins. Genes Dev. 19, 1294–1305.
Willis, S.N., Fletcher, J.I., Kaufmann, T., Van Delft, M.F., Chen, L., Czabotar, P.E., Ierino, H., Lee, E.F., Fairlie, W.D., Bouillet, P., Strasser, A., Kluck, R.M., Adams, J.M., Huang, D.C., 2007. Apoptosis initiated when BH3 ligands engage multiple Bcl-2 homologs, not Bax or Bak. Science 315, 856–859.
Wirawan, E., Vande Walle, L., Kersse, K., Cornelis, S., Claerhout, S., Vanoverberghe, I., Roelandt, R., De Rycke, R., Verspurten, J., Declercq, W., Agostinis, P., Vanden Berghe, T., Lippens, S., Vandenabeele, P., 2010. Caspase-mediated cleavage of Beclin-1 inactivates Beclin-1-induced autophagy and enhances apoptosis by promoting the release of proapoptotic factors from mitochondria. Cell Death Dis. 1, e18.

Yamada, Y., Suzuki, N.N., Hanada, T., Ichimura, Y., Kumeta, H., Fujioka, Y., Ohsumi, Y., Inagaki, F., 2007. The crystal structure of Atg3, an autophagy-related ubiquitin carrier protein (E2) enzyme that mediates Atg8 lipidation. J. Biol. Chem. 282, 8036–8043.
Yan, X.Y., Zhong, X.R., Yu, S.H., Zhang, L.C., Liu, Y.N., Zhang, Y., Sun, L.K., Su, J., 2019. p62 aggregates mediated Caspase 8 activation is responsible for progression of ovarian cancer. J. Cell. Mol. Med. 23, 4030–4042.
Yao, S., Lee, E.F., Pettikiriarachchi, A., Evangelista, M., Keizer, D.W., Fairlie, W.D., 2016. Characterisation of the conformational preference and dynamics of the intrinsically disordered N-terminal region of Beclin 1 by NMR spectroscopy. Biochim. Biophys. Acta 1864, 1128–1137.
Young, M.M., Takahashi, Y., Khan, O., Park, S., Hori, T., Yun, J., Sharma, A.K., Amin, S., Hu, C.D., Zhang, J., Kester, M., Wang, H.G., 2012. Autophagosomal membrane serves as platform for intracellular death-inducing signaling complex (iDISC)-mediated caspase-8 activation and apoptosis. J. Biol. Chem. 287, 12455–12468.
Young, L.N., Cho, K., Lawrence, R., Zoncu, R., Hurley, J.H., 2016. Dynamics and architecture of the NRBF2-containing phosphatidylinositol 3-kinase complex I of autophagy. Proc. Natl. Acad. Sci. U. S. A. 113, 8224–8229.
Young, L.N., Goerdeler, F., Hurley, J.H., 2019. Structural pathway for allosteric activation of the autophagic PI 3-kinase complex I. Proc. Natl. Acad. Sci. U. S. A. 116, 21508–21513.
Yousefi, S., Perozzo, R., Schmid, I., Ziemiecki, A., Schaffner, T., Scapozza, L., Brunner, T., Simon, H.U., 2006. Calpain-mediated cleavage of Atg5 switches autophagy to apoptosis. Nat. Cell Biol. 8, 1124–1132.
Yuan, H.X., Russell, R.C., Guan, K.L., 2013. Regulation of PIK3C3/VPS34 complexes by MTOR in nutrient stress-induced autophagy. Autophagy 9, 1983–1995.
Zalckvar, E., Berissi, H., Eisenstein, M., Kimchi, A., 2009a. Phosphorylation of Beclin 1 by DAP-kinase promotes autophagy by weakening its interactions with Bcl-2 and Bcl-XL. Autophagy 5, 720–722.
Zalckvar, E., Berissi, H., Mizrachy, L., Idelchuk, Y., Koren, I., Eisenstein, M., Sabanay, H., Pinkas-Kramarski, R., Kimchi, A., 2009b. DAP-kinase-mediated phosphorylation on the BH3 domain of beclin 1 promotes dissociation of beclin 1 from Bcl-XL and induction of autophagy. EMBO Rep. 10, 285–292.
Zhong, Y., Wang, Q.J., Li, X., Yan, Y., Backer, J.M., Chait, B.T., Heintz, N., Yue, Z., 2009. Distinct regulation of autophagic activity by Atg14L and Rubicon associated with Beclin 1-phosphatidylinositol-3-kinase complex. Nat. Cell Biol. 11, 468–476.
Zhou, C., Ma, K., Gao, R., Mu, C., Chen, L., Liu, Q., Luo, Q., Feng, D., Zhu, Y., Chen, Q., 2017. Regulation of mATG9 trafficking by Src- and ULK1-mediated phosphorylation in basal and starvation-induced autophagy. Cell Res. 27, 184–201.
Zhu, Y., Zhao, L., Liu, L., Gao, P., Tian, W., Wang, X., Jin, H., Xu, H., Chen, Q., 2010. Beclin 1 cleavage by caspase-3 inactivates autophagy and promotes apoptosis. Protein Cell 1, 468–477.

CHAPTER FIVE

The interplay of autophagy and non-apoptotic cell death pathways

Dannah R. Miller, Scott D. Cramer, Andrew Thorburn*
Department of Pharmacology, University of Colorado Anschutz Medical Campus, Aurora, CO, United States
*Corresponding author: e-mail address: andrew.thorburn@cuanschutz.edu

Contents

Abstract

Autophagy, the process of macromolecular degradation through the lysosome, has been extensively studied for the past decade or two. Autophagy can regulate cell death, especially apoptosis, through selective degradation of both positive and negative apoptosis regulators. However, multiple other programmed cell death pathways exist. As knowledge of these other types of cell death expand, it has been suggested that they also interact with autophagy. In this review, we discuss the molecular mechanisms that comprise three non-apoptotic forms of cell death (necroptosis, pyroptosis and ferroptosis) focusing on how the autophagy machinery regulates these different cell death mechanisms through (i) its degradative functions, i.e., true autophagy, and (ii) other non-degradative functions of the autophagy machinery such as serving as a signaling scaffold or by participating in other autophagy-independent cellular processes.

International Review of Cell and Molecular Biology, Volume 352
ISSN 1937-6448
https://doi.org/10.1016/bs.ircmb.2019.12.004

1. Introduction

Autophagy is the process by which a cell recycles nutrients through targeted macromolecule degradation by the lysosome. Typically, autophagy provides energy and metabolites under conditions of starvation and alleviates cellular stress resulting from damaged proteins or organelles (Codogno and Meijer, 2005; Dikic and Elazar, 2018; He and Klionsky, 2009; Marino et al., 2014). Autophagy is executed by several different protein complexes. The Unc-51-like kinase (ULK) kinase complex, containing ULK1 or ULK2, FAK-family kinase-interacting protein at 200 kDa (FIP200), autophagy-related protein 13 (ATG13), and ATG101 (Hosokawa et al., 2009a,b), promotes autophagosome initiation and elongation through recruitment and activation of a lipid kinase complex, which includes VPS34, BECLIN-1, VPS15, and ATG14L. This complex promotes production of phosphatidylinositol 3-phosphate (PI3P) (Russell et al., 2013) leading to the recruitment of other autophagy regulators including several proteins that function in protein conjugations including ATG3, ATG4, ATG5, ATG7, ATG10 and ATG16L1. Together these activities lead to the conjugation of microtubule-associated proteins 1A/1B light chain 3B (LC3) and gamma-aminobutyric acid receptor-associated protein (GABARAP) family members to phosphatidylethanolamine (PE), which allows the LC3 family members to be recruited to autophagosome membranes (Dikic and Elazar, 2018). After formation of the complete autophagosome and sequestering of proteins and other cytoplasmic cargo, the autophagosome fuses with the lysosome leading to degradation of the engulfed molecules and organelles.

Autophagy has been extensively studied in relation to its ability to regulate apoptosis and is known to both positively and, more commonly, negatively regulate apoptosis. For example, CD95/FAS-driven extrinsic apoptosis is enhanced by the autophagic degradation of protein tyrosine phosphatase, non-receptor type 13 (PTPN13) while in the same cells autophagy inhibited apoptosis by the very similar TRAIL death receptors (Gump et al., 2014). These studies and their implications for normal physiology and disease have been extensively reviewed (Anding and Baehrecke, 2015; Denton et al., 2015; Fitzwalter and Thorburn, 2015; Green and Levine, 2014; Marino et al., 2014; Zhang and Baehrecke, 2015). Although the role of autophagy has been well characterized in apoptosis, its role in other forms of cell death is less well understood. We will focus on the role that autophagy plays in necroptosis, ferroptosis, and pyroptosis through autophagy-mediated degradation and recruitment of cell death components to the necrosome. Studies show that the autophagy machinery is vital for both degradation of cell death components and regulators, while it also functions as a scaffold for the necrosome and inhibition of lysosome fusion can exacerbate cell death. Importantly, these two functions can lead to competing effects, i.e., to inhibit or enhance the likelihood that a cell will die.

2. Non-apoptotic forms of cell death

2.1 Necroptosis

Necroptosis is a form of programmed cell necrosis characterized by rupture of the plasma membrane, enhanced cell size, and swollen organelles (Li et al., 2018; Nikseresht et al., 2019; Ofengeim and Yuan, 2013). Necroptosis is caspase-independent and is best understood after it is initiated by the activation of death receptors such as tumor necrosis factor receptor 1 (TNFR1) (Degterev et al., 2005, 2008; Galluzzi et al., 2014, 2017b) or pathogen recognition receptors (PRRs), including toll-like receptor 3 (TLR3), TLR4, and Z-DNA binding protein 1 (ZBP1) (Kaiser et al., 2013; Upton et al., 2010, 2012). During necroptosis, receptor-interacting serine/threonine kinase 3 (RIPK3) and its substrate mixed lineage kinase domain-like (MLKL) play significant roles in the activation and execution of cell death, respectively (Hadifar et al., 2019). In its inactivate state, RIPK1 is highly ubiquitated, which allows for NFκB cell survival signaling (Christofferson et al., 2014; Kondylis et al., 2017). After initiation of TNF-induced necroptosis, RIPK3 is activated by RIPK1 through the physical interaction between their respective RIP homotypic interaction motif (RHIM) domains and RIPK1 catalytic activity (Cho et al., 2009; Li et al., 2012; Vandenabeele et al., 2010). RIPK1 and RIPK3 can also be directly activated by Toll/IL-1 receptor domain-containing adaptor protein inducing interferon-β (TRIF) upon death signaling by TRLs (He et al., 2011; Kaiser et al., 2013; Lim et al., 2019). This kinase complex (the necrosome) recruits MLKL and can also contain other molecules including the well-known apoptosis proteins Fas-associated protein with death domain (FADD) and caspase-8 (Oberst et al., 2011). Upon phosphorylation by RIPK3, MLKL is oligomerized and translocated to the plasma membrane, where it binds specific acidic phosphoinositide species to promote plasma membrane permeabilization (Cai et al., 2014; Dovey et al., 2018; Murphy et al., 2013; Remijsen et al., 2014; Sun et al., 2012; Wang et al., 2014; Wu et al., 2013; Zhao

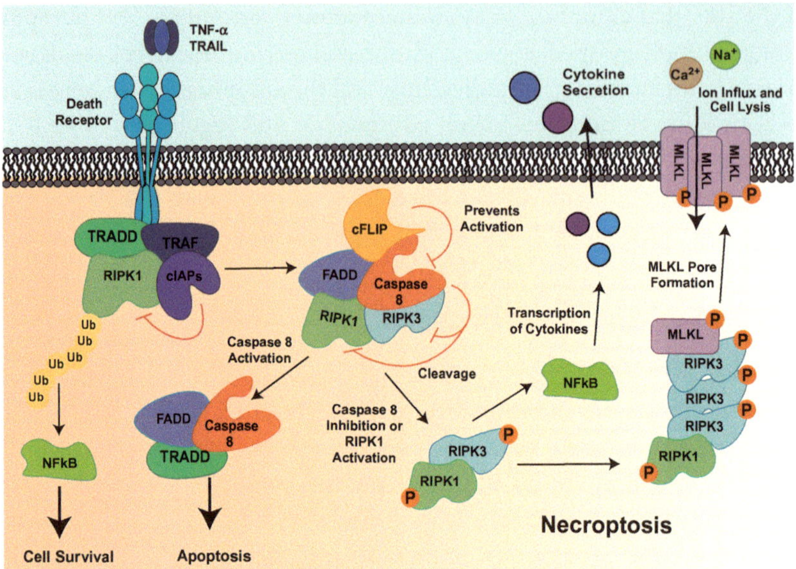

Fig. 1 The necroptosis cell death pathway forms pores in the plasma membrane. Necroptosis is a caspase-independent form of cell death. In a pro-survival state, RIPK1 is localized to a death receptor along with TRADD, TRAF, and cIAPs. cIAPs inhibit RIPK1 function through polyubiquitination. Upon the introduction of cell death stimuli, such as TNF-α or TRAIL, RIPK1 undergoes deubiquitination by CYLD to form the ripoptosome. cIAP inhibition can also form this complex, which is composed of FADD, Caspase 8, cFLIP, RIPK1, and RIPK3. Caspase 8 can cleave RIPK1 or RIPK3 to prevent necroptosis, while cFLIP is an inhibitor of Caspase 8 activity. If Caspase 8 is activated, the cell will undergo apoptosis. If Caspase 8 remains inactive or RIPK3 is phosphorylated, the necrosome complex will form to execute necroptosis. Phosphorylation of RIPK1 and RIPK3 leads to MLKL phosphorylation and activation. Upon phosphorylation, MLKL forms oligomers that can create pores in the plasma membrane for Ca^{2+} or Na^{+} influx that will eventually lead to the cell to burst and leakage of cellular contents.

et al., 2012) as shown in Fig. 1. This plasma membrane permeabilization step early in the death process is significantly different from apoptosis, where plasma membrane rupture is a relatively late event. This, along with specific signaling events (Aaes et al., 2016; Snyder et al., 2019; Yatim et al., 2015), means that necroptosis, unlike apoptosis, releases cytokines and chemokines, resulting in inflammation and immune response (Pasparakis and Vandenabeele, 2015). Necroptosis is involved in various pathological conditions such as ischemia/reperfusion (I/R) injury, non-alcoholic steatohepatitis, sepsis, and atherosclerosis (Shan et al., 2018). Additionally, necroptosis can also be activated by diverse anti-cancer agents (if the tumor cells express RIPK3, which many do not) (Koo et al., 2015) and is much more efficient at activating an adaptive immune response to dying tumor cells than apoptosis (Aaes et al., 2016; Snyder et al., 2019; Yatim et al., 2015). The expression of cytokines and the eventual effect on the immune responses are largely driven by NFκB through RIPK1/RIPK3 activity (Snyder et al., 2019; Yatim et al., 2015). Although cells can still undergo cell death with NFκB inhibition (Snyder et al., 2019), this suggests that cytokine production in necroptosis may be more therapeutically relevant than cell lysis by itself.

The vital function of MLKL in necroptosis has been well documented (Murphy et al., 2013; Sun et al., 2012), but the mechanism through which MLKL executes necroptosis is still incompletely understood. Initial studies suggested that MLKL-mediated PGAM5 activation was vital for necroptosis execution. The activation of PGAM5 then leads to recruitment and activation of mitochondrial fission factor DRP1 to fragment mitochondria which can reduce energy generation and increase ROS levels (Wang et al., 2012). Other studies have suggested that MLKL binding to phosphatidylinositol phosphates (PIPs) on the plasma membrane can induce leakage of PIP-containing liposomes and forms a pore to permeabilize the plasma membrane (Dondelinger et al., 2014). It has since been reported that MLKL oligomerization promotes Ca^{2+} influx mediated by the transient receptor potential cation channel subfamily M member 7 (TRPM7) (Cai et al., 2014). Na^{+} influx from MLKL pore formation has also been attributed to membrane rupture (Chen et al., 2014). Further evidence for MLKL-mediated execution of necroptotic cell death comes from the ability of ESCRT-III to antagonize this mechanism of cell death (Gong et al., 2017; Yoon et al., 2017). In this context, ESCRT-III delayed death by repairing damaged membranes to increase cytokine production or allow for cell survival (Gong et al., 2017). Additionally, MLKL also activates cell surface proteases of the ADAM family at the plasma membrane to enhance shedding of plasma membrane-associated proteins (Cai et al., 2016) or form Mg^{2+}permeant channels (Xia et al., 2016). An important issue for the field is that studies should incorporate utilization of MLKL deficient cells or mice in addition to MLKL inhibitors instead of just RIPK1/RIPK3 inhibition to tease out the role of necroptosis in the effects that are being studied.

2.2 Ferroptosis

Ferroptosis is an iron-dependent cell death mechanism triggered by the inhibition of glutathione (GSH) biosynthesis or glutathione peroxidase (GPX4)

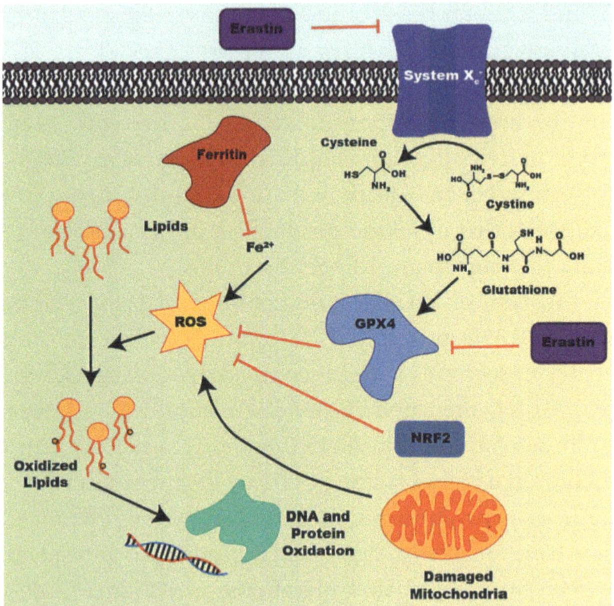

Fig. 2 Ferroptotic cell death is promoted by increased ROS. Ferroptotic cell death occurs due to increased free iron that promotes the generation of toxic ROS molecules that oxidize lipids and other vital molecules. GPX4, NRF2 and iron metabolism proteins such as Ferritin protect the cells from ferroptosis through ROS scavenging and iron storage, respectively. Transporter protein System x_c^- is vital to GPX4 function as it converts cystine to cysteine, which is incorporated into antioxidant GSH. Small molecule erastin induces ferroptosis through inhibition of GPX4 and System x_c^-.

(Dixon et al., 2012; Stockwell et al., 2017) that results from excessive membrane lipid peroxidation and depletion of polyunsaturated fatty acids from the plasma membrane (Martin-Sanchez et al., 2017; Muller et al., 2017; Xie et al., 2016), as shown in Fig. 2. System x_c^- is an important amino acid transporter involved in ferroptotic cell death due to its ability to reduce cystine to cysteine for GSH synthesis (Fig. 2). Small molecule inhibitors such as erastin can inhibit system x_c^- to induce ferroptosis (Dixon et al., 2012; Sato et al., 2018). Because iron is crucial to this cell death pathway, ferritin, transferrin, divalent metal transporter 1 (DMT1) and other similar iron metabolism-related proteins are therefore vital for preventing ferroptosis. Ferroptosis also can be inhibited by iron chelators DFO, 311, ciclopirox, and 2,2-bipyridyl or lipophilic antioxidants trolox, butylated hydroxyanisole, butylated hydroxytoluene, and synthetic antioxidant ferrostatin-1 (Yang and

Stockwell, 2008). An antioxidant response by NRF2 can also inhibit ferroptosis by promoting the transcription of genes quinone oxidoreductase 1 and heme oxygenase 1 (Sun et al., 2016). Ferroptosis can contribute to a multitude of diseases, including periventricular leukomalacia, Huntington's disease, iron-induced kidney tubule damage, and I/R (Stockwell et al., 2017). A multitude of agents including buthioniesulfoximine (BSO)(Yang et al., 2014), acetaminophen (Lorincz et al., 2015), surafenib (Louandre et al., 2015), etc., can induce ferroptosis, as outline in Xie et al. (2016).

2.3 Pyroptosis

Pyroptosis is a regulated form of cell death that relies on the activation of Caspase 1. Pyroptosis is an inflammatory form of cell death triggered by microbial infections or host factors. As such, pyroptosis can be triggered by bacteria, pathogens, or their endotoxins (Cheng et al., 2017). Pyroptosis is also involved in several diseases characterized by inflammation and cell death including autoinflammation syndromes (Simon and Van Der Meer, 2007), neurodegenerative diseases (Ona et al., 1999), inflammatory bowel disease (Siegmund et al., 2001), and cerebral ischemia (Schielke et al., 1998). As demonstrated in Fig. 3, Caspase 1 (or Caspase 3/4/5/11) activation leads to maturation of proinflammatory cytokines such as IL-1β and IL-18 (He et al., 2015). Pyroptosis is characterized by rapid rupture of the plasma membrane that releases proinflammatory molecules mediated by Caspase 1-dependent pore formation (Fink and Cookson, 2006; Fink et al., 2008). When a danger signal arises in the host cell from an infection, Nod-like receptors (NLRs) recognize these signals and initiate pyroptosis (Kufer and Sansonetti, 2007). The NLRs, usually NLRP3 or NLRC4, promote the formation of the inflammasome and Caspase 1 activation through inflammasome adaptor protein ASC (Broz and Dixit, 2016; Faustin et al., 2007; Fernandes-Alnemri et al., 2007; Martinon and Tschopp, 2007) (Fig. 3). Cleavage of the pore forming gasdermin D (GSDMD) proteins by caspases is crucial for pyroptosis-mediated cytokine release and ion and water influx, which eventually leads to osmotic lysis (Evavold et al., 2018; He et al., 2015; Heilig et al., 2018; Sborgi et al., 2016; Shi et al., 2015, 2017). Upon cleavage, GSDMD selectively binds to membranes containing phosphatidylserine, cardiolipin, or PIPs to generate oligomeric pores to trigger pyroptosis (Liu et al., 2016; Ruan et al., 2018). Importantly, the lipid-binding properties of GSDMD allow for less damage to neighboring tissues when cells die of pyroptosis (Liu et al., 2016).

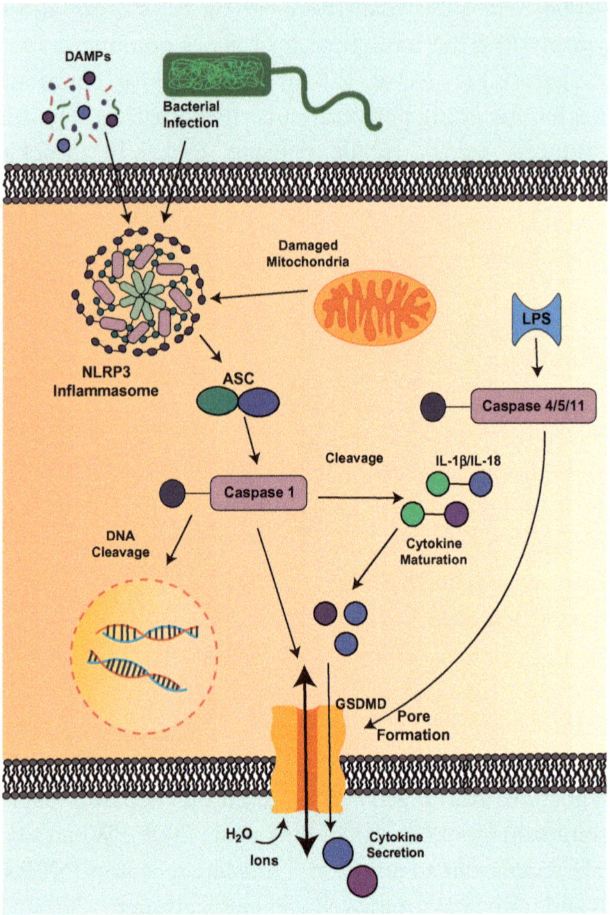

Fig. 3 DAMPs and bacterial infections promote Caspase 1-dependent pyroptosis. The presence of DAMPs or bacterial infections promote a caspase-dependent form of cell death called pyroptosis. Pyroptosis is characterized by the generation of an NLRP3 inflammasome that promotes Caspase 1 activation. Upon cleavage, Caspase 1 promotes GSDMD pore formation, in addition to DNA and cytokine cleavage that results in cell death and an immunogenic response.

In the presence of LPS, murine inflammatory Caspase-11, or human inflammatory Caspases 4 and 5 can also bind to this molecule to promote GSDMD processing and consequent pyroptotic cell death (Hagar et al., 2013; Kayagaki et al., 2015; Shi et al., 2014). Caspase 1-dependent DNA cleavage also occurs during pyroptosis (Bergsbaken and Cookson, 2007; Fink and Cookson, 2006), but the mechanism by which this occurs is still unclear.

3. Autophagy regulation of cell death through degradation

3.1 Autophagic degradation of necroptosis pathway proteins

Various reports have suggested that autophagy flux or lysosomal dysfunction is key to the propagation of necroptosis (Fig. 4; Table 1). Necroptosis often occurs concurrently with the activation of autophagy (Degterev et al., 2005; Ye et al., 2011). Early evidence for the interdependent relationship between autophagy and necroptosis came from reports showing that knockdown of ATG5, ATG7, or Beclin 1 prevent what has subsequently been recognized as a type of necroptotic cell death (Shimizu et al., 2004; Yu et al., 2004). A more recent study by Frank et al. demonstrates that the phosphorylation of MLKL correlates with the accumulation of LC3-II due to phosphorylated MLKL causing a block in autophagic flux, thus suggesting that autophagy is specifically halted during necroptosis (Frank et al., 2019). Consistent with the idea that autophagy needs to be blocked in order to increase the likelihood of necroptosis being completed, depletion of ATG16L1 or ATG7, which reduces autophagic flux can promote necroptosis (Matsuzawa-Ishimoto et al., 2017; Zhou et al., 2017). Further, impaired autophagosome and lysosome fusion through the loss of Lamp-2 also resulted in both apoptotic and necroptotic cell death (Fortunato et al., 2009). The inhibition of autophagy flux through lysosomal damage leads to accumulation of autophagosomes and subsequently increased LC3 and p62 protein levels. The accrual of these proteins correlates with RIPK1 and RIPK3 protein levels and enhanced necroptosis. This is at least in part because RIPK1 and RIPK3 are degraded by the lysosome; therefore, inhibition of lysosomal fusion sensitizes cells to necroptosis (Bray et al., 2012; Liu et al., 2018). More recently, autophagy was shown to be involved in the turnover of complexes containing TRIF, RIPK1, RIPK3, and ZBP1 (Lim et al., 2019). The underlying molecular mechanisms of autophagy regulation of necroptosis are complicated. For example, the autophagic receptor optineurin (OPTN) actively regulates proteasomal turnover of RIPK1, meaning that loss of OPTN induces RIPK1-dependent necroptosis (Galluzzi et al., 2017a; Ito et al., 2016). Similarly, rapamycin treatment and a subsequent increase in autophagy flux protected photoreceptor cells against zVAD-induced necroptosis through reduced RIPK1 expression (Ding et al., 2019). Such studies have been interpreted as meaning that necroptosis and, therefore, inflammatory

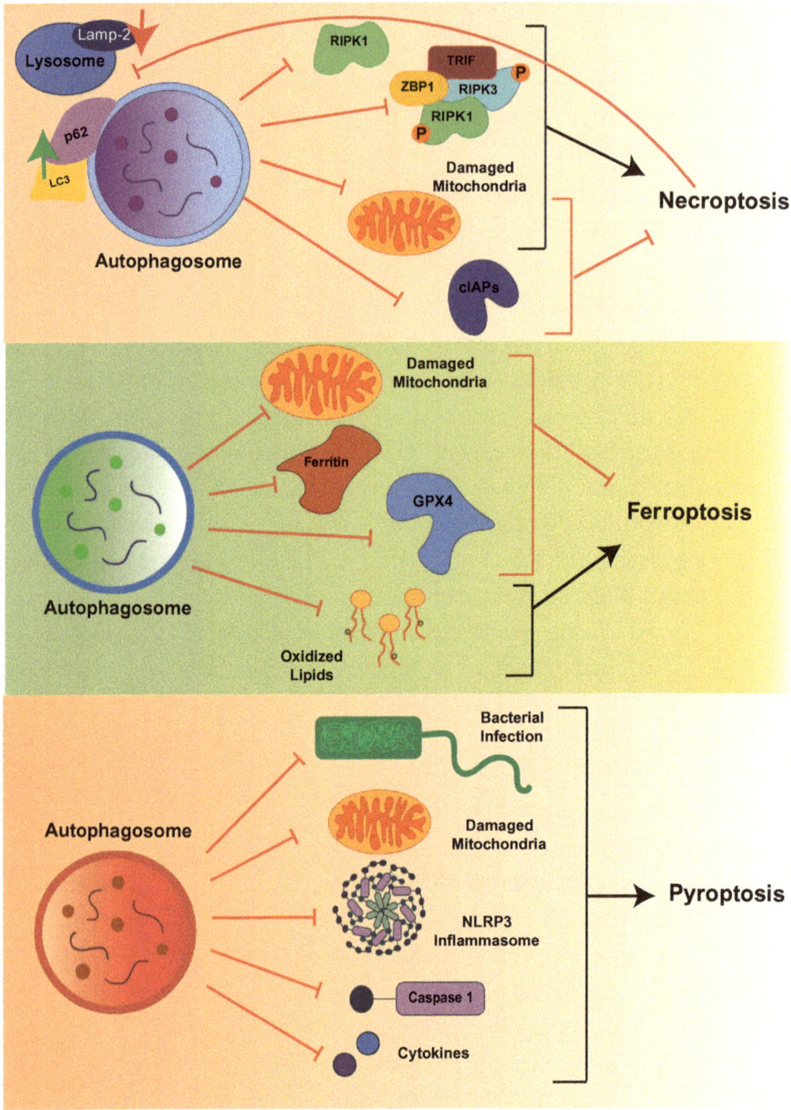

Fig. 4 Selective degradation by autophagy contributes to cell death. Autophagy can contribute to or inhibit necroptosis, ferroptosis, and pyroptosis in numerous ways. Autophagy can promote the selective degradation of RIPK proteins or the necrosome to prevent cell death, while degradation of cIAPs can occur to promote necroptosis. Mitophagy can remove damaged mitochondria that would induce necroptotic cell death. Conversely, increased mitophagy can also contribute to increased ROS levels that promote necroptosis. Importantly, necroptosis can also halt autophagy flux through accumulation of LC3 and p62 molecules or reduction of Lamp-2. In the context of ferroptosis, autophagy can promote cell death through selective degradation of GPX4 or Ferritin in addition to mitophagy. Autophagy inhibits ferroptosis through degradation of the toxic oxidized lipids that contribute to cellular DNA and protein damaged. Autophagy flux often increases as a pro-survival and anti-pyroptosis response. Autophagy can inhibit pyroptosis through degradation of Caspase 1, cytokines, the NLRP3 inflammasome, damaged mitochondria, or the bacteria itself.

Table 1 Effects of autophagy on non-apoptotic cell death.

Cell death pathway	Cell death molecule	Autophagy molecule	Positive or negative regulation	Mechanism of regulation	Reference
Necroptosis		ATG16L1	Negative		Matsuzawa-Ishimoto et al. (2017)
		ATG7	Negative		Zhou et al. (2017)
		Lamp-2	Negative		Fortunato et al. (2009)
	RIPK1		Negative	Autophagosomal degradation of RIPK1	Bray et al. (2012)
	TRIF, RIPK1, RIPK3, ZBP1		Negative	Autophagosomal degradation of TRIF, RIPK1, RIPK3, and ZBP1	Lim et al. (2019)
	MLKL	LC3	Positive	Phosphorylation of MLKL correlates with LC3 protein levels	Frank et al. (2019)
	RIPK1, RIPK3	LC3 and p62	Negative	Correlating protein levels of RIPK and LC3/p62, lysosomal damage leads to accumulation of RIPK proteins	Liu et al. (2018)
	RIPK1	OPTN	Negative	Autophagosomal degradation of RIPK1	Galluzzi et al. (2017a) and Ito et al. (2016)
	cIAP1		Positive	Autophagosomal degradation of cIAP1	He et al. (2014)
		ATG5, PINK1, Drp1	Positive	Mitophagy	Basit et al. (2017) and Mizumura et al. (2014)
	Damaged mitochondria, ROS		Neutral	Increased ROS due to mitochondrial damage is a side effect	Murphy et al. (2013) and Tait et al. (2013)
	PGAM5	PINK1	Negative	Mitophagy removes damaged mitochondria	Liao et al. (2019), Lu et al. (2016), and Sulkshane and Teni (2017)
	RIPK1	p62	Positive	Necrosome assembly and activation	Goodall et al. (2016) and Kharaziha et al. (2015)
	RIPK1, RIPK3, FADD	ATG5, p62	Positive	Necrosome assembly and activation	Basit et al. (2013) and Sulkshane and Teni (2017)
	RIPK1, RIPK3	ATG5, LC3	Positive	Necrosome assembly and activation	Arya et al. (2018)

Continued

Table 1 Effects of autophagy on non-apoptotic cell death.—cont'd

Cell death pathway	Cell death molecule	Autophagy molecule	Positive or negative regulation	Mechanism of regulation	Reference
Pyroptosis	Caspase-1, cytokines	Atg16L1, Atg7	Negative	Reduces caspase-1 activity and cytokine secretion	Pu et al. (2017) and Saitoh et al. (2008)
	Caspase-1, NLRP3, cytokines	Beclin-1, LC3	Negative	Removal of dysfunctional mitochondrial that would promote caspase activity and cytokine production and secretion	Nakahira et al. (2011) and Zhou et al. (2011)
	Cytokines, NLRP3		Negative	Selective degradation of cytokines or inflammasome	Harris et al. (2011)
	IL-1β	Atg5	Positive	Promotes cytokine secretion	Dupont et al. (2011)
Ferroptosis	NCOA4, FTH1	ATG3, ATG5, ATG7, ATG13	Positive	Degradation of FTH1 to allow for accumulation of iron	Gao et al. (2016) and Hou et al. (2016)
	System x_c^-	Beclin-1	Positive	Reduce system x_c^- activity by Beclin-1 binding to SLC7A11	Song et al. (2018)
	GPX4	Lamp-2	Positive	Autophagosomal degradation of GPX4	Wu et al. (2019)
	GSH		Positive	Autophagosomal degradation of GSH	Mancilla et al. (2015)
	Oxidized lipids		Negative	Degradation of toxic oxidized lipids	Haberzettl and Hill (2013) and Hill et al. (2008)

cytokine production, which would be associated with the necroptosis (Snyder et al., 2019; Yatim et al., 2015), can be prevented through increased autophagy due to degradation of vital necrosome components. However, it is important to note that many such studies have not directly tested if these responses truly are due to necroptosis and are, for example, dependent on MLKL.

Autophagic turnover of more general necroptosis regulators also serves as a mechanism by which autophagy can control necroptosis sensitivity (Gyrd-Hansen, 2017; Lim et al., 2019; Seo et al., 2016). For example, autophagosome-mediated degradation of antioxidant protein catalase can promote necroptosis (Yu et al., 2006). Necroptosis is inhibited by inhibitor of apoptosis proteins (IAPs), due to their ability to ubiquitinate RIPK1 which prevents necrosome formation to promote cell survival (Mccomb et al., 2012; Moulin et al., 2012; Vanlangenakker et al., 2011; Yabal et al., 2014) as shown in Fig. 1. As such, necroptosis can be promoted through the elimination of IAPs (Feoktistova et al., 2012; Geserick et al., 2009; Tenev et al., 2011) or the autophagic digestion of cIAP1 through reduced RIPK1 ubiquitination (He et al., 2014; Xu et al., 2016) (Fig. 4). Notably, Caspase 8 activity in the necrosome complex is crucial for determining the particular mechanism by which the cell dies (Oberst et al., 2011). Active Caspase 8 can cleave RIPK1 and RIPK3 to block necroptosis and instead promotes apoptosis (Feng et al., 2007; O'donnell et al., 2011; Rebe et al., 2007). Caspase 8 can be degraded by autophagy upon a cell death stimuli such as TRAIL (Hou et al., 2010), which could therefore also potentially enhance the likelihood that a cell will undergo necroptosis. Similarly, apoptosis modulator cellular FLICE-like inhibitory protein (cFLIP) can be selectively degraded by autophagy to initiate cell death (Xu et al., 2016). In the context of necroptosis, splice variants of cFLIP can regulate necroptosis through altering caspase-8 activation at the DISC complex. $cFLIP_L$ inhibits necroptosis jointly with caspase-8. The formation of $cFLIP_L$/Caspase-8 heterodimers results in localized Caspase-8 activity and inactivation of RIPK1 and RIPK3 (Dillon et al., 2014; Geserick et al., 2009; Kaiser et al., 2011; Krueger et al., 2001; Lemke et al., 2014; Oberst et al., 2011; Weinlich et al., 2013; Zhao et al., 2017). When the $cFLIP_L$ complex localizes to the ripoptosome complex, it blocks cell death and maintains cell survival (Schilling et al., 2014; Tenev et al., 2011). In summary, there are several negative regulators of necroptosis that have been shown to be selectively degraded through autophagy and these mechanisms may thus also affect the likelihood of a cell dying by necroptosis.

It is thought that the functional status of mitochondria can influence necroptotic cell death as well. It has been suggested that mitochondrial dysfunction, particularly through increased ROS generation contributes to necroptosis (Sulkshane and Teni, 2017). Therefore, mitophagy can prevent necroptosis, as seen in a study by Lu et al. They demonstrate that the loss of mitochondrial PGAM5 exacerbates necroptosis due to the inability of cells to undergo PINK1-dependent mitophagy. Thus, in this context, the accumulation of defective mitochondria initiates necroptosis (Lu et al., 2016). Studies by Mizumura et al. showed that mitochondrial dysfunction promotes mitophagy and cell death. Inhibition of mitophagy through Mdivi-1 or PINK1 knockdown dramatically reduces MLKL phosphorylation and cell death, suggesting that mitophagy of dysfunctional mitochondria promotes cell death (Mizumura et al., 2014). Similarly, inhibition of complex I activity of the electron transport chain by BAY 87-2243 increased ROS levels through mitophagy that induced necroptosis (Basit et al., 2017). Depletion of ATG5, mitochondrial protective proteins PINK1 or Drp1, RIPK1, or MLKL reduced necroptotic and ferroptotic cell death, again supporting the pro-death role of mitophagy (Basit et al., 2017). However, other groups found that the depletion of mitochondria or scavenging of ROS does not alter the cell's fate upon a TNF-α death signal suggesting that in some contexts damaged mitochondria and increased ROS are merely a side effect of necroptosis (Murphy et al., 2013; Tait et al., 2013). Thus the exact mechanism by which mitophagy regulates necroptosis is unclear at this time.

3.2 Autophagy regulation of ferroptosis proteins

Ferroptosis is also regulated by autophagy, as the selective degradation of various molecules can initiate or augment this cell death process. Autophagy-dependent degradation of ferritin is termed ferritinophagy. This process is mediated by nuclear receptor coactivator 4 (NCOA4), which binds to ferritin heavy chain 1 (FTH1) to target this molecule to autophagosomes for degradation (Mancias et al., 2014). Increased ferritinophagy thus allows for accumulation of free iron and promotes ferroptosis (Hou et al., 2016) while knockdown of NCOA4, or general regulators of autophagy such as ATG3, ATG5, ATG7 or ATG13 can suppress ferroptosis (Gao et al., 2016; Hou et al., 2016). One study by Zhang et al. noted that the kinase inhibitor sorafenib promoted an increase in autophagic flux and ferritinophagy that promoted ferroptosis (Zhang et al., 2018), while

Gao et al. showed that inhibition of autophagy could prevent ferroptosis (Gao et al., 2016). Inhibition of autophagy through Bafilomycin A1 treatment also greatly reduced erastin-induced ferroptosis (Gao et al., 2018; Sun et al., 2018). GPX4 can also be selectively degraded by chaperone-mediated autophagy to promote ferroptosis (Wu et al., 2019). Similarly, degradation of circadian clock regulator ARNTL/BMAL1 can promote ferroptosis in vitro and in vivo through promotion of lipid oxidation (Liu et al., 2019). In a pro-survival attempt by the cell, autophagy-dependent degradation of oxidized lipids thus controls the accumulation of these molecules and can prevent ferroptosis (Haberzettl and Hill, 2013; Hill et al., 2008). Therefore, the role of autophagy in ferroptosis more often aids in the initiation of cell death, although it can prevent death through the removal of toxic, damaged molecules resulting from increased ROS levels, as well.

3.3 Autophagy degradation of pro-pyroptosis molecules

Various studies have reported interdependent relationships between autophagy and pyroptosis, as well. In particular, autophagy reduces endotoxin-mediated secretion of proinflammatory cytokinase IL-1β and IL-18. Upon deletion of *atg16l1* or *atg7*, Caspase 1 activity and cytokine secretion was enhanced with lipopolysaccharide (LPS) treatment (Pu et al., 2017; Saitoh et al., 2008). Consistently, *atg16l1*-null mice were more susceptible to colitis, which could be improved by anti-IL-18 therapy. Additionally, loss of autophagy proteins LC3-II or Beclin-1 increased Caspase 1 activity and secretion of IL-1β and IL-18 cytokines upon treatment with the endotoxin LPS. In this study, Nakahira et al. show that autophagy or mitophagy prevented the LPS-induced activation of pyroptosis in response to mtDNA released by dysfunctional mitochondria (Nakahira et al., 2011). As such, activation of autophagy was proposed to delay pyroptosis by removal of damaged molecules within the cell (Byrne et al., 2013) and targeting cytokines for selective degradation in addition to regulation of the inflammasome (Harris et al., 2011) (Fig. 4). A unique feature of pyroptosis regulation by autophagy is through degradation of the NLRP3 inflammasome and bacterial pathogens, not just signaling molecules. The NLRP3 inflammasome acts as a sensor for mitochondrial dysfunction, which demonstrates the link between mitochondrial damage, autophagy/mitophagy, and inflammation (Zhou et al., 2011). The NLRP3 inflammasome can also be selectively degraded by autophagy, limiting cytokine secretion and reduced pyroptosis

(Chuang et al., 2013; Mehto et al., 2019; Shi et al., 2012). Conversely, the NRLP3 inflammasome can also inhibit autophagy to promote pyroptosis (Wei et al., 2019). Increased autophagic flux upon host cell infection can selectively degrade invading bacteria to prevent cell death altogether (Amer and Swanson, 2005; Gutierrez et al., 2004; Nakagawa et al., 2004). Consistent with this, several reports have concluded that activation of autophagy can prevent pyroptosis (Li et al., 2019; Tu et al., 2019; Wang et al., 2019; Zi et al., 2019) or that the inhibition of autophagy can enhance pyroptotic cell death (Qiu et al., 2018; Yu et al., 2019). It is important to note that in order to prevent pyroptosis, the autophagosome often targets large protein complexes or organelles for degradation, not just individual signaling molecules, and thus creates a rather different feature of pyroptosis regulation by autophagy compared to ferroptosis and necroptosis.

4. Regulation of cell death through scaffolding and degradation-independent functions of the autophagy machinery

Execution of necroptosis can also be achieved through recruitment of necrosome components to the autophagy machinery, as shown in Fig. 1. This function is distinct from the degradative function of autophagy and exemplifies an important concept: the autophagy machinery can elicit biological effects both by its canonical function of degrading proteins and other cellular material or by serving as a scaffold to enhance cell signaling events. In a specific, genetically-defined, subtype of prostate cancer with loss of MAP3K7 (encoding TGF-β activating kinase 1), ATG5 or ATG7 knockdown reduces necroptosis in response to TRAIL, and alters the mechanism of cell death by controlling whether the cell dies by apoptosis or necroptosis (Goodall et al., 2016). The underlying mechanism is that the autophagy adaptor protein SQSTM1/p62 recruits RIPK1 to autophagosomal membranes, which serve as a signaling scaffold for the necrosome to execute necroptosis. Upon TRAIL treatment, RIPK1 is recruited to the necrosome by p62 in Map3k7-null mouse prostate epithelial cells and this was found to lead to the phosphorylation of MLKL on autophagosome-like membranes by immuno-electron microscopy (Goodall et al., 2016). Importantly, loss of p62 in this model switches the mechanism of death to apoptosis rather than necroptosis. This provides an example where the form of cell death is tunable and can be controlled by the autophagy machinery (Goodall et al., 2016). A similar scaffolding function was described for the BH3 mimetic

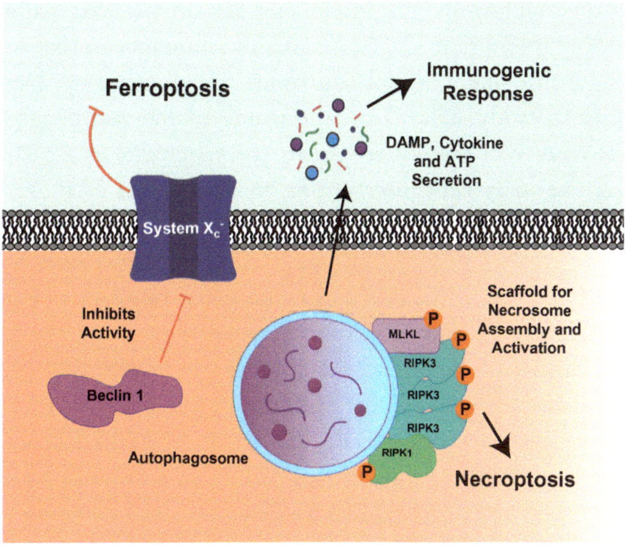

Fig. 5 Autophagy-independent functions of the autophagy machinery that promote cell death. The autophagosome can serve as a scaffold for the assembly and activation of the necrosome for execution of necroptosis which requires p62, Atg5, and LC3. Through an non-canonical pathway, autophagy also promotes the secretion of DAMPs and ATP to be recognized by immune cells during necroptosis and pyroptosis. Beclin-1 can block System x_c^- activity through interaction with SLC7A11 to induce ferroptosis by reduced GSH levels in an autophagy-independent manner.

Obatoclax (Basit et al., 2013; Sulkshane and Teni, 2017) and graphene oxide conjugated to CQ (Arya et al., 2018) (Fig. 5). The kinase inhibitor sorafenib has also been reported to facilitate interaction between p62 and RIPK1 to promote necroptosis in a manner that was dependent on the expression of ATG5 but was unaffected by ATG7 knockdown (Kharaziha et al., 2015). In that study, the authors did not test if other components of the necroptosis pathway were required. It is therefore unclear if this is another example of the necrosome being activated on autophagosomes.

Similar scaffolding by the autophagosome can also affect apoptosis as shown by studies where autophagosome formation was found to activate Caspase 8 in a complex with FADD and ATG5 for initiation of cell death (Tang et al., 2017; Young et al., 2012). Scaffolding functions of the autophagy machinery may also regulate broader signaling pathways in addition to the death pathways. For example, the autophagosome can promote ERK phosphorylation through recruitment of MEK in a manner independent of canonical autophagy (Martinez-Lopez et al., 2013). Further analysis is required to understand what molecules are on the external side of the autophagosome membrane since presumably, interactions that occur inside the autophagosome likely do not contribute to cell signaling. Development of techniques to study what is on the inside of the autophagosome have greatly improved over the last few years (Le Guerroue et al., 2017); however, to date the analysis of interactions on the outside of autophagosomes has been more limited.

In addition to scaffolding, other degradation-independent functions of the autophagy machinery contribute to cell death regulation. For example, an important feature of necroptosis is that it promotes immune activation that requires secretion of DAMPs and cytokines into the environment (Aaes et al., 2016; Bottcher et al., 2006; Kaczmarek et al., 2013; Pasparakis and Vandenabeele, 2015; Yatim et al., 2015). Necroptotic DAMPs include HMGB1 and ATP (He and Wang, 2018; Kaczmarek et al., 2013). Additionally, even after the cell has undergone necroptosis-mediated cell lysis, necroptotic cells continue to generate cytokines through mRNA translation dependent on the endoplasmic reticulum integrity (Orozco et al., 2019). Such activities can also involve the autophagy machinery. For example, secretion of IL-1β occurs in an autophagy machinery-dependent but degradation-independent manner through an unconventional secretory pathway which requires ATG5, the inflammasome, and Golgi stacking paralogs (Dupont et al., 2011). HMGB1 can also be secreted from dying cells in an autophagy machinery-dependent manner (Thorburn et al., 2009). And the Chinese herbal medicine shikonin has been reported to act via an ATG5- and Beclin 1-dependent surface DAMP upregulation to enhance immunosurveillance (Fig. 5). In this context, blockage of autophagosome-lysosome fusion with chloroquine (CQ) was reported to enhance shikonin-mediated surface DAMP expression and promotes immunogenic phenotype (Lin et al., 2018). Additionally, ferroptosis can also be regulated by components of the autophagy machinery in what appears to be an autophagy-independent manner. For example, Beclin-1 can bind to system x_c^- core component SLC7A11 to reduce its activity and promote ferroptosis (Aaes et al., 2016; Song et al., 2018).

5. Conclusions

The role of autophagy in cell death has been well documented for years. In this review, we highlighted non-apoptotic mechanisms of programmed cell death (Figs. 1–3) and some of the proposed ways that mutual regulation of autophagy and these cell death mechanisms occurs (Table 1). Not only is autophagy involved in the selective degradation of regulatory molecules that are involved in controlling death pathways (Fig. 4), but the autophagosome (or autophagosome-like structures) can have scaffolding functions that promote protein interactions that regulate necroptosis and perhaps other forms of death (Fig. 5).

It is important to note however that many, if not all, components of the autophagy machinery have critical roles in other cellular processes (Galluzzi and Green, 2019). In particular, vital autophagy proteins ATG5 and ATG7 are critical for other endocytic pathways, such as LC3-associated phagocytosis (LAP). LAP processes extracellular molecules including, but not limited to, dead cells, pathogens, amyloid structures, cell surface receptors, etc. (Heckmann and Green, 2019; Heckmann et al., 2017). The distinction between autophagy and LAP and other autophagy-independent functions of the ATG proteins is vital. Unfortunately, this has not been made clear in the publications discussed in this review (or in most other cases where autophagy is proposed to have a certain biological role). Most of the literature cited above where a conclusion was made that autophagy affected the non-apoptotic death processes came to this conclusion by targeting one or two autophagy regulators (most commonly ATG5, ATG7 or BECN1) then seeing the noted effect on induction of necroptosis, ferroptosis or pyroptosis. However, such an experimental design cannot exclude the possibility that the effect seen was actually due to an autophagy-independent function, such as LAP, of the targeted gene. Targeted knockout of genes that are more distinct to autophagy itself vs other pathways (Galluzzi and Green, 2019), such as FIP200 (required for autophagy but not LAP) or RUBCN (a negative regulator of autophagy but a positive regulator of LAP) can offer more information on whether any specific response is in fact due to autophagy or something else. Additionally, the recognition that the autophagy machinery can participate in creating scaffolds that are used to promote signaling provides an added level of complexity. This needs to be addressed by experiments that directly examine proposed scaffolding functions, e.g., by examining interactions in cells with the proposed scaffold.

More rigorous studies could be designed to address these issues. For example, we propose that before the conclusion is made that autophagy was important for promoting or inhibiting a particular type of cell death, that the following considerations be adopted. Autophagy is defined as the process by which cellular material is delivered to the lysosome and degraded via double-membraned autophagosomes. Thus to make the claim that autophagy was involved in a given response one should show the following:

1. Targeted inhibition of both autophagy-specific regulators (e.g., FIP200) as well as components of the autophagy machinery that are involved in other processes (such as ATG7) should be shown to have the same effects on the output that is being measured. If an effect is seen upon loss of the LC3 conjugation machinery, the investigators should directly test if other processes like LAP are actually responsible for the effect seen.
2. In order to claim that autophagy is important in regulating the response, one should identify the component whose degradation is thought to be responsible for the effect. This molecule should then be shown to be delivered to lysosomes in a manner that requires the autophagy machinery and inhibition of lysosomal degradation should be shown to cause reduced turnover of the proposed target.
3. The turnover of the proposed target via the lysosomes should be demonstrated to explain why overall manipulation of autophagy (e.g., with autophagy inhibitors) elicits the biological effect that is being studied (more or less cell death by the pathway under investigation).
4. Alternative non-degradative mechanistic explanations such as scaffolding by autophagosomes or autophagosome-like membranes should be considered, tested and excluded, if necessary.

As far as we are aware, none of the proposed mechanisms for autophagy as a mechanism of regulating non-apoptotic cell death has, as yet, achieved this level of analysis. Indeed even the much more widely studied roles of autophagy in controlling apoptosis has rarely been done in this kind of way. One of our own recent studies did go part way to address these issues (Fitzwalter et al., 2018). In this study it was shown that for a particular anti-cancer agent (Nutlin 3, a MDM2 inhibitor), one could enhance tumor cell apoptosis by inhibiting autophagy. This was due to basal autophagy causing the lysosomal turnover of a transcription factor, FOXO3a, which was prevented from being delivered to the lysosome by both pharmacological and genetic inhibition of the autophagy machinery. This activity required the LC3 conjugation machinery and specifically ATG5 and ATG7. Since FOXO3a is a cellular protein, we proposed that this was true autophagy although we did not fully exclude that this might involve other LC3 dependent degradation processes. The mechanism of apoptosis sensitization was due to enhanced basal transcription of the PUMA/BBC3 gene, a pro-apoptotic member of the BCL family of proteins. Importantly we were also able to demonstrate that the enhanced apoptosis by autophagy inhibitors was abolished by inactivation of PUMA, and, more important, also by inactivation of just the single DNA binding site in the PUMA gene locus that binds FOXO3a. Thus the effect on apoptosis in this case seems to be entirely due to autophagic turnover of a specific protein (FOXO3a), which works through a specific mechanism (regulation of basal transcription of just one of its target genes (PUMA/BBC3)). By understanding how both degradative and non-degradative functions of the autophagy machinery as well as different degradative processes like LAP that involve components of the autophagy machinery elicit effects on non-apoptotic cell death pathways using similarly detailed mechanistic approaches, it may be possible to better influence these processes to obtain outcomes that we want.

Acknowledgments

This work was supported by the National Institutes of Health under Ruth L. Kirschstein National Research Service Award T32CA190216-04 and R01s CA150925, CA190170 and CA199741. Its contents are solely the responsibility of the authors and do not necessarily represent the official views of the NIH.

References

Aaes, T.L., Kaczmarek, A., Delvaeye, T., et al., 2016. Vaccination with necroptotic cancer cells induces efficient anti-tumor immunity. Cell Rep. 15, 274–287.

Amer, A.O., Swanson, M.S., 2005. Autophagy is an immediate macrophage response to Legionella pneumophila. Cell. Microbiol. 7, 765–778.

Anding, A.L., Baehrecke, E.H., 2015. Vps15 is required for stress induced and developmentally triggered autophagy and salivary gland protein secretion in Drosophila. Cell Death Differ. 22, 457–464.

Arya, B.D., Mittal, S., Joshi, P., Pandey, A.K., Ramirez-Vick, J.E., Singh, S.P., 2018. Graphene oxide-chloroquine nanoconjugate induce necroptotic death in A549 cancer cells through autophagy modulation. Nanomedicine (Lond.) 13, 2261–2282.

Basit, F., Cristofanon, S., Fulda, S., 2013. Obatoclax (GX15-070) triggers necroptosis by promoting the assembly of the necrosome on autophagosomal membranes. Cell Death Differ. 20, 1161–1173.

Basit, F., Van Oppen, L.M., Schockel, L., et al., 2017. Mitochondrial complex I inhibition triggers a mitophagy-dependent ROS increase leading to necroptosis and ferroptosis in melanoma cells. Cell Death Dis. 8, e2716.

Bergsbaken, T., Cookson, B.T., 2007. Macrophage activation redirects yersinia-infected host cell death from apoptosis to caspase-1-dependent pyroptosis. PLoS Pathog. 3, e161.

Bottcher, A., Gaipl, U.S., Furnrohr, B.G., et al., 2006. Involvement of phosphatidylserine, alphavbeta3, CD14, CD36, and complement C1q in the phagocytosis of primary necrotic lymphocytes by macrophages. Arthritis Rheum. 54, 927–938.

Bray, K., Mathew, R., Lau, A., et al., 2012. Autophagy suppresses RIP kinase-dependent necrosis enabling survival to mTOR inhibition. PLoS One 7, e41831.

Broz, P., Dixit, V.M., 2016. Inflammasomes: mechanism of assembly, regulation and signalling. Nat. Rev. Immunol. 16, 407–420.

Byrne, B.G., Dubuisson, J.F., Joshi, A.D., Persson, J.J., Swanson, M.S., 2013. Inflammasome components coordinate autophagy and pyroptosis as macrophage responses to infection. MBio 4, e00620-12.

Cai, Z., Jitkaew, S., Zhao, J., et al., 2014. Plasma membrane translocation of trimerized MLKL protein is required for TNF-induced necroptosis. Nat. Cell Biol. 16, 55–65.

Cai, Z., Zhang, A., Choksi, S., et al., 2016. Activation of cell-surface proteases promotes necroptosis, inflammation and cell migration. Cell Res. 26, 886–900.

Chen, X., Li, W., Ren, J., et al., 2014. Translocation of mixed lineage kinase domain-like protein to plasma membrane leads to necrotic cell death. Cell Res. 24, 105–121.

Cheng, K.T., Xiong, S., Ye, Z., et al., 2017. Caspase-11-mediated endothelial pyroptosis underlies endotoxemia-induced lung injury. J. Clin. Invest. 127, 4124–4135.

Cho, Y.S., Challa, S., Moquin, D., et al., 2009. Phosphorylation-driven assembly of the RIP1-RIP3 complex regulates programmed necrosis and virus-induced inflammation. Cell 137, 1112–1123.

Christofferson, D.E., Li, Y., Yuan, J., 2014. Control of life-or-death decisions by RIP1 kinase. Annu. Rev. Physiol. 76, 129–150.

Chuang, S.Y., Yang, C.H., Chou, C.C., Chiang, Y.P., Chuang, T.H., Hsu, L.C., 2013. TLR-induced PAI-2 expression suppresses IL-1beta processing via increasing autophagy and NLRP3 degradation. Proc. Natl. Acad. Sci. U. S. A. 110, 16079–16084.

Codogno, P., Meijer, A.J., 2005. Autophagy and signaling: their role in cell survival and cell death. Cell Death Differ. 12 (Suppl. 2), 1509–1518.

Degterev, A., Huang, Z., Boyce, M., et al., 2005. Chemical inhibitor of nonapoptotic cell death with therapeutic potential for ischemic brain injury. Nat. Chem. Biol. 1, 112–119.

Degterev, A., Hitomi, J., Germscheid, M., et al., 2008. Identification of RIP1 kinase as a specific cellular target of necrostatins. Nat. Chem. Biol. 4, 313–321.

Denton, D., Xu, T., Kumar, S., 2015. Autophagy as a pro-death pathway. Immunol. Cell Biol. 93, 35–42.

Dikic, I., Elazar, Z., 2018. Mechanism and medical implications of mammalian autophagy. Nat. Rev. Mol. Cell Biol. 19, 349–364.

Dillon, C.P., Weinlich, R., Rodriguez, D.A., et al., 2014. RIPK1 blocks early postnatal lethality mediated by caspase-8 and RIPK3. Cell 157, 1189–1202.

Ding, J., Yang, N., Yan, Y., et al., 2019. Rapamycin inhibited photoreceptor necroptosis and protected the retina by activation of autophagy in experimental retinal detachment. Curr. Eye Res. 44, 739–745.

Dixon, S.J., Lemberg, K.M., Lamprecht, M.R., et al., 2012. Ferroptosis: an iron-dependent form of nonapoptotic cell death. Cell 149, 1060–1072.

Dondelinger, Y., Declercq, W., Montessuit, S., et al., 2014. MLKL compromises plasma membrane integrity by binding to phosphatidylinositol phosphates. Cell Rep. 7, 971–981.

Dovey, C.M., Diep, J., Clarke, B.P., et al., 2018. MLKL requires the inositol phosphate code to execute necroptosis. Mol. Cell 70, 936-48.e7.

Dupont, N., Jiang, S., Pilli, M., Ornatowski, W., Bhattacharya, D., Deretic, V., 2011. Autophagy-based unconventional secretory pathway for extracellular delivery of IL-1beta. EMBO J. 30, 4701–4711.

Evavold, C.L., Ruan, J., Tan, Y., Xia, S., Wu, H., Kagan, J.C., 2018. The pore-forming protein gasdermin D regulates interleukin-1 secretion from living macrophages. Immunity 48, 35-44.e6.

Faustin, B., Lartigue, L., Bruey, J.M., et al., 2007. Reconstituted NALP1 inflammasome reveals two-step mechanism of caspase-1 activation. Mol. Cell 25, 713–724.

Feng, S., Yang, Y., Mei, Y., et al., 2007. Cleavage of RIP3 inactivates its caspase-independent apoptosis pathway by removal of kinase domain. Cell. Signal. 19, 2056–2067.

Feoktistova, M., Geserick, P., Panayotova-Dimitrova, D., Leverkus, M., 2012. Pick your poison: the Ripoptosome, a cell death platform regulating apoptosis and necroptosis. Cell Cycle 11, 460–467.

Fernandes-Alnemri, T., Wu, J., Yu, J.W., et al., 2007. The pyroptosome: a supramolecular assembly of ASC dimers mediating inflammatory cell death via caspase-1 activation. Cell Death Differ. 14, 1590–1604.
Fink, S.L., Cookson, B.T., 2006. Caspase-1-dependent pore formation during pyroptosis leads to osmotic lysis of infected host macrophages. Cell. Microbiol. 8, 1812–1825.
Fink, S.L., Bergsbaken, T., Cookson, B.T., 2008. Anthrax lethal toxin and salmonella elicit the common cell death pathway of caspase-1-dependent pyroptosis via distinct mechanisms. Proc. Natl. Acad. Sci. U. S. A. 105, 4312–4317.
Fitzwalter, B.E., Thorburn, A., 2015. Recent insights into cell death and autophagy. FEBS J. 282, 4279–4288.
Fitzwalter, B.E., Towers, C.G., Sullivan, K.D., et al., 2018. Autophagy inhibition mediates apoptosis sensitization in cancer therapy by relieving FOXO3a turnover. Dev. Cell 44, 555-65.e3.
Fortunato F, Burgers H, Bergmann F, et al., 2009. Impaired autolysosome formation correlates with Lamp-2 depletion: role of apoptosis, autophagy, and necrosis in pancreatitis. Gastroenterology 137, 350–60, 60.e1-5.
Frank, D., Vaux, D.L., Murphy, J.M., Vince, J.E., Lindqvist, L.M., 2019. Activated MLKL attenuates autophagy following its translocation to intracellular membranes. J. Cell Sci. 132, jcs220996.
Galluzzi, L., Green, D.R., 2019. Autophagy-independent functions of the autophagy machinery. Cell 177, 1682–1699.
Galluzzi, L., Kepp, O., Krautwald, S., Kroemer, G., Linkermann, A., 2014. Molecular mechanisms of regulated necrosis. Semin. Cell Dev. Biol. 35, 24–32.
Galluzzi, L., Baehrecke, E.H., Ballabio, A., et al., 2017a. Molecular definitions of autophagy and related processes. EMBO J. 36, 1811–1836.
Galluzzi, L., Kepp, O., Chan, F.K., Kroemer, G., 2017b. Necroptosis: mechanisms and relevance to disease. Annu. Rev. Pathol. 12, 103–130.
Gao, M., Monian, P., Pan, Q., Zhang, W., Xiang, J., Jiang, X., 2016. Ferroptosis is an autophagic cell death process. Cell Res. 26, 1021–1032.
Gao, H., Bai, Y., Jia, Y., et al., 2018. Ferroptosis is a lysosomal cell death process. Biochem. Biophys. Res. Commun. 503, 1550–1556.
Geserick, P., Hupe, M., Moulin, M., et al., 2009. Cellular IAPs inhibit a cryptic CD95-induced cell death by limiting RIP1 kinase recruitment. J. Cell Biol. 187, 1037–1054.
Gong, Y.N., Guy, C., Crawford, J.C., Green, D.R., 2017. Biological events and molecular signaling following MLKL activation during necroptosis. Cell Cycle 16, 1748–1760.
Goodall, M.L., Fitzwalter, B.E., Zahedi, S., et al., 2016. The autophagy machinery controls cell death switching between apoptosis and necroptosis. Dev. Cell 37, 337–349.
Green, D.R., Levine, B., 2014. To be or not to be? How selective autophagy and cell death govern cell fate. Cell 157, 65–75.
Gump, J.M., Staskiewicz, L., Morgan, M.J., Bamberg, A., Riches, D.W., Thorburn, A., 2014. Autophagy variation within a cell population determines cell fate through selective degradation of Fap-1. Nat. Cell Biol. 16, 47–54.
Gutierrez, M.G., Master, S.S., Singh, S.B., Taylor, G.A., Colombo, M.I., Deretic, V., 2004. Autophagy is a defense mechanism inhibiting BCG and mycobacterium tuberculosis survival in infected macrophages. Cell 119, 753–766.
Gyrd-Hansen, M., 2017. All roads lead to ubiquitin. Cell Death Differ. 24, 1135–1136.
Haberzettl, P., Hill, B.G., 2013. Oxidized lipids activate autophagy in a JNK-dependent manner by stimulating the endoplasmic reticulum stress response. Redox Biol. 1, 56–64.
Hadifar, S., Behrouzi, A., Fateh, A., et al., 2019. Comparative study of interruption of signaling pathways in lung epithelial cell by two different mycobacterium tuberculosis lineages. J. Cell. Physiol. 234, 4739–4753.

Hagar, J.A., Powell, D.A., Aachoui, Y., Ernst, R.K., Miao, E.A., 2013. Cytoplasmic LPS activates caspase-11: implications in TLR4-independent endotoxic shock. Science 341, 1250–1253.

Harris, J., Hartman, M., Roche, C., et al., 2011. Autophagy controls IL-1beta secretion by targeting pro-IL-1beta for degradation. J. Biol. Chem. 286, 9587–9597.

He, C., Klionsky, D.J., 2009. Regulation mechanisms and signaling pathways of autophagy. Annu. Rev. Genet. 43, 67–93.

He, S., Wang, X., 2018. RIP kinases as modulators of inflammation and immunity. Nat. Immunol. 19, 912–922.

He, S., Liang, Y., Shao, F., Wang, X., 2011. Toll-like receptors activate programmed necrosis in macrophages through a receptor-interacting kinase-3-mediated pathway. Proc. Natl. Acad. Sci. U. S. A. 108, 20054–20059.

He, W., Wang, Q., Srinivasan, B., et al., 2014. A JNK-mediated autophagy pathway that triggers c-IAP degradation and necroptosis for anticancer chemotherapy. Oncogene 33, 3004–3013.

He, W.T., Wan, H., Hu, L., et al., 2015. Gasdermin D is an executor of pyroptosis and required for interleukin-1beta secretion. Cell Res. 25, 1285–1298.

Heckmann, B.L., Green, D.R., 2019. LC3-associated phagocytosis at a glance. J. Cell Sci. 132, jcs222984.

Heckmann, B.L., Boada-Romero, E., Cunha, L.D., Magne, J., Green, D.R., 2017. LC3-associated phagocytosis and inflammation. J. Mol. Biol. 429, 3561–3576.

Heilig, R., Dick, M.S., Sborgi, L., Meunier, E., Hiller, S., Broz, P., 2018. The gasdermin-D pore acts as a conduit for IL-1beta secretion in mice. Eur. J. Immunol. 48, 584–592.

Hill, B.G., Haberzettl, P., Ahmed, Y., Srivastava, S., Bhatnagar, A., 2008. Unsaturated lipid peroxidation-derived aldehydes activate autophagy in vascular smooth-muscle cells. Biochem. J. 410, 525–534.

Hosokawa, N., Sasaki, T., Iemura, S., Natsume, T., Hara, T., Mizushima, N., 2009a. Atg101, a novel mammalian autophagy protein interacting with Atg13. Autophagy 5, 973–979.

Hosokawa, N., Hara, T., Kaizuka, T., et al., 2009b. Nutrient-dependent mTORC1 association with the ULK1-Atg13-FIP200 complex required for autophagy. Mol. Biol. Cell 20, 1981–1991.

Hou, W., Han, J., Lu, C., Goldstein, L.A., Rabinowich, H., 2010. Autophagic degradation of active caspase-8: a crosstalk mechanism between autophagy and apoptosis. Autophagy 6, 891–900.

Hou, W., Xie, Y., Song, X., et al., 2016. Autophagy promotes ferroptosis by degradation of ferritin. Autophagy 12, 1425–1428.

Ito, Y., Ofengeim, D., Najafov, A., et al., 2016. RIPK1 mediates axonal degeneration by promoting inflammation and necroptosis in ALS. Science 353, 603–608.

Kaczmarek, A., Vandenabeele, P., Krysko, D.V., 2013. Necroptosis: the release of damage-associated molecular patterns and its physiological relevance. Immunity 38, 209–223.

Kaiser, W.J., Upton, J.W., Long, A.B., et al., 2011. RIP3 mediates the embryonic lethality of caspase-8-deficient mice. Nature 471, 368–372.

Kaiser, W.J., Sridharan, H., Huang, C., et al., 2013. Toll-like receptor 3-mediated necrosis via TRIF, RIP3, and MLKL. J. Biol. Chem. 288, 31268–31279.

Kayagaki, N., Stowe, I.B., Lee, B.L., et al., 2015. Caspase-11 cleaves gasdermin D for non-canonical inflammasome signalling. Nature 526, 666–671.

Kharaziha, P., Chioureas, D., Baltatzis, G., et al., 2015. Sorafenib-induced defective autophagy promotes cell death by necroptosis. Oncotarget 6, 37066–37082.

Kondylis, V., Kumari, S., Vlantis, K., Pasparakis, M., 2017. The interplay of IKK, NF-kappaB and RIPK1 signaling in the regulation of cell death, tissue homeostasis and inflammation. Immunol. Rev. 277, 113–127.

Koo, G.-B., Morgan, M.J., Lee, D.-G., et al., 2015. Methylation-dependent loss of RIP3 expression in cancer represses programmed necrosis in response to chemotherapeutics. Cell Res. 25, 707–725.

Krueger, A., Schmitz, I., Baumann, S., Krammer, P.H., Kirchhoff, S., 2001. Cellular FLICE-inhibitory protein splice variants inhibit different steps of caspase-8 activation at the CD95 death-inducing signaling complex. J. Biol. Chem. 276, 20633–20640.

Kufer, T.A., Sansonetti, P.J., 2007. Sensing of bacteria: NOD a lonely job. Curr. Opin. Microbiol. 10, 62–69.

Le Guerroue, F., Eck, F., Jung, J., et al., 2017. Autophagosomal content profiling reveals an LC3C-dependent piecemeal mitophagy pathway. Mol. Cell 68, 786-96.e6.

Lemke, J., Von Karstedt, S., Abd El Hay, M., et al., 2014. Selective CDK9 inhibition overcomes TRAIL resistance by concomitant suppression of cFlip and Mcl-1. Cell Death Differ. 21, 491–502.

Li, J., Mcquade, T., Siemer, A.B., et al., 2012. The RIP1/RIP3 necrosome forms a functional amyloid signaling complex required for programmed necrosis. Cell 150, 339–350.

Li, S., Ning, L.G., Lou, X.H., Xu, G.Q., 2018. Necroptosis in inflammatory bowel disease and other intestinal diseases. World J. Clin. Cases 6, 745–752.

Li, M.Y., Zhu, X.L., Zhao, B.X., et al., 2019. Adrenomedullin alleviates the pyroptosis of Leydig cells by promoting autophagy via the ROS-AMPK-mTOR axis. Cell Death Dis. 10, 489.

Liao, Z., Dai, Z., Cai, C., et al., 2019. Knockout of Atg5 inhibits proliferation and promotes apoptosis of DF-1 cells. In Vitro Cell. Dev. Biol. Anim. 55, 341–348.

Lim, J., Park, H., Heisler, J., et al., 2019. Autophagy regulates inflammatory programmed cell death via turnover of RHIM-domain proteins. Elife 8, e44452.

Lin, S.Y., Hsieh, S.Y., Fan, Y.T., et al., 2018. Necroptosis promotes autophagy-dependent upregulation of DAMP and results in immunosurveillance. Autophagy 14, 778–795.

Liu, X., Zhang, Z., Ruan, J., et al., 2016. Inflammasome-activated gasdermin D causes pyroptosis by forming membrane pores. Nature 535, 153–158.

Liu, S., Li, Y., Choi, H.M.C., et al., 2018. Lysosomal damage after spinal cord injury causes accumulation of RIPK1 and RIPK3 proteins and potentiation of necroptosis. Cell Death Dis. 9, 476.

Liu, J., Yang, M., Kang, R., Klionsky, D.J., Tang, D., 2019. Autophagic degradation of the circadian clock regulator promotes ferroptosis. Autophagy 15, 2033–2035.

Lorincz, T., Jemnitz, K., Kardon, T., Mandl, J., Szarka, A., 2015. Ferroptosis is involved in acetaminophen induced cell death. Pathol. Oncol. Res. 21, 1115–1121.

Louandre, C., Marcq, I., Bouhlal, H., et al., 2015. The retinoblastoma (Rb) protein regulates ferroptosis induced by sorafenib in human hepatocellular carcinoma cells. Cancer Lett. 356, 971–977.

Lu, W., Sun, J., Yoon, J.S., et al., 2016. Mitochondrial protein PGAM5 regulates mitophagic protection against cell necroptosis. PLoS One 11, e0147792.

Mancias, J.D., Wang, X., Gygi, S.P., Harper, J.W., Kimmelman, A.C., 2014. Quantitative proteomics identifies NCOA4 as the cargo receptor mediating ferritinophagy. Nature 509, 105–109.

Mancilla, H., Maldonado, R., Cereceda, K., et al., 2015. Glutathione depletion induces spermatogonial cell autophagy. J. Cell. Biochem. 116, 2283–2292.

Marino, G., Niso-Santano, M., Baehrecke, E.H., Kroemer, G., 2014. Self-consumption: the interplay of autophagy and apoptosis. Nat. Rev. Mol. Cell Biol. 15, 81–94.

Martinez-Lopez, N., Athonvarangkul, D., Mishall, P., Sahu, S., Singh, R., 2013. Autophagy proteins regulate ERK phosphorylation. Nat. Commun. 4, 2799.

Martinon, F., Tschopp, J., 2007. Inflammatory caspases and inflammasomes: master switches of inflammation. Cell Death Differ. 14, 10–22.

Martin-Sanchez, D., Ruiz-Andres, O., Poveda, J., et al., 2017. Ferroptosis, but not necroptosis, is important in nephrotoxic folic acid-induced AKI. J. Am. Soc. Nephrol. 28, 218–229.
Matsuzawa-Ishimoto, Y., Shono, Y., Gomez, L.E., et al., 2017. Autophagy protein ATG16L1 prevents necroptosis in the intestinal epithelium. J. Exp. Med. 214, 3687–3705.
Mccomb, S., Cheung, H.H., Korneluk, R.G., Wang, S., Krishnan, L., Sad, S., 2012. cIAP1 and cIAP2 limit macrophage necroptosis by inhibiting Rip1 and Rip3 activation. Cell Death Differ. 19, 1791–1801.
Mehto, S., Jena, K.K., Nath, P., et al., 2019. The Crohn's disease risk factor IRGM limits NLRP3 inflammasome activation by impeding its assembly and by mediating its selective autophagy. Mol. Cell 73, 429-45.e7.
Mizumura, K., Cloonan, S.M., Nakahira, K., et al., 2014. Mitophagy-dependent necroptosis contributes to the pathogenesis of COPD. J. Clin. Invest. 124, 3987–4003.
Moulin, M., Anderton, H., Voss, A.K., et al., 2012. IAPs limit activation of RIP kinases by TNF receptor 1 during development. EMBO J. 31, 1679–1691.
Muller, T., Dewitz, C., Schmitz, J., et al., 2017. Necroptosis and ferroptosis are alternative cell death pathways that operate in acute kidney failure. Cell. Mol. Life Sci. 74, 3631–3645.
Murphy, J.M., Czabotar, P.E., Hildebrand, J.M., et al., 2013. The pseudokinase MLKL mediates necroptosis via a molecular switch mechanism. Immunity 39, 443–453.
Nakagawa, I., Amano, A., Mizushima, N., et al., 2004. Autophagy defends cells against invading group A streptococcus. Science 306, 1037–1040.
Nakahira, K., Haspel, J.A., Rathinam, V.A., et al., 2011. Autophagy proteins regulate innate immune responses by inhibiting the release of mitochondrial DNA mediated by the NALP3 inflammasome. Nat. Immunol. 12, 222–230.
Nikseresht, S., Khodagholi, F., Ahmadiani, A., 2019. Protective effects of ex-527 on cerebral ischemia-reperfusion injury through necroptosis signaling pathway attenuation. J. Cell. Physiol. 234, 1816–1826.
Oberst, A., Dillon, C.P., Weinlich, R., et al., 2011. Catalytic activity of the caspase-8-FLIP (L) complex inhibits RIPK3-dependent necrosis. Nature 471, 363–367.
O'donnell, M.A., Perez-Jimenez, E., Oberst, A., et al., 2011. Caspase 8 inhibits programmed necrosis by processing CYLD. Nat. Cell Biol. 13, 1437–1442.
Ofengeim, D., Yuan, J., 2013. Regulation of RIP1 kinase signalling at the crossroads of inflammation and cell death. Nat. Rev. Mol. Cell Biol. 14, 727–736.
Ona, V.O., Li, M., Vonsattel, J.P., et al., 1999. Inhibition of caspase-1 slows disease progression in a mouse model of Huntington's disease. Nature 399, 263–267.
Orozco, S.L., Daniels, B.P., Yatim, N., et al., 2019. RIPK3 activation leads to cytokine synthesis that continues after loss of cell membrane integrity. Cell. Rep. 28, 2275-87.e5.
Pasparakis, M., Vandenabeele, P., 2015. Necroptosis and its role in inflammation. Nature 517, 311–320.
Pu, Q., Gan, C., Li, R., et al., 2017. Atg7 deficiency intensifies inflammasome activation and pyroptosis in pseudomonas sepsis. J. Immunol. 198, 3205–3213.
Qiu, T., Pei, P., Yao, X., et al., 2018. Taurine attenuates arsenic-induced pyroptosis and nonalcoholic steatohepatitis by inhibiting the autophagic-inflammasomal pathway. Cell Death Dis. 9, 946.
Rebe, C., Cathelin, S., Launay, S., et al., 2007. Caspase-8 prevents sustained activation of NF-kappaB in monocytes undergoing macrophagic differentiation. Blood 109, 1442–1450.
Remijsen, Q., Goossens, V., Grootjans, S., et al., 2014. Depletion of RIPK3 or MLKL blocks TNF-driven necroptosis and switches towards a delayed RIPK1 kinase-dependent apoptosis. Cell Death Dis. 5, e1004.

Ruan, J., Xia, S., Liu, X., Lieberman, J., Wu, H., 2018. Cryo-EM structure of the gasdermin A3 membrane pore. Nature 557, 62–67.

Russell, R.C., Tian, Y., Yuan, H., et al., 2013. ULK1 induces autophagy by phosphorylating Beclin-1 and activating VPS34 lipid kinase. Nat. Cell Biol. 15, 741–750.

Saitoh, T., Fujita, N., Jang, M.H., et al., 2008. Loss of the autophagy protein Atg16L1 enhances endotoxin-induced IL-1beta production. Nature 456, 264–268.

Sato, M., Kusumi, R., Hamashima, S., et al., 2018. The ferroptosis inducer erastin irreversibly inhibits system xc − and synergizes with cisplatin to increase cisplatin's cytotoxicity in cancer cells. Sci. Rep. 8, 968.

Sborgi, L., Ruhl, S., Mulvihill, E., et al., 2016. GSDMD membrane pore formation constitutes the mechanism of pyroptotic cell death. EMBO J. 35, 1766–1778.

Schielke, G.P., Yang, G.Y., Shivers, B.D., Betz, A.L., 1998. Reduced ischemic brain injury in interleukin-1 beta converting enzyme-deficient mice. J. Cereb. Blood Flow Metab. 18, 180–185.

Schilling, R., Geserick, P., Leverkus, M., 2014. Characterization of the ripoptosome and its components: implications for anti-inflammatory and cancer therapy. Methods Enzymol. 545, 83–102.

Seo, J., Lee, E.W., Sung, H., et al., 2016. CHIP controls necroptosis through ubiquitylation- and lysosome-dependent degradation of RIPK3. Nat. Cell Biol. 18, 291–302.

Shan, B., Pan, H., Najafov, A., Yuan, J., 2018. Necroptosis in development and diseases. Genes Dev. 32, 327–340.

Shi, C.S., Shenderov, K., Huang, N.N., et al., 2012. Activation of autophagy by inflammatory signals limits IL-1beta production by targeting ubiquitinated inflammasomes for destruction. Nat. Immunol. 13, 255–263.

Shi, J., Zhao, Y., Wang, Y., et al., 2014. Inflammatory caspases are innate immune receptors for intracellular LPS. Nature 514, 187–192.

Shi, J., Zhao, Y., Wang, K., et al., 2015. Cleavage of GSDMD by inflammatory caspases determines pyroptotic cell death. Nature 526, 660–665.

Shi, J., Gao, W., Shao, F., 2017. Pyroptosis: gasdermin-mediated programmed necrotic cell death. Trends Biochem. Sci. 42, 245–254.

Shimizu, S., Kanaseki, T., Mizushima, N., et al., 2004. Role of Bcl-2 family proteins in a non-apoptotic programmed cell death dependent on autophagy genes. Nat. Cell Biol. 6, 1221–1228.

Siegmund, B., Lehr, H.A., Fantuzzi, G., Dinarello, C.A., 2001. IL-1 beta-converting enzyme (caspase-1) in intestinal inflammation. Proc. Natl. Acad. Sci. U. S. A. 98, 13249–13254.

Simon, A., Van Der Meer, J.W., 2007. Pathogenesis of familial periodic fever syndromes or hereditary autoinflammatory syndromes. Am. J. Physiol. Regul. Integr. Comp. Physiol. 292, R86–R98.

Snyder, A.G., Hubbard, N.W., Messmer, M.N., et al., 2019. Intratumoral activation of the necroptotic pathway components RIPK1 and RIPK3 potentiates antitumor immunity. Sci. Immunol. 4, eaaw2004.

Song, X., Zhu, S., Chen, P., et al., 2018. AMPK-mediated BECN1 phosphorylation promotes ferroptosis by directly blocking system xc(−) activity. Curr. Biol. 28, 2388-99.e5.

Stockwell, B.R., Friedmann Angeli, J.P., Bayir, H., et al., 2017. Ferroptosis: a regulated cell death nexus linking metabolism, redox biology, and disease. Cell 171, 273–285.

Sulkshane, P., Teni, T., 2017. BH3 mimetic Obatoclax (GX15-070) mediates mitochondrial stress predominantly via MCL-1 inhibition and induces autophagy-dependent necroptosis in human oral cancer cells. Oncotarget 8, 60060–60079.

Sun, L., Wang, H., Wang, Z., et al., 2012. Mixed lineage kinase domain-like protein mediates necrosis signaling downstream of RIP3 kinase. Cell 148, 213–227.

Sun, X., Ou, Z., Chen, R., et al., 2016. Activation of the p62-Keap1-NRF2 pathway protects against ferroptosis in hepatocellular carcinoma cells. Hepatology 63, 173–184.

Sun, Y., Zheng, Y., Wang, C., Liu, Y., 2018. Glutathione depletion induces ferroptosis, autophagy, and premature cell senescence in retinal pigment epithelial cells. Cell Death Dis. 9, 753.

Tait, S.W., Oberst, A., Quarato, G., et al., 2013. Widespread mitochondrial depletion via mitophagy does not compromise necroptosis. Cell Rep. 5, 878–885.

Tang, Z., Takahashi, Y., Chen, C., et al., 2017. Atg2A/B deficiency switches cytoprotective autophagy to non-canonical caspase-8 activation and apoptosis. Cell Death Differ. 24, 2127–2138.

Tenev, T., Bianchi, K., Darding, M., et al., 2011. The Ripoptosome, a signaling platform that assembles in response to genotoxic stress and loss of IAPs. Mol. Cell 43, 432–448.

Thorburn, J., Horita, H., Redzic, J., Hansen, K., Frankel, A.E., Thorburn, A., 2009. Autophagy regulates selective HMGB1 release in tumor cells that are destined to die. Cell Death Differ. 16, 175–183.

Tu, Y., Guo, C., Song, F., et al., 2019. Mild hypothermia alleviates diabetes aggravated cerebral ischemic injury via activating autophagy and inhibiting pyroptosis. Brain Res. Bull. 150, 1–12.

Upton, J.W., Kaiser, W.J., Mocarski, E.S., 2010. Virus inhibition of RIP3-dependent necrosis. Cell Host Microbe 7, 302–313.

Upton, J.W., Kaiser, W.J., Mocarski, E.S., 2012. DAI/ZBP1/DLM-1 complexes with RIP3 to mediate virus-induced programmed necrosis that is targeted by murine cytomegalovirus vIRA. Cell Host Microbe 11, 290–297.

Vandenabeele, P., Declercq, W., Van Herreweghe, F., Vanden Berghe, T., 2010. The role of the kinases RIP1 and RIP3 in TNF-induced necrosis. Sci. Signal. 3, re4.

Vanlangenakker, N., Vanden Berghe, T., Bogaert, P., et al., 2011. cIAP1 and TAK1 protect cells from TNF-induced necrosis by preventing RIP1/RIP3-dependent reactive oxygen species production. Cell Death Differ. 18, 656–665.

Wang, Z., Jiang, H., Chen, S., Du, F., Wang, X., 2012. The mitochondrial phosphatase PGAM5 functions at the convergence point of multiple necrotic death pathways. Cell 148, 228–243.

Wang, H., Sun, L., Su, L., et al., 2014. Mixed lineage kinase domain-like protein MLKL causes necrotic membrane disruption upon phosphorylation by RIP3. Mol. Cell 54, 133–146.

Wang, Z., Li, Y., Yang, X., et al., 2019. Protective effects of rapamycin induced autophagy on CLP septic mice. Comp. Immunol. Microbiol. Infect. Dis. 64, 47–52.

Wei, Q., Zhu, R., Zhu, J., Zhao, R., Li, M., 2019. E2-induced activation of the NLRP3 inflammasome triggers pyroptosis and inhibits autophagy in HCC cells. Oncol. Res. 27, 827–834.

Weinlich, R., Oberst, A., Dillon, C.P., et al., 2013. Protective roles for caspase-8 and cFLIP in adult homeostasis. Cell Rep. 5, 340–348.

Wu, J., Huang, Z., Ren, J., et al., 2013. Mlkl knockout mice demonstrate the indispensable role of Mlkl in necroptosis. Cell Res. 23, 994–1006.

Wu, Z., Geng, Y., Lu, X., et al., 2019. Chaperone-mediated autophagy is involved in the execution of ferroptosis. Proc. Natl. Acad. Sci. U. S. A. 116, 2996–3005.

Xia, B., Fang, S., Chen, X., et al., 2016. MLKL forms cation channels. Cell Res. 26, 517–528.

Xie, Y., Hou, W., Song, X., et al., 2016. Ferroptosis: process and function. Cell Death Differ. 23, 369–379.

Xu, J., Xu, X., Shi, S., et al., 2016. Autophagy-mediated degradation of IAPs and c-FLIP(L) potentiates apoptosis induced by combination of TRAIL and Chal-24. J. Cell. Biochem. 117, 1136–1144.

Yabal, M., Muller, N., Adler, H., et al., 2014. XIAP restricts TNF- and RIP3-dependent cell death and inflammasome activation. Cell Rep. 7, 1796–1808.

Yang, W.S., Stockwell, B.R., 2008. Synthetic lethal screening identifies compounds activating iron-dependent, nonapoptotic cell death in oncogenic-RAS-harboring cancer cells. Chem. Biol. 15, 234–245.

Yang, W.S., Sriramaratnam, R., Welsch, M.E., et al., 2014. Regulation of ferroptotic cancer cell death by GPX4. Cell 156, 317–331.

Yatim, N., Jusforgues-Saklani, H., Orozco, S., et al., 2015. RIPK1 and NF-kappaB signaling in dying cells determines cross-priming of CD8(+) T cells. Science 350, 328–334.

Ye, Y.C., Yu, L., Wang, H.J., Tashiro, S., Onodera, S., Ikejima, T., 2011. TNFalpha-induced necroptosis and autophagy via supression of the p38-NF-kappaB survival pathway in L929 cells. J. Pharmacol. Sci. 117, 160–169.

Yoon, S., Kovalenko, A., Bogdanov, K., Wallach, D., 2017. MLKL, the protein that mediates necroptosis, also regulates endosomal trafficking and extracellular vesicle generation. Immunity 47, 51-65.e7.

Young, M.M., Takahashi, Y., Khan, O., et al., 2012. Autophagosomal membrane serves as platform for intracellular death-inducing signaling complex (iDISC)-mediated caspase-8 activation and apoptosis. J. Biol. Chem. 287, 12455–12468.

Yu, L., Alva, A., Su, H., et al., 2004. Regulation of an ATG7-beclin 1 program of autophagic cell death by caspase-8. Science 304, 1500–1502.

Yu, L., Wan, F., Dutta, S., et al., 2006. Autophagic programmed cell death by selective catalase degradation. Proc. Natl. Acad. Sci. U. S. A. 103, 4952–4957.

Yu, P., Wang, H.Y., Tian, M., et al., 2019. Eukaryotic elongation factor-2 kinase regulates the cross-talk between autophagy and pyroptosis in doxorubicin-treated human melanoma cells in vitro. Acta Pharmacol. Sin. 40, 1237–1244.

Zhang, H., Baehrecke, E.H., 2015. Eaten alive: novel insights into autophagy from multicellular model systems. Trends Cell Biol. 25, 376–387.

Zhang, Z., Yao, Z., Wang, L., et al., 2018. Activation of ferritinophagy is required for the RNA-binding protein ELAVL1/HuR to regulate ferroptosis in hepatic stellate cells. Autophagy 14, 2083–2103.

Zhao, J., Jitkaew, S., Cai, Z., et al., 2012. Mixed lineage kinase domain-like is a key receptor interacting protein 3 downstream component of TNF-induced necrosis. Proc. Natl. Acad. Sci. U. S. A. 109, 5322–5327.

Zhao, Q., Yu, X., Zhang, H., et al., 2017. RIPK3 mediates necroptosis during embryonic development and postnatal inflammation in Fadd-deficient mice. Cell Rep. 19, 798–808.

Zhou, R., Yazdi, A.S., Menu, P., Tschopp, J., 2011. A role for mitochondria in NLRP3 inflammasome activation. Nature 469, 221–225.

Zhou, X., Xie, L., Xia, L., et al., 2017. RIP3 attenuates the pancreatic damage induced by deletion of ATG7. Cell Death Dis. 8, e2918.

Zi, Y., Yi-An, Y., Bing, J., et al., 2019. Sirt6-induced autophagy restricted TREM-1-mediated pyroptosis in ox-LDL-treated endothelial cells: relevance to prognostication of patients with acute myocardial infarction. Cell Death Dis. 5, 88.

CHAPTER SIX

Redox signaling in the pathogenesis of human disease and the regulatory role of autophagy

Shazib Pervaiz[a,b,c], Gregory L. Bellot[d], Antoinette Lemoine[e], Catherine Brenner[f,]*

[a]Department of Physiology and Medical Science Cluster Cancer Program, Yong Loo Lin School of Medicine, National University of Singapore, Singapore
[b]NUS Graduate School for Integrative Sciences and Engineering, National University of Singapore, Singapore
[c]National University Cancer Institute, National University Health System, Singapore
[d]Department of Hand & Reconstructive Microsurgery, University Orthopedic, Hand & Reconstructive Microsurgery Cluster, National University Health System, Singapore
[e]Departement of Oncogenetics, APHP, GH Paris-Sud, Hôpital Paul Brousse, Inserm UMR-S 1193, Université Paris-Saclay, Villejuif, France
[f]Université Paris-Saclay, Institut Gustave Roussy, CNRS, Metabolic and Systemic Aspects of Oncogenesis, Villejuif, France
*Corresponding author: e-mail address: catherine.brenner@universite-paris-saclay.fr

Contents

Abstract

Aberrant cell death signaling and oxidative stress are implicated in myriad of human pathological states such as neurodegenerative, cardiovascular, metabolic and liver diseases, as well as drug-induced toxicities. While regulated cell death and mild oxidative stress are essential during normal tissue homeostasis, deregulated signaling can trigger

International Review of Cell and Molecular Biology, Volume 352
ISSN 1937-6448
https://doi.org/10.1016/bs.ircmb.2020.03.002

massive depletion in a particular cell type and/or damage tissues and impair organ function with deleterious consequences that manifest as disease states. If regeneration cannot restore tissue homeostasis, the severity of the disease correlates with the extent of cell loss. Cell death can be executed via multiple modalities such as apoptosis, necrosis, pyroptosis, necroptosis and ferroptosis, depending on cell autonomous mechanisms (e.g., reactive oxygen species production, calcium overload and altered proteostasis) and/or non-cell autonomous processes (e.g., environmental stress, irradiation, chemotherapeutic agents, inflammation and pathogens). Accordingly, the inhibition of aberrant cell death and oxidative stress together with activation of autophagy, a regulated self-degradation process, are progressively emerging as relevant cytoprotective strategies to sustain homeostasis. In this review, we summarize the current literature on the crosstalk between cellular redox state and cell fate signaling, specifically from the standpoint of autophagy and its role in the maintenance of tissue/organ homeostasis via regulating oxidative stress and the potential implications for the design of novel therapeutic strategies.

Abbreviations

AD	Alzheimer's disease
AMPK	5′-AMP-activated protein kinase
ETC	electron transport chain
GPx	glutathione peroxidase
GPX4	glutathione peroxidase 4
GR	glutathione reductase
GSH	reduced glutathione
H_20_2	hydrogen peroxide
HO	heme oxygenase
mTORC1	mammalian target of rapamycin complex 1
NO	nitric oxide
NOX	nitric oxide synthase
PD	Parkinson's disease
PS1	presenilin 1
PTP	permeability transition pore
RNS	reactive nitrogen species
ROS	reactive oxygen species
Sirt1	sirtuin 1
SOD	superoxide dismutase
VDAC	voltage dependent anion channel

1. Introduction

Accumulated evidences obtained over the past decades indicate that numerous human pathologies involve a deregulation of the balance between cell life and death associated with harmful oxidative stress and cell demise

(Carson and Ribeiro, 1993). This applies notably to neurodegenerative pathologies, cardiovascular diseases (e.g., myocardial infarction, heart failure, ischemia-reperfusion), liver diseases and drug-induced toxicities (Carson and Ribeiro, 1993). Whereas cell loss prevention is a recognized goal in therapeutic innovation, identification of new druggable cell death inhibitor compounds is still an unmet need.

Until recently, programmed cell death or apoptosis and accidental necrosis were believed to be the sole causes of cell loss observed in human degenerative diseases (Carson and Ribeiro, 1993). This paradigm evolved in the last few years with the discovery that various cell death modalities can co-exist. For example, multiple modalities of cell death execution such as regulated necrosis, pyroptosis, necroptosis and ferroptosis have been detected in various pathophysiological states, sometimes revealed upon postmortem examination (Galluzzi et al., 2018). Whereas, these forms of cell death are different in terms of morphological criteria and biochemical mechanisms, they are interconnected and display a common link in the form of reactive oxygen species (ROS) production, rendering their specific identification complex. Moreover, as these forms are still incompletely described, there is a need for reliable biomarkers to allow for the accurate identification in vivo and design of personalized treatments.

Due to the multiplicity of signaling pathways, it appears highly unlikely that targeting a single node (effector pathway) via pharmacological or genetic approaches would suffice as a strategy to alleviate disease etiology and prevent tissue and organ damage. As a matter of fact, it might be more judicious to activate a signaling pathway to prevent the deleterious effects of cell death and ROS, well before the induction of irreversible cellular damages and organ dysfunction. To that end, the induction of autophagy that promotes the self-renewal of cellular components has been proposed as an attractive cytoprotective strategy. Indeed, autophagy could have potential applications in the management of disease states or toxin-induced tissue/organ damage that directly or indirectly involve redox stress and excessive cell death signaling (Galluzzi et al., 2017; Khandia et al., 2019; Rubinsztein et al., 2012; Yin et al., 2016).

2. Cellular redox state and its regulation

2.1 Intracellular sources of reactive oxygen species (ROS)

The cellular redox status is an outcome of the tight balance between the generation of intracellular ROS and their regulation by the multiple anti-oxidant systems. The primary sources of intracellular ROS generation are the membrane NADPH oxidase (NOX) family (NOX 1-5, DUOX 1/2), and the mitochondrial electron transport chain (ETC). While NOX 2 was first described in phagocytic cells, similar oxidase complexes have been described in other cell types (Lambeth, 2004), thus underscoring the importance of ROS in myriad cellular processes. Thus, the physiological NOX function being ROS production, ROS may have a key signaling role in the cell and may not be viewed only as provoking accidental insults. For the ETC, there is a general consensus that the prime sites for the initial generation of ROS are complexes I and III; however there is evidence now to implicate complexes II and IV as well (Chong et al., 2014). The two most extensively studied ROS under physiological and pathological states are superoxide ($O_2^{\bullet}$) and hydrogen peroxide (H_2O_2). Superoxide, generated by one electron reduction of molecular oxygen, is a radical that has low diffusibility across membranes and relatively low reactivity with biomolecules (Fogg et al., 2011; Halliwell, 2007; Han et al., 2003; Imlay, 1995, 2003; Madesh and Hajnoczky, 2001). However, selective reactivity with proteins containing iron-sulfur [4Fe—4S] clusters has been described$^{\bullet}$, which results in the oxidation and degradation of the cluster and release of excess Fe^{++} (Flint et al., 1993; Gardner and Fridovich, 1991a,b; Liochev and Fridovich, 1993). Intracellular $O_2^{\bullet -}$ is readily converted to the non-radical product H_2O_2 by the action of the various superoxide dismutase (cytosolic Cu/ZnSOD and mitochondrial MnSOD). Hydrogen peroxide has a much longer lifespan and ability to diffuse through membranes and as such most of the secondary messenger function attributed to intracellular ROS is believed to be mediated by the oxidizing activity of H_2O_2, such as observed with the post-translational modification(s) of kinases (Akt/PKB, MAPK) and phosphatases (PTPase, PTEN) (Bae et al., 1997; Chiarugi, 2003; Denu and Dixon, 1998; Finkel, 2012; Heffetz et al., 1992; Lo and Cruz, 1995; Sullivan et al., 1994; Sundaresan et al., 1995; Ushio-Fukai et al., 2002). It is also noteworthy that a second critical factor in the overall readout of redox signaling is the involvement of reactive nitrogen species (RNS), in particular nitric oxide ($^{\bullet}NO$), and more importantly its reactivity with $O_2^{\bullet -}$ to generate peroxynitrite ($OONO^-$), which has a better and more selective reactivity with biomolecules (Beckman and Koppenol, 1996). In fact, peroxynitrite modifies relatively slowly biological molecules such as DNA and proteins, making peroxynitrite a selective oxidant (see below). Thus, notably, the rate constant (k_{NO}) for the reaction between $O_2^{\bullet -}$ and$^{\cdot}$NO is significantly higher than the dismutation of $O_2^{\bullet -}$ to

H_2O_2 (function of SODs) (Beckman and Koppenol, 1996). As such, the intracellular fate of $O_2^{\bullet -}$ and its biological outcome could be determined by the ratio between SODs and nitric oxide synthases (inducible NOS (iNOS) and endothelial NOS (eNOS)) that generate $^{\bullet}NO$ and by its subcellular localization.

2.2 ROS-induced modifications and their cellular regulation

Chemically, redox-induced changes fall in two broad categories, reversible and irreversible (reviewed in (Cai and Yan, 2013)). The reversible changes are associated with modifications at the cysteine residues within proteins brought about by H_2O_2, $^{\bullet}NO$, $OONO^-$ or glutathione, such as S-nitrosylation (S-NO), S-glutathionylation, disulfide formation and s-sulfonation. The irreversible changes include the most commonly observed protein carbonylation that targets a number of amino acids (arginine, histidine, lysine, proline, threonine and cysteine) and the more specific modifications such as 3-nitrotyrosine (3-NT) formation ($OONO^-$ mediated). For instance, it has been shown in vivo that peroxynitrite modifies tyrosine in proteins creating detectable nitrotyrosines (Beckman and Koppenol, 1996). In general, these modifications induced by different redox milieu can either have beneficial or detrimental effect(s) on cell function and/or survival.

The biological outcome of the diverse effects of ROS is primarily dependent upon the amount of ROS generation and its regulation by the cellular anti-oxidant defenses. In this regard, a battery of redox defense systems ensures tight regulation of intracellular oxidative stress. These include the major anti-oxidant proteins such as glutathione, various SODs, catalase, and peroxiredoxins. The glutathione system deserves a special mention here as it consists of two enzymes [glutathione peroxidase (GPx) and glutathione reductase (GR)] that catalyze the reduction of H_2O_2 while redox recycling of the entire system (Marengo et al., 2016). This redox recycling results in a stable but flexible intracellular level of oxidized (GSSG) to reduced glutathione (GSH), which can be called upon to buffer the redox milieu under states of overt oxidative stress.

The intricate crosstalk between the aforementioned ROS and RNS play a critical role in cell fate decisions (Chen and Pervaiz, 2007; Clement and Pervaiz, 2001; Pervaiz et al., 2001; Pervaiz and Clement, 2007). Although the dogmatic view held for a long time associated an increase in intracellular ROS/RNS with cellular toxicity and tissue injury/damage, there is a general consensus now that a mild "pro-oxidant" milieu is linked to signaling networks involved in normal cell growth, proliferation and survival (Chen and Pervaiz, 2007; Janssen-Heininger et al., 2008; Pervaiz and Clement, 2007). This dichotomy of redox-mediated effects from the standpoint of homeostasis and its deregulation is a function of the various modifications of biomolecules, such as DNA, lipids and proteins. In turn, these modifications alter the course and/or outcome of cellular processes governing cell survival or execution, such as apoptosis, necrosis, necroptosis and autophagy.

3. Autophagy and cell fate signaling

An evolutionary conserved process, autophagy, is a mean to deliver damaged organelles or cellular components to the lysosomes for digestion and removal. As such, the process is critically important during periods of nutrient crisis as "self-eating" can generate the metabolic substrates for sustaining cell function and survival (Mizushima et al., 2008; Yorimitsu and Klionsky, 2005). Autophagy is a highly coordinated and genetically programmed process under the control of a set of genes termed as autophagy-related genes (*Atg*), which are conserved in yeast, flies, and mammals (Klionsky et al., 2008). Briefly the process is a sequential execution of membrane-linked events: phagophore (isolated membrane) to autophagosome formation, its fusion with the lysosome to form the autolysophagsome and finally lysosomal degradation. Phagophore formation is orchestrated by ULK1/Atg1 via regulation of mammalian target of rapamycin (mTOR) (Kuroyanagi et al., 1998), which promotes association of Atg1 with dephosphorylated Atg13 to kick start a series of Atg-mediated events, culminating in autophagy (Simonsen and Tooze, 2009). These include, recruitment of Beclin 1 by Vps34, activation of class III phosphatidylinositol 3-kinase (PtdIns3K) and formation of the Atg12-Atg5 and LC3/Atg8 ubiquitin-conjugation complexes. It's important to note that nascent LC3/Atg8 is processed by the cysteine protease Atg4, (Nakatogawa and Ohsumi, 2008) and subsequently conjugated with phosphotidylethanolamine through the action of Atg7 and Atg3; lipidated LC3 (LC3-II) is a bona fide marker for autophagy detection (Ichimura et al., 2000). Finally, autophagic flux is completed by the degradation of contents delivered to the lysosomes (Maiuri et al., 2007). As we shall see in the proceeding discussion, the cysteine residue within Atg4 is an attractive target for redox-dependent regulation of autophagy.

4. Crosstalk between autophagy and cellular redox status

Based on the seemingly similar cellular effects associated with alterations in cellular redox milieu and the autophagic process, there has been a long-standing debate *vis-à-vis* the crosstalk between redox-dependent signaling networks and autophagy. While, the initial observations suggested that autophagy was a redox regulatory process and as such the association between an increase in cellular ROS and autophagy was actually an indicator of stalled autophagy or in other words incomplete autophagic flux. However, over the past decade or so, there is ample experimental evidence to indicate that alterations in cellular redox state could be a trigger for or regulator of autophagy, and in contrast autophagy could be a mechanism for regulating mitochondrial redox metabolism (elegantly reviewed in Filomeni et al., 2015; Roca-Agujetas et al., 2019; Sanchez-Alvarez et al., 2019). In the following section we will attempt to briefly review the literature on the crosstalk between ROS and autophagy and mitophagy as a response to clear damaged mitochondria and manage toxic levels of ROS.

4.1 Redox-dependent modification of autophagic process

It is to be noted that a number of cellular stimuli that activate autophagy, such as nutrient deprivation, also result in an increase in intracellular ROS, and importantly anti-oxidants seem to inhibit autophagy induction under such scenarios. These observations point to an upstream role for oxidative stress in autophagy induction, and most of the evidence implicates mitochondrial ETC as the primary source of ROS. The rapid induction of autophagy in response to mitochondrial ROS has been linked to oxidative modification of proteins. In this regard, redox-mediated glutathionylation (and activation) of AMPK (Filomeni et al., 2015) has been documented as a driver of autophagy via phosphorylation of its substrate ULK1. Furthermore, the redox-sensitive kinase $p66^{Shc}$ upon phosphorylation translocates to the mitochondria, binds and oxidizes cytochrome C, which drives AMPK-mediated autophagy and mitophagy (Roca-Agujetas et al., 2019). More importantly, proper elongation of the autophagosome is a function of oxidative modification of cysteine 81 residue on Atg4 (Scherz-Shouval et al., 2007).

4.2 Role of GSH efflux on mitochondrial ROS and cellular response

Nutrient deprivation has been shown to trigger glutathione efflux and redox-mediated thiol modification also targets the glutathione redox status within the cell, rendering the ratio between oxidized (GSSG) to reduced glutathione (GSH) in favor of GSSG (Filomeni et al., 2015) Thus, oxidative modification and degradation of Keap1 activates the transcriptional activity of Nrf2, influencing notably the nuclear genes transcription as an anti-oxidant response. In parallel, as a matter of fact, the *cause and effect* relationship between thiol modification and autophagy is strong and direct even in the absence of an extrinsic or intrinsic stimulus (for autophagy). A brief summary illustrating some of the main relationships between cellular redox status and autophagy is presented in Fig. 1.

The involvement of intracellular ROS/RNS has also been reported upon drug-induced execution of cells via autophagy. Under such conditions, redox-dependent inhibition of mTOR has been observed. In a recent report, small molecule induced targeting of mutant KRAS driven cancer cells was shown to involve MAPK and Akt-dependent mitochondrial ROS production with the simultaneous induction of autophagy and apoptosis (Iskandar et al., 2016). Redox-mediated activation of MAPK has also been shown to trigger phosphorylation of Bcl-2, which disrupts its interaction with Beclin 1, thus enabling Beclin 1-induced autophagy (Wei et al., 2008a,b). Along similar lines, ROS-induced NF-kB mediated transcription of Beclin 1 as well as JNK-dependent upregulation of *ATG* genes positively influence autophagy output (Roca-Agujetas et al., 2019). Mitochondrial ROS via modulation of lysosomal Ca^{2+} is also implicated in calcineurin-dependent activation (nuclear localization) of the lysosomal transcription factor TFEB, important in the transcription of ATG and lysosomal genes (Zhang et al., 2016a); however, contrasting results have been reported by a different study (Vlahakis et al., 2016). Interestingly, autophagy is also a mechanism for the removal of damaged mitochondria that result in a significant increase in oxidative stress. Under such states, autophagy serves to alleviate redox stress by taking out the source of aberrant ROS production, i.e., dysfunctional mitochondria (Filomeni et al., 2015).

It is worth mentioning that oxidative stress results in the activation of ROS responsive genes/transcription factors, namely, HIF-1/2, p53 and FOXOs. These factors work upstream of a redox responsive gene family, the *sestrins SESN1, 2 and 3* (Sanchez-Alvarez et al., 2019). Although, there

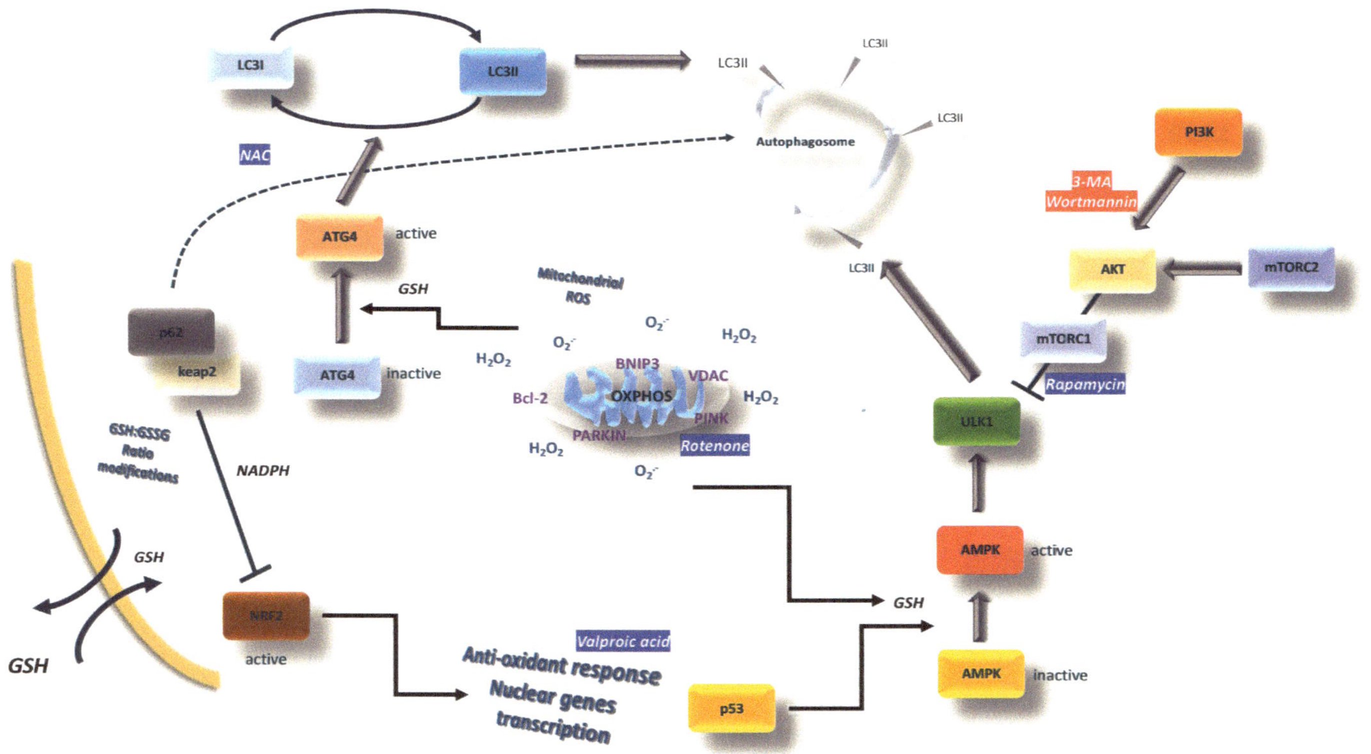

Fig. 1 Main crosstalks between autophagy regulation and redox homeostasis modification. GSH:GSSG ratio and mitochondrial ROS production have been shown to be major hubs involved in this interaction between autophagy induction and cellular redox status. Various arrows indicate ROS and RNS-induced proteins modifications such as S-nitrotyrosine modification, glutathionylation, disulfide bridge formation, as well as phosphorylation (see text for more details). Examples of molecules regulating autophagy process are highlighted in blue (anti-oxidant and/or autophagy inhibitor) or red (autophagy inducer). 3-MA, 3-methyladenine; NAC, *N*-acetylcysteine.

are some distinct differences in the functional biology of the various sestrin proteins, overall they all converge to trigger activation of anti-oxidant factors, such as Nrf-2 transcriptional targets (Bae et al., 2013), and importantly block mTORC1 signaling to promote autophagy (Lee et al., 2010). SESN2 is also a substrate of ULK1 and its phosphorylation (during leucine deprivation) mediates mTORC1 inhibition (Kimball et al., 2016), thus highlighting the intricate crosstalk between the sestrin family and autophagy in regulating cellular redox metabolism. From the therapeutic standpoint, a host of small molecule compounds as well as natural products have been shown to activate sestrins, such as HDAC inhibitors, bortezomib, quercetin, resveratrol, arsenic, and others (Sanchez-Alvarez et al., 2019).

Notably, under specific conditions, ROS and RNS have also been shown to block autophagy via redox modification of proteins such as Bcl-2 and JNK, leading to inactivation of JNK phosphorylation of Bcl-2, increasing its interaction with beclin-1 or by activating mTORC1 (Dodson et al., 2013; Sarkar et al., 2011). The latter has also specifically been shown with NO. donors that target proteins for nitrosylation, thus blocking autophagy (Fig. 1).

5. Interplay between cell death, ROS and autophagy in pathological states

5.1 Neurodegenerative disorders

Neurodegenerative disorders are a group of diseases affecting the central nervous system (e.g., Alzheimer's disease (AD), Parkinson's disease (PD), Huntington's disease, prion disease, spinocerebellar ataxia types 1–3). These pathologies are associated with several forms of cell death (e.g., apoptosis, necrosis and necroptosis) as well as mitochondrial dysfunction, ROS accumulation and deregulation of autophagy at various stages of the disease (Nixon, 2013). Notably, loss of autophagy promotes neurodegeneration in mice (Hara et al., 2008; Komatsu et al., 2006). Along similar lines, elevated autophagy has been reported as a pro-survival and protective response to damaging insults, and genetic studies have identified numerous mutations in autophagy-related genes and linked these mutations to human pathology (Levine and Kroemer, 2008). The stage at which the autophagic dysfunction might be pharmacologically modulated can be the determinant for controlling the pathogenesis of the disease. In this regard, various strategies are currently being evaluated to enhance autophagy induction and sequestration, increase lysosome efficiency and promote selective autophagy (Nixon, 2013; Vidal et al., 2014).

5.1.1 Alzheimer's disease

As early as the 1980s, neuronal cell death has been described as an important contributor in AD pathogenesis, long before the identification of the underlying mechanisms of cell death (Mann, 1985). Since human brain autopsies revealed the loss of neurons at the proximity of neurofibrillary tangles and amyloid plaques, the occurrence of apoptosis was strongly suspected. Indeed, the involvement of apoptosis in the loss of neurons and astrocytes was clearly demonstrated by TUNEL labeling of human hippocampal tissue (Smale et al., 1995). In addition, peroxynitrite has been evidenced to alter the nitration and oxidation of Tau protein, decreasing its ability to polymerize into neurofibrillary tangles and also, to cause DNA damage by activating PARP and depleting NAD+, in turn slowing rates of glycolysis and OXPHOS (Pacher, 2008; Vana et al., 2011). Recently, necroptosis has also been identified upon postmortem examination of human AD brains (Caccamo et al., 2017). This form of cell death requires the activation of multiple kinases. Mechanistically, receptor-interacting serine/threonine-protein kinase (RIPK) 1 and 3 form the "necrosome" and subsequently phosphorylate mixed lineage kinase domain-like protein (MLKL), resulting in the manifestation of the necroptotic phenotype in various types of nerve cells (Vandenabeele et al., 2010).

Regarding autophagy during the AD progression, several steps of the autophagic process have been described as defective with disruption of substrate proteolysis within autolysosomes by mutant Presenilin 1 (PS1), as the main alteration. PS1, a ubiquitous transmembrane protein, has diverse biological roles in cell adhesion, apoptosis, neurite outgrowth, calcium homeostasis and synaptic plasticity (Shen and Kelleher, 2007). Importantly, PS1 mutations are the most common cause of early-onset familial AD (Korovaitseva et al., 1997; Rogaev et al., 1997). Mitophagy, the selective process leading to the elimination of dysfunctional mitochondria involving a role for proteins such as PARKIN, PINK1, voltage dependent anion channel (VDAC) and BNIP3 has also proved to be dysfunctional in AD (Kerr et al., 2017) (Fig. 1). The loss of maintenance of a healthy mitochondrial pool affecting collectively neuronal development, function and survival (Kerr et al., 2017).

Recently, a role of mitochondrial dynamics in AD pathogenesis has been proposed. While studying the anti-oxidant and mitochondrial response to exogenous Abeta1-42 oligomers (AbetaO) mimicking amyloid beta peptide, Cieslik et al. found significant downregulation of sirtuin1 (Sirt1), mitofusin 1 (Mfn1) and NADH-ubiquinone oxidoreductase chain 1 (mt-Nd1) together with upregulation of Dynamin-1-like Protein (Dnm1L) in the cortical regions of Tg-AD mice (Cieslik et al., 2019). Consistent with these findings, the group also reported changes in microRNA profile (miRNA-9, miRNA-34a, miRNA-146a and miRNA-155), responsible for the regulation of *Sirt1* expression, in human AD brains. These studies suggest that targeting processes/pathways that modulate mitochondrial dynamics and Sirt1 expression may represent a putative strategy for AD therapy (Cieslik et al., 2019).

5.1.2 Parkinson's disease

In PD, the exact mechanisms of cell death pathogenesis are still under debate, despite decades of intense research in this domain. The pathological hallmarks of PD are a progressive depletion of dopaminergic neurons within the substantia nigra pars compacta, which results in marked dopamine deficit in the striatum and accumulation of α-synuclein in the form of Lewy bodies in the vulnerable populations of neurons (Bhat et al., 2018). Notably, a complex interplay between caspase activation, oxidative and nitrosative stress, mitochondrial dysfunction, inflammation, excitotoxicity, proteolytic stress and apoptosis have been proposed as causal determinants in PD progression (Olanow, 2007). More precisely, postmortem studies revealed a depletion of reduced glutathione and an increase in MnSOD (SOD2) activity, thereby suggesting the direct or indirect involvement of oxidative stress in the underlying etiology associated with substantia nigra damage (Jenner, 1998). In addition, selective impairment of complex I of the mitochondrial ETC and iron accumulation are observed in SNPC in PD patients. More recent evidence points to a prominent early role of impaired mitochondria dynamics, biogenesis and mitophagy in the pathogenesis of PD, which provides food for thought in the targeting of these nodes as potential therapeutic strategy against PD (Rani and Mondal, 2019).

Intriguingly, several indirect hallmarks in AD, PD or other neurological disorders such as amyotrophic lateral sclerosis suggest that ferroptosis, a lethal cell death pathway marked by glutathione peroxidase 4 (GPX4) inhibition, mitochondrial shrinkage, lipid peroxidation and iron deregulation is triggered in these diseases (Stockwell et al., 2017; Weiland et al., 2019). However, this remains to be better documented, which is particularly challenging using in vivo models.

5.1.3 Cardiomyocyte injury and death

The heart is a high energy-demanding organ composed of several cell types devoted to specific functions. For example, while some cell types within the cardiovascular system are actively proliferating, the cardiomyocytes (contractile cells) are post-mitotic and do not proliferate (Whelan et al., 2010). Since cardiomyocytes are non-dividing cells, the basal rate of cell death is very low and as such the regenerative capacity of the heart is limited. Therefore, any damage to cardiomyocytes that results in cell death, as evidenced during ischemia and myocardial infarction, is usually accompanied by fibroblast activation (resulting in tissue fibrosis) and immune cell infiltration that severely compromise the contractile function of the myocardium (Whelan et al., 2010). While a large majority of literature favors the association of necrosis with ischemia-reperfusion and/or infarction, there is also evidence for the involvement of apoptosis, albeit the percentage of apoptotic cells is currently being debated and influenced by the method used. In this regard, a study by Olivetti et al. used tissue histopathology, DNA electrophoresis analysis, fluorescence microscopy and western blot analysis of heart failure tissue and reported relatively low incidence of apoptosis ($\sim$0.23%) compared to a significant appearance of necrosis ($\sim$10%) of the cardiomyocytes (Olivetti et al., 1997). It should also be pointed out that cardiomyocytes injury/death is a major side effect of some anticancer drugs (e.g., doxorubicin or adriamycin, tyrosine kinase inhibitors) and radiotherapy, acutely following the treatment and many years following the cure (Carvalho et al., 2014; Kerkelä et al., 2006). In addition to the classical apoptosis and necrosis, there is also experimental evidence to implicate necroptosis in the death of cardiomyocytes, such as during doxorubicin treatment and upon ischemia/reperfusion (Zhang et al., 2016b). Interestingly, necroptosis in this setting was independent of RIPK1 and mixed lineage kinase domain-like protein (MLKL), but driven by calmodulin-dependent kinase II (CaMKII), a RIP3 substrate (Zhang et al., 2016b). Thus, CaMKII appears to be a nodal signal for multiple programmed cell death pathways, at least in the myocardium (Feng and Anderson, 2017).

The autophagy machinery is particularly active in the cardiovascular system in order to adapt to changes in physiological or pathological demand and to prevent cardiac failure and (irreversible) damages (Lavandero et al., 2013). Thus, several pathways of regulation more or less specific to the heart and vessels have been well documented and shown to involve mTORC1, AMPK, IP_3 and IP_3R, PKA, GSK3B, HDAC, TP53 and the so-called NAD+ salvage pathway (for review: Lavandero et al., 2013).

Nevertheless, in ischemic pathologies, cardiac cell death generally involves mitochondrial dysfunction and ROS production (Chiong et al., 2011; Galluzzi et al., 2018; Le Bras et al., 2005). Mitochondria undergo an irreversible mitochondrial membrane permeabilization affecting either the outer membrane and implicating proteins such as VDAC and BAX/BCL-2 family members or both the outer and the inner mitochondrial membranes via the opening of the permeability transition pore (PTP) (Brenner and Moulin, 2012). Mitochondrial membrane permeabilization leads to calcium overload and arrest of ATP production by the respiratory chain, and as cardiomyocytes require a significantly higher pool of ATP to sustain their function the cessation of ATP generation invariably triggers cell death. Moreover, the aberrant generation of ROS, induced upon ischemia and/or ischemia/reperfusion, shifts the balance between ROS production and anti-oxidant defense, thereby resulting in abnormal ROS accumulation with deleterious impact on the structure and function of DNA, lipids and proteins (Le Bras et al., 2005). Notably, as ROS generation in the presence of iron (Fe^{3+}) promotes the Fenton reaction (for the generation of highly reactive and toxic hydroxyl radical), a form of cell death dependent on iron, Ferroptosis, has also been linked to cardiomyocyte death, and as such targeting ferroptosis seems a plausible therapeutic/preventive strategy for cardiomyopathy (Fang et al., 2019). Furthermore, it reports also indicate that mitochondrial ROS production by damaged mitochondria that have not been eliminated by mitophagy can activate the NRLP3 inflammasome, leading to interleukin-1 (IL-1) production and inflammation (Liu et al., 2014; Zhou et al., 2011), thus suggesting a possible role for preventing inflammation to abate cardiomyocyte injury and prevent cell death. Along similar lines, as mitochondria are involved in an intricate crosstalk with the endoplasmic reticulum (ER) and nucleus, therapeutic strategies designed to disturb or derail these networks are also been proposed; however, the precise targets and ways to modulate them still remain elusive.

5.1.4 Hepatic disorders

The liver is an organ composed of resident cells (e.g., Kupffer cells, hepatic stellate cells, sinusoidal endothelial cells and hepatocytes) and cells that are recruited in response to injury (e.g., monocytes, macrophages, dendritic cells, natural killer cells). In response to stress, both type of cells emit pro-inflammatory signals such as cytokines, chemokines, lipid messengers, and ROS that ultimately trigger apoptotic or necrotic elimination of hepatocytes by autonomous as well as non-autonomous mechanisms (Brenner et al., 2013). While inflammation can be beneficial to the liver, when it reaches excessive intensity and duration, it may induce a massive loss of hepatocytes and exacerbate the severity of various hepatic pathological conditions, including ischemia-reperfusion injury, systemic metabolic alterations (e.g., obesity, diabetes, non-alcoholic fatty liver disorders), alcoholic hepatitis and infection, de facto being associated with irreversible liver damage, fibrosis, and carcinogenesis. Additionally, dying hepatocytes release damage-associated molecular proteins (DAMP) that can themselves stimulate inflammatory stress responses, hence stimulating a vicious cycle of inflammation and cell death (Vénéreau et al., 2015).

Autophagy is deeply involved in liver function with beneficial aspects allowing hepatocytes to cope with nutrient deprivation and anoxia during liver ischemia-reperfusion and protection from hepatitis B or C viral infection (Czaja et al., 2013; Rautou et al., 2010). In contrast, in hepatocellular carcinoma, the autophagy level is decreased and associated with modulation of Beclin-1 (Rautou et al., 2010). This has been exploited to reduce chemotherapy-induced necrosis trough induction of autophagy by ischemic pre-conditioning of patient livers (Esposti et al., 2010) as well as to decrease the incidence of rejection episodes of fatty livers (Degli Esposti et al., 2011). Similarly, following alcohol consumption, autophagy has been shown to be defective in liver cells. This failure may result from a decrease in 5′-AMP-activated protein kinase (AMPK) and due to an alteration in vesicle transport in hepatocytes as in pancreatic cells (da Silva Xavier et al., 2003; Tsuboi et al., 2003). The decrease in autophagy contributes to the formation of Mallory-Denk bodies, damaged intermediate filaments within hepatocytes, and to liver cell death (Zatloukal et al., 2007). Finally, hepatic autophagy is defective in obesity and linked to insulin resistance (Codogno and Meijer, 2010; Yang et al., 2010). Accordingly, autophagy stimulation and oxidative stress reduction by adiponectin, a protein hormone or adipokine secreted by adipose tissue to regulate glucose levels and fatty acids oxidation, restores insulin sensitivity (Liu et al., 2015). Therefore, adiponectin corrected SOD activity, the GSH to GSSG ratio, and lipid peroxidation through the modulation of NOX4 as well as several anti-oxidant enzymes, including GPXs, Prdx and SOD (Liu et al., 2015).

Among others, a promising molecular target in sepsis would be heme oxygenase 1 (HO-1), because it has been demonstrated that HO-1 mediated autophagy protects against hepatocyte cell death and hepatic injury from infection/sepsis in mice (Carchman et al., 2011).

5.1.5 Drug-induced cytotoxicity

Many xenobiotics such as commonly used drugs are responsible for side effects that are directly or indirectly associated with oxidative stress and cell death and manifest as cell/tissue damage and organ dysfunction. This has been extensively described for drugs affecting the cardiovascular system in patients cured from cancer and particularly, cardiomyocytes that are post-mitotic cells and thus, do not proliferate. These cardiotoxic effects can be observed in older cancer patients treated during childhood by anthracyclines such as adriamycin/doxorubicin (Mancilla et al., 2011). Tyrosine kinase inhibitors (TKIs) constitute another class of anticancer agents with marked cardiovascular deleterious effects ranging from heart failure, left ventricle dysfunction, conduction abnormalities, delayed ventricular repolarization (i.e., QT prolongation), acute coronary syndromes, myocardial injury, arterial thrombosis and hypertension (Chen et al., 2008; Kerkelä et al., 2006). Depending on the agent, cellular toxicity mechanisms can involve DNA damage, ER stress, cytochrome C release, mitochondrial dysfunction, ROS production, bioenergetic catastrophe and apoptosis (Carvalho et al., 2014; Chen et al., 2008). Moreover, these deleterious effects of chemotherapy are accumulative with those of associated upon radiotherapy. To prevent cardiotoxicity, an interesting approach has been to develop combination therapy with cardioprotective molecules. Thus, dexrazoxane was approved by American Food and Drug administration (FDA) and administrated to patients as a potential cardioprotective agent; however, a major limitation to its use is the reported association with higher rate of secondary malignant neoplasms (Shaikh et al., 2016).

Another well-known example of drug-induced cell death associated syndrome comes from the study of acetaminophen on the liver. Acetaminophen-induced liver injury is the combined result of drug toxicity and a potent innate immune response. Imaeda and colleagues reported that DNA released from dying hepatocytes is a key stimulus of innate immune activation in the acetaminophen-treated mouse liver (Imaeda et al., 2009). They demonstrated that hepatocyte DNA promotes immune activation by acting as damage-associated molecular pattern (DAMP) factors that stimulates cytokine production in neighboring sinusoidal endothelial cells via Tlr9 and the Nalp3 inflammasome (Imaeda et al., 2009).

6. Pro-autophagic molecules and therapeutic potential

Many compounds have been previously shown to target mTORC1 activity and strongly activate autophagic pathway (Heras-Sandoval et al., 2020). Some of these compounds efficiently modulate PI3K, or SHIP-1, such as theophylline (selective for PI3Kδ subunit), whereas others directly target mTORC1 such as rapamycin (Yang et al., 2013). Rapamycin was isolated from *Streptomyces hygroscopicus* during the 70s. It is a macrocyclic lactone harboring antifungal and antibiotic properties. The study of rapamycin favored the discovery of TOR (Target Of Rapamycin) and its mammalian analogue, mTOR in yeast. Thus, rapamycin forms a ternary complex with the FK506-binding protein 12 (FKBP12) and the FRB domain of mTOR via an allosteric mechanism, blocking substrate recruitment. First, pharmacological studies allowed its classification as a potent immunosuppressive drug based on its properties to inhibit proliferation of activated T lymphocytes. Later, rapamycin was evaluated for its clinical use in cardiovascular area and in oncology as well (Chan, 2004; Kahan and Camardo, 2001; Mancini et al., 2003).

Rapamycin studies prompted chemists and biologist researchers to launch high throughput screening campaigns of large chemical libraries to identify small molecules that activate autophagy and could be useful against AD, PD and Huntington disease (Sarkar et al., 2007; Zhang et al., 2007). For instance, using H4 neuroglioma cells, some screens took advantage of the localization of light chain 3 coupled to GFP (LC3-GFP) to the autophagosomal membrane upon induction of autophagy and its detection by high content imagery (Zhang et al., 2007). Other screens were based on the ability of small molecules to (indirectly) modulate the levels of PtdIns(3)*P*, to degrade long lived-proteins, the accumulation of mutant polyglutamine as well as the activation of mTOR (Zhang et al., 2007). In summary, these screens identified eight FDA-approved compounds that can be divided into two classes: three antipsychotic drugs (fluspirilene, trifluoperazine, and pimozide) and three cardiovascular drugs (niguldipine, nicardipine and amiodarone, that inhibit intracellular Ca^{2+} currents, loperamide, a FDA-approved drug for diarrhea, and penitrem A, an indole-diterpenoid mycotoxin) (Zhang et al., 2007). Amiodarone was further confirmed to induce autophagy and inhibit cell death in respiratory epithelial cells carrying the ΔF508 mutation of CFTR and liver cells (Lan et al., 2014; Lin et al., 2015; Zhang et al., 2018). Trifluoroperazine was also found in a zebrafish screen and found to protect human dopaminergic cells (Haddad and Nakamura, 2017).

Interestingly, a structure relationship analysis identified a particular three cycles-containing scaffold as able to induce autophagy in neurons and protective in a model of Huntington disease model (Tsvetkov et al., 2010).

Regarding chemotherapy-induced cardiotoxicity, fewer examples have been reported so far to exploit autophagy induction and cell death inhibition. However, proofs of concept have been obtained with rapamycin and resveratrol in vitro as well as in vivo (Danz et al., 2009). In both cases, protective effects involved also mitochondrial stabilization and ROS inhibition. Moreover, two high throughput screening campaigns have been undertaken to identify Beclin1 mimetics and disruptors of the interaction of Beclin1-BCL-2 susceptible to affect autophagosome formation (Yu et al., 2017). Some of the compounds identified are devoid of cytotoxicity while others inhibit cell growth. However, it is not known if these compounds can prevent cell death and decrease ROS-induced damage.

Trehalose is a disaccharide present in bacteria, yeast, fungi, insects, invertebrates, lower and higher plants, but not in humans (Elbein et al., 2003). It may serve as a source of energy and carbon and also as a signaling molecule to direct or control certain metabolic pathways or even to affect growth in yeast and plants (Elbein et al., 2003). Interestingly, trehalose, which is known to induce autophagy, reduces aggregate formation and apoptosis and delays pathology in a transgenic mouse model of oculopharyngeal muscular dystrophy (Davies et al., 2005). Trehalose also protects from aggravation of amyloid pathology induced by isoflurane anesthesia in *APPswe* mutant mice. Thus, co-treatment with trehalose reduces the levels of beta-amyloid peptide aggregates, tau plaques and levels of phospho-tau (Perucho et al., 2012).

Carbon monoxide (CO) is an endogenous gaso-transmitter that under low concentrations prevents cell death and inflammation. Endogenously, CO can be generated from the degradation of biliverdin-Ixα by HO-1. Recently, the role of autophagy in CO-mediated cytoprotection against oxidative stress has been investigated in the model yeast *Saccharomyces cerevisiae* and confirmed in astrocytes (Figueiredo-Pereira et al., 2019), opening the door to novel therapeutic applications of CO and CO-releasing molecules such as CORM-A1 (Ryter and Choi, 2010).

7. Conclusions/perspectives

Pathologies involving an excess of cell death represent an important threat for public health worldwide. This applies to neurodegenerative diseases, cardiac and liver pathologies, drug-induced toxicities that are believed to reach higher incidence in the future, notably due to the aging of population and cure of certain cancers. Since some mechanisms of cell death and ROS signaling are necessary for organism development and life, we believe that the development of innovative strategies based on the prevention of cell loss through autophagy induction hold promise (Fig. 2). However, the fact

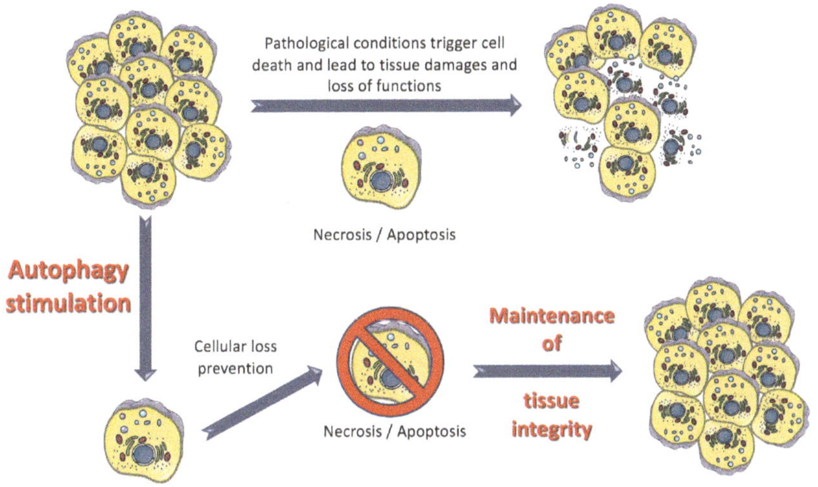

Fig. 2 Graphic of the potential of autophagy induction for cell death inhibition in a therapeutic perspective. Some drugs have been shown to promote autophagy and inhibit cell death, exhibiting cytoprotective properties.

that autophagy is associated with cell death under certain circumstances have to be acknowledged and kept in mind while designing and testing autophagy induction as a cytoprotective mechanism. As briefly highlighted in this review, an area of immense importance could be the repurposing of already approved and in use drugs for identifying novel molecules that elicit autophagy-mediated cytoprotection. This will be aided in the coming thanks to the presence of excellent high throughput screening platforms.

Acknowledgments

C.B. would like to thank Centre National de la Recherche Scientifique (CNRS), the Investment for the Future program ANR-11-IDEX-0003-01 within the LABEX ANR-10-LABX-0033 and Institut National du Cancer (INCa) for the program 2017-1-PLBIO-08-IN.

References

Bae, Y.S., Kang, S.W., Seo, M.S., Baines, I.C., Tekle, E., Chock, P.B., Rhee, S.G., 1997. Epidermal growth factor (EGF)-induced generation of hydrogen peroxide. Role in EGF receptor-mediated tyrosine phosphorylation. J. Biol. Chem. 272, 217.

Bae, S.H., Sung, S.H., Oh, S.Y., Lim, J.M., Lee, S.K., Park, Y.N., Lee, H.E., Kang, D., Rhee, S.G., 2013. Sestrins activate Nrf2 by promoting p62-dependent autophagic degradation of Keap1 and prevent oxidative liver damage. Cell Metab. 17, 73.

Beckman, J.S., Koppenol, W.H., 1996. Nitric oxide, superoxide, and peroxynitrite: the good, the bad, and ugly. Am. J. Physiol. 271, C1424.

Bhat, S., Acharya, U.R., Hagiwara, Y., Dadmehr, N., Adeli, H., 2018. Parkinson's disease: cause factors, measurable indicators, and early diagnosis. Comput. Biol. Med. 102, 234.
Brenner, C., Moulin, M., 2012. Physiological roles of the permeability transition pore. Circ. Res. 111, 1237.
Brenner, C., Galluzzi, L., Kepp, O., Kroemer, G., 2013. Decoding cell death signals in liver inflammation. J. Hepatol. 59, 583.
Caccamo, A., Branca, C., Piras, I.S., Ferreira, E., Huentelman, M.J., Liang, W.S., Readhead, B., Dudley, J.T., Spangenberg, E.E., Green, K.N., 2017. Necroptosis activation in Alzheimer's disease. Nat. Neurosci. 20, 1236.
Cai, Z., Yan, L.J., 2013. Protein oxidative modifications: beneficial roles in disease and health. J. Biochem. Pharmacol. Res. 1, 15.
Carchman, E.H., Rao, J., Loughran, P.A., Rosengart, M.R., Zuckerbraun, B.S., 2011. Heme oxygenase-1-mediated autophagy protects against hepatocyte cell death and hepatic injury from infection/sepsis in mice. Hepatology 53, 2053.
Carson, D.A., Ribeiro, J.M., 1993. Apoptosis and disease. Lancet 341, 1251.
Carvalho, F.S., Burgeiro, A., Garcia, R., Moreno, A.J., Carvalho, R.A., Oliveira, P.J., 2014. Doxorubicin-induced cardiotoxicity: from bioenergetic failure and cell death to cardiomyopathy. Med. Res. Rev. 34, 106.
Chan, S., 2004. Targeting the mammalian target of rapamycin (mTOR): a new approach to treating cancer. Br. J. Cancer 91, 1420.
Chen, Z.X., Pervaiz, S., 2007. Bcl-2 induces pro-oxidant state by engaging mitochondrial respiration in tumor cells. Cell Death Differ. 14, 1617.
Chen, M.H., Kerkelä, R., Force, T., 2008. Mechanisms of cardiac dysfunction associated with tyrosine kinase inhibitor cancer therapeutics. Circulation 118, 84.
Chiarugi, P., 2003. Reactive oxygen species as mediators of cell adhesion. Ital. J. Biochem. 52, 28.
Chiong, M., Wang, Z., Pedrozo, Z., Cao, D., Troncoso, R., Ibacache, M., Criollo, A., Nemchenko, A., Hill, J., Lavandero, S., 2011. Cardiomyocyte death: mechanisms and translational implications. Cell Death Dis. 2, e244.
Chong, S.J., Low, I.C., Pervaiz, S., 2014. Mitochondrial ROS and involvement of Bcl-2 as a mitochondrial ROS regulator. Mitochondrion 19 (Pt A), 39.
Cieslik, M., Czapski, G.A., Wojtowicz, S., Wieczorek, I., Wencel, P.L., Strosznajder, R.P., Jaber, V., Lukiw, W.J., Strosznajder, J.B., 2019. Alterations of transcription of genes coding anti-oxidative and mitochondria-related proteins in amyloid beta toxicity: relevance to alzheimer's disease. Mol. Neurobiol. 57, 1374–1388. https://doi.org/10.1007/s12035-019-01819-y.
Clement, M.V., Pervaiz, S., 2001. Intracellular superoxide and hydrogen peroxide concentrations: a critical balance that determines survival or death. Redox Rep. 6, 211.
Codogno, P., Meijer, A.J., 2010. Autophagy: a potential link between obesity and insulin resistance. Cell Metab. 11, 449.
Czaja, M.J., Ding, W.-X., Donohue, T.M., Friedman, S.L., Kim, J.-S., Komatsu, M., Lemasters, J.J., Lemoine, A., Lin, J.D., Ou, J.-H.J., 2013. Functions of autophagy in normal and diseased liver. Autophagy 9, 1131.
da Silva Xavier, G., Leclerc, I., Varadi, A., Tsuboi, T., Moule, S.K., Rutter, G.A., 2003. Role for AMP-activated protein kinase in glucose-stimulated insulin secretion and preproinsulin gene expression. Biochem. J. 371, 761.
Danz, E.D.B., Skramsted, J., Henry, N., Bennett, J.A., Keller, R.S., 2009. Resveratrol prevents doxorubicin cardiotoxicity through mitochondrial stabilization and the Sirt1 pathway. Free Radic. Biol. Med. 46, 1589.
Davies, J.E., Sarkar, S., Rubinsztein, D.C., 2005. Trehalose reduces aggregate formation and delays pathology in a transgenic mouse model of oculopharyngeal muscular dystrophy. Hum. Mol. Genet. 15, 23.

Degli Esposti, D., Sebagh, M., Pham, P., Reffas, M., Poüs, C., Brenner, C., Azoulay, D., Lemoine, A., 2011. Ischemic preconditioning induces autophagy and limits necrosis in human recipients of fatty liver grafts, decreasing the incidence of rejection episodes. Cell Death Dis. 2, e111.

Denu, J.M., Dixon, J.E., 1998. Protein tyrosine phosphatases: mechanisms of catalysis and regulation. Curr. Opin. Chem. Biol. 2, 633.

Dodson, M., Darley-Usmar, V., Zhang, J., 2013. Cellular metabolic and autophagic pathways: traffic control by redox signaling. Free Radic. Biol. Med. 63, 207.

Elbein, A.D., Pan, Y., Pastuszak, I., Carroll, D., 2003. New insights on trehalose: a multifunctional molecule. Glycobiology 13, 17R.

Esposti, D.D., Domart, M.-C., Sebagh, M., Harper, F., Pierron, G., Brenner, C., Lemoine, A., 2010. Autophagy is induced by ischemic preconditioning in human livers formerly treated by chemotherapy to limit necrosis. Autophagy 6, 172.

Fang, X., Wang, H., Han, D., Xie, E., Yang, X., Wei, J., Gu, S., Gao, F., Zhu, N., Yin, X., 2019. Ferroptosis as a target for protection against cardiomyopathy. Proc. Natl. Acad. Sci. 116, 2672.

Feng, N., Anderson, M.E., 2017. CaMKII is a nodal signal for multiple programmed cell death pathways in heart. J. Mol. Cell. Cardiol. 103, 102.

Figueiredo-Pereira, C., Menezes, R., Ferreira, S., Santos, C.N., Vieira, H.L.A., 2019. Carbon monoxide released by CORM-A1 prevents yeast cell death via autophagy stimulation. FEMS Yeast Res. 19 (5), pii: foz051. https://doi.org/10.1093/femsyr/foz051.

Filomeni, G., De Zio, D., Cecconi, F., 2015. Oxidative stress and autophagy: the clash between damage and metabolic needs. Cell Death Differ. 22, 377.

Finkel, T., 2012. Relief with rapamycin: mTOR inhibition protects against radiation-induced mucositis. Cell Stem Cell 11, 287.

Flint, D.H., Emptage, M.H., Finnegan, M.G., Fu, W., Johnson, M.K., 1993. The role and properties of the iron-sulfur cluster in Escherichia coli dihydroxy-acid dehydratase. J. Biol. Chem. 268, 14732.

Fogg, V.C., Lanning, N.J., Mackeigan, J.P., 2011. Mitochondria in cancer: at the crossroads of life and death. Chin. J. Cancer 30, 526.

Galluzzi, L., Bravo-San Pedro, J.M., Levine, B., Green, D.R., Kroemer, G., 2017. Pharmacological modulation of autophagy: therapeutic potential and persisting obstacles. Nat. Rev. Drug Discov. 16, 487.

Galluzzi, L., Vitale, I., Aaronson, S.A., Abrams, J.M., Adam, D., Agostinis, P., Alnemri, E.S., Altucci, L., Amelio, I., Andrews, D.W., 2018. Molecular mechanisms of cell death: recommendations of the nomenclature committee on cell death 2018. Cell Death Differ. 25, 486.

Gardner, P.R., Fridovich, I., 1991a. Quinolinate synthetase: the oxygen-sensitive site of de novo NAD(P)+ biosynthesis. Arch. Biochem. Biophys. 284, 106.

Gardner, P.R., Fridovich, I., 1991b. Superoxide sensitivity of the Escherichia coli aconitase. J. Biol. Chem. 266, 19328.

Haddad, D., Nakamura, K., 2017. PINK1-based screen shines light on autophagy enhancers for Parkinson's disease. Cell Chem. Biol. 24, 429.

Halliwell, B., 2007. Biochemistry of oxidative stress. Biochem. Soc. Trans. 35, 1147.

Han, D., Antunes, F., Canali, R., Rettori, D., Cadenas, E., 2003. Voltage-dependent anion channels control the release of the superoxide anion from mitochondria to cytosol. J. Biol. Chem. 278, 5557.

Hara, T., Takamura, A., Kishi, C., Iemura, S., Natsume, T., Guan, J.L., Mizushima, N., 2008. FIP200, a ULK-interacting protein, is required for autophagosome formation in mammalian cells. J. Cell Biol. 181, 497.

Heffetz, D., Rutter, W.J., Zick, Y., 1992. The insulinomimetic agents H2O2 and vanadate stimulate tyrosine phosphorylation of potential target proteins for the insulin receptor kinase in intact cells. Biochem. J. 288 (Pt 2), 631.

Heras-Sandoval, D., Pérez-Rojas, J.M., Pedraza-Chaverri, J., 2020. Novel compounds for the modulation of mTOR and autophagy to treat neurodegenerative diseases. Cell. Signal. 65, 109442. https://doi.org/10.1016/j.cellsig.2019.109442.
Ichimura, Y., Kirisako, T., Takao, T., Satomi, Y., Shimonishi, Y., Ishihara, N., Mizushima, N., Tanida, I., Kominami, E., Ohsumi, M., et al., 2000. A ubiquitin-like system mediates protein lipidation. Nature 408, 488.
Imaeda, A.B., Watanabe, A., Sohail, M.A., Mahmood, S., Mohamadnejad, M., Sutterwala, F.S., Flavell, R.A., Mehal, W.Z., 2009. Acetaminophen-induced hepatotoxicity in mice is dependent on Tlr9 and the Nalp3 inflammasome. J. Clin. Invest. 119, 305.
Imlay, J.A., 1995. A metabolic enzyme that rapidly produces superoxide, fumarate reductase of Escherichia coli. J. Biol. Chem. 270, 19767.
Imlay, J.A., 2003. Pathways of oxidative damage. Annu. Rev. Microbiol. 57, 395.
Iskandar, K., Rezlan, M., Yadav, S.K., Foo, C.H., Sethi, G., Qiang, Y., Bellot, G.L., Pervaiz, S., 2016. Synthetic lethality of a novel small molecule against mutant KRAS-expressing cancer cells involves AKT-dependent ROS production. Antioxid. Redox Signal. 24, 781.
Janssen-Heininger, Y.M., Mossman, B.T., Heintz, N.H., Forman, H.J., Kalyanaraman, B., Finkel, T., Stamler, J.S., Rhee, S.G., van der Vliet, A., 2008. Redox-based regulation of signal transduction: principles, pitfalls, and promises. Free Radic. Biol. Med. 45, 1.
Jenner, P., 1998. Oxidative mechanisms in nigral cell death in Parkinson's disease. Mov. Disord. 13, 24.
Kahan, B.D., Camardo, J.S., 2001. Rapamycin: clinical results and future opportunities. Transplantation 72, 1181.
Kerkelä, R., Grazette, L., Yacobi, R., Iliescu, C., Patten, R., Beahm, C., Walters, B., Shevtsov, S., Pesant, S., Clubb, F.J., 2006. Cardiotoxicity of the cancer therapeutic agent imatinib mesylate. Nat. Med. 12, 908.
Kerr, J.S., Adriaanse, B.A., Greig, N.H., Mattson, M.P., Cader, M.Z., Bohr, V.A., Fang, E.F., 2017. Mitophagy and Alzheimer's disease: cellular and molecular mechanisms. Trends Neurosci. 40, 151.
Khandia, R., Dadar, M., Munjal, A., Dhama, K., Karthik, K., Tiwari, R., Yatoo, M., Iqbal, H., Singh, K.P., Joshi, S.K., 2019. A comprehensive review of autophagy and its various roles in infectious, non-infectious, and lifestyle diseases: current knowledge and prospects for disease prevention, novel drug design, and therapy. Cell 8, 674.
Kimball, S.R., Gordon, B.S., Moyer, J.E., Dennis, M.D., Jefferson, L.S., 2016. Leucine induced dephosphorylation of Sestrin2 promotes mTORC1 activation. Cell. Signal. 28, 896.
Klionsky, D.J., Abeliovich, H., Agostinis, P., Agrawal, D.K., Aliev, G., Askew, D.S., Baba, M., Baehrecke, E.H., Bahr, B.A., Ballabio, A., et al., 2008. Guidelines for the use and interpretation of assays for monitoring autophagy in higher eukaryotes. Autophagy 4, 151.
Komatsu, M., Waguri, S., Chiba, T., Murata, S., Iwata, J.-I., Tanida, I., Ueno, T., Koike, M., Uchiyama, Y., Kominami, E., 2006. Loss of autophagy in the central nervous system causes neurodegeneration in mice. Nature 441, 880.
Korovaitseva, G.I., Bukina, A., Farrer, L.A., Rogaev, E.I., 1997. Presenilin polymorphisms in Alzheimer's disease. Lancet 350, 959.
Kuroyanagi, H., Yan, J., Seki, N., Yamanouchi, Y., Suzuki, Y., Takano, T., Muramatsu, M., Shirasawa, T., 1998. Human ULK1, a novel serine/threonine kinase related to UNC-51 kinase of Caenorhabditis elegans: cDNA cloning, expression, and chromosomal assignment. Genomics 51, 76–85.
Lambeth, J.D., 2004. NOX enzymes and the biology of reactive oxygen. Nat. Rev. Immunol. 4, 181–189.

Lan, S.-H., Wu, S.-Y., Zuchini, R., Lin, X.-Z., Su, I.-J., Tsai, T.-F., Lin, Y.-J., Wu, C.-T., Liu, H.-S., 2014. Autophagy-preferential degradation of MIR224 participates in hepatocellular carcinoma tumorigenesis. Autophagy 10, 1687.

Lavandero, S., Troncoso, R., Rothermel, B.A., Martinet, W., Sadoshima, J., Hill, J.A., 2013. Cardiovascular autophagy: concepts, controversies, and perspectives. Autophagy 9, 1455.

Le Bras, M., Clement, M., Pervaiz, S., Brenner, C., 2005. Reactive oxygen species and the mitochondrial signaling pathway of cell death. Histol. Histopathol. 20, 15.

Lee, Y.K., Park, S.Y., Kim, Y.M., Kim, D.C., Lee, W.S., Surh, Y.J., Park, O.J., 2010. Suppression of mTOR via Akt-dependent and -independent mechanisms in selenium-treated colon cancer cells: involvement of AMPKalpha1. Carcinogenesis 31, 1092.

Levine, B., Kroemer, G., 2008. Autophagy in the pathogenesis of disease. Cell 132, 27–42.

Lin, C.-W., Chen, Y.-S., Lin, C.-C., Chen, Y.-J., Lo, G.-H., Lee, P.-H., Kuo, P.-L., Dai, C.-Y., Huang, J.-F., Chung, W.-L., 2015. Amiodarone as an autophagy promoter reduces liver injury and enhances liver regeneration and survival in mice after partial hepatectomy. Sci. Rep. 5, 15807.

Liochev, S.I., Fridovich, I., 1993. Modulation of the fumarases of Escherichia coli in response to oxidative stress. Arch. Biochem. Biophys. 301, 379.

Liu, W., Yin, Y., Zhou, Z., He, M., Dai, Y., 2014. OxLDL-induced IL-1beta secretion promoting foam cells formation was mainly via CD36 mediated ROS production leading to NLRP3 inflammasome activation. Inflamm. Res. 63, 33.

Liu, X.H., Yang, Y.W., Dai, H.T., Cai, S.W., Chen, R.H., Ye, Z.Q., 2015. Protective role of adiponectin in a rat model of intestinal ischemia reperfusion injury. World J. Gastroenterol. 21 (47), 13250–13258. PMID: 26715807. https://doi.org/10.3748/wjg.v21.i47.13250.

Lo, Y.Y., Cruz, T.F., 1995. Involvement of reactive oxygen species in cytokine and growth factor induction of c-fos expression in chondrocytes. J. Biol. Chem. 270, 11727.

Madesh, M., Hajnoczky, G., 2001. VDAC-dependent permeabilization of the outer mitochondrial membrane by superoxide induces rapid and massive cytochrome c release. J. Cell Biol. 155, 1003.

Maiuri, M.C., Zalckvar, E., Kimchi, A., Kroemer, G., 2007. Self-eating and self-killing: crosstalk between autophagy and apoptosis. Nat. Rev. Mol. Cell Biol. 8, 741.

Mancilla, T.R., Iskra, B., Aune, G.J., 2011. Doxorubicin-induced cardiomyopathy in children. Compr. Physiol. 9, 905.

Mancini, D., Pinney, S., Burkhoff, D., LaManca, J., Itescu, S., Burke, E., Edwards, N., Oz, M., Marks, A.R., 2003. Use of rapamycin slows progression of cardiac transplantation vasculopathy. Circulation 108, 4.

Mann, D.M., 1985. The neuropathology of Alzheimer's disease: a review with pathogenetic, aetiological and therapeutic considerations. Mech. Ageing Dev. 31, 213.

Marengo, B., Nitti, M., Furfaro, A.L., Colla, R., Ciucis, C.D., Marinari, U.M., Pronzato, M.A., Traverso, N., Domenicotti, C., 2016. Redox homeostasis and cellular antioxidant systems: crucial players in cancer growth and therapy. Oxid. Med. Cell. Longev. 2016, 6235641.

Mizushima, N., Tsukamoto, S., Kuma, A., 2008. Autophagy in embryogenesis and cell differentiation. Tanpakushitsu Kakusan Koso 53, 2170.

Nakatogawa, H., Ohsumi, Y., 2008. Molecular mechanisms of autophagy in yeast. Tanpakushitsu Kakusan Koso 53, 2099.

Nixon, R.A., 2013. The role of autophagy in neurodegenerative disease. Nat. Med. 19, 983.

Olanow, C.W., 2007. The pathogenesis of cell death in Parkinson's disease–2007. Mov. Disord. 22, S33.

Olivetti, G., Abbi, R., Quaini, F., Kajstura, J., Cheng, W., Nitahara, J.A., Quaini, E., Di Loreto, C., Beltrami, C.A., Krajewski, S., Reed, J.C., Anversa, P., 1997. Apoptosis in the failing human heart. N. Engl. J. Med. 336 (16), 1131–1141.

Pacher, P., 2008. Poly(ADP-ribose) polymerase inhibition as a novel therapeutic approach against intraepidermal nerve fiber loss and neuropathic pain associated with advanced diabetic neuropathy: a commentary on "PARP inhibition or gene deficiency counteracts intraepidermal nerve fiber loss and neuropathic pain in advanced diabetic neuropathy" Free Radic. Biol. Med. 44, 969.

Perucho, J., Casarejos, M.J., Gomez, A., Solano, R.M., Garcia de Yebenes, J., Mena, M.A., 2012. Trehalose protects from aggravation of amyloid pathology induced by isoflurane anesthesia in APPswe mutant mice. Curr. Alzheimer Res. 9, 334.

Pervaiz, S., Clement, M.V., 2007. Superoxide anion: oncogenic reactive oxygen species? Int. J. Biochem. Cell Biol. 39, 1297.

Pervaiz, S., Cao, J., Chao, O.S., Chin, Y.Y., Clement, M.V., 2001. Activation of the RacGTPase inhibits apoptosis in human tumor cells. Oncogene 20, 6263.

Rani, L., Mondal, A.C., 2019. Emerging concepts of mitochondrial dysfunction in Parkinson's disease progression: pathogenic and therapeutic implications. Mitochondrion 50, 25.

Rautou, P.-E., Mansouri, A., Lebrec, D., Durand, F., Valla, D., Moreau, R., 2010. Autophagy in liver diseases. J. Hepatol. 53, 1123.

Roca-Agujetas, V., de Dios, C., Leston, L., Mari, M., Morales, A., Colell, A., 2019. Recent insights into the mitochondrial role in autophagy and its regulation by oxidative stress. Oxid. Med. Cell. Longev. 2019, 3809308.

Rogaev, E.I., Sherrington, R., Wu, C., Levesque, G., Liang, Y., Rogaeva, E.A., Ikeda, M., Holman, K., Lin, C., Lukiw, W.J., et al., 1997. Analysis of the 5′ sequence, genomic structure, and alternative splicing of the presenilin-1 gene (PSEN1) associated with early onset Alzheimer disease. Genomics 40, 415.

Rubinsztein, D.C., Codogno, P., Levine, B., 2012. Autophagy modulation as a potential therapeutic target for diverse diseases. Nat. Rev. Drug Discov. 11, 709.

Ryter, S.W., Choi, A.M., 2010. Heme oxygenase-1/carbon monoxide: novel therapeutic strategies in critical care medicine. Curr. Drug Targets 11, 1485.

Sanchez-Alvarez, M., Strippoli, R., Donadelli, M., Bazhin, A.V., Cordani, M., 2019. Sestrins as a therapeutic bridge between ROS and autophagy in cancer. Cancers (Basel) 11, 371.

Sarkar, S., Perlstein, E.O., Imarisio, S., Pineau, S., Cordenier, A., Maglathlin, R.L., Webster, J.A., Lewis, T.A., O'Kane, C.J., Schreiber, S.L., 2007. Small molecules enhance autophagy and reduce toxicity in Huntington's disease models. Nat. Chem. Biol. 3, 331.

Sarkar, S., Korolchuk, V.I., Renna, M., Imarisio, S., Fleming, A., Williams, A., Garcia-Arencibia, M., Rose, C., Luo, S., Underwood, B.R., et al., 2011. Complex inhibitory effects of nitric oxide on autophagy. Mol. Cell 43, 19.

Scherz-Shouval, R., Shvets, E., Elazar, Z., 2007. Oxidation as a post-translational modification that regulates autophagy. Autophagy 3, 371.

Shaikh, F., Dupuis, L.L., Alexander, S., Gupta, A., Mertens, L., Nathan, P.C., 2016. Cardioprotection and second malignant neoplasms associated with dexrazoxane in children receiving anthracycline chemotherapy: A systematic review and meta-analysis. J. Natl. Cancer Inst. 108, djv357.

Shen, J., Kelleher 3rd, R.J., 2007. The presenilin hypothesis of Alzheimer's disease: evidence for a loss-of-function pathogenic mechanism. Proc. Natl. Acad. Sci. U. S. A. 104, 403–409.

Simonsen, A., Tooze, S.A., 2009. Coordination of membrane events during autophagy by multiple class III PI3-kinase complexes. J. Cell Biol. 186, 773.

Smale, G., Nichols, N.R., Brady, D.R., Finch, C.E., Horton Jr., W.E., 1995. Evidence for apoptotic cell death in Alzheimer's disease. Exp. Neurol. 133, 225.

Stockwell, B.R., Angeli, J.P.F., Bayir, H., Bush, A.I., Conrad, M., Dixon, S.J., Fulda, S., Gascón, S., Hatzios, S.K., Kagan, V.E., 2017. Ferroptosis: a regulated cell death nexus linking metabolism, redox biology, and disease. Cell 171, 273.

Sullivan, S.G., Chiu, D.T., Errasfa, M., Wang, J.M., Qi, J.S., Stern, A., 1994. Effects of H2O2 on protein tyrosine phosphatase activity in HER14 cells. Free Radic. Biol. Med. 16, 399.
Sundaresan, M., Yu, Z.X., Ferrans, V.J., Irani, K., Finkel, T., 1995. Requirement for generation of H2O2 for platelet-derived growth factor signal transduction. Science 270, 296.
Tsuboi, T., da Silva Xavier, G., Leclerc, I., Rutter, G.A., 2003. 5′-AMP-activated protein kinase controls insulin-containing secretory vesicle dynamics. J. Biol. Chem. 278, 52042.
Tsvetkov, A.S., Miller, J., Arrasate, M., Wong, J.S., Pleiss, M.A., Finkbeiner, S., 2010. A small-molecule scaffold induces autophagy in primary neurons and protects against toxicity in a Huntington disease model. Proc. Natl. Acad. Sci. 107, 16982.
Ushio-Fukai, M., Tang, Y., Fukai, T., Dikalov, S.I., Ma, Y., Fujimoto, M., Quinn, M.T., Pagano, P.J., Johnson, C., Alexander, R.W., 2002. Novel role of gp91(phox)-containing NAD(P)H oxidase in vascular endothelial growth factor-induced signaling and angiogenesis. Circ. Res. 91, 1160.
Vana, L., Kanaan, N.M., Ugwu, I.C., Wuu, J., Mufson, E.J., Binder, L.I., 2011. Progression of tau pathology in cholinergic basal forebrain neurons in mild cognitive impairment and Alzheimer's disease. Am. J. Pathol. 179, 2533.
Vandenabeele, P., Galluzzi, L., Berghe, T.V., Kroemer, G., 2010. Molecular mechanisms of necroptosis: an ordered cellular explosion. Nat. Rev. Mol. Cell Biol. 11, 700.
Vénéreau, E., Ceriotti, C., Bianchi, M.E., 2015. DAMPs from cell death to new life. Front. Immunol. 6, 422.
Vidal, R.L., Matus, S., Bargsted, L., Hetz, C., 2014. Targeting autophagy in neurodegenerative diseases. Trends Pharmacol. Sci. 35, 583–591.
Vlahakis, A., Lopez Muniozguren, N., Powers, T., 2016. Calcium channel regulator Mid1 links TORC2-mediated changes in mitochondrial respiration to autophagy. J. Cell Biol. 215, 779.
Wei, Y., Pattingre, S., Sinha, S., Bassik, M., Levine, B., 2008a. JNK1-mediated phosphorylation of Bcl-2 regulates starvation-induced autophagy. Mol. Cell 30, 678.
Wei, Y., Sinha, S., Levine, B., 2008b. Dual role of JNK1-mediated phosphorylation of Bcl-2 in autophagy and apoptosis regulation. Autophagy 4, 949.
Weiland, A., Wang, Y., Wu, W., Lan, X., Han, X., Li, Q., Wang, J., 2019. Ferroptosis and its role in diverse brain diseases. Mol. Neurobiol. 56, 4880.
Whelan, R.S., Kaplinskiy, V., Kitsis, R.N., 2010. Cell death in the pathogenesis of heart disease: mechanisms and significance. Annu. Rev. Physiol. 72, 19.
Yang, L., Li, P., Fu, S., Calay, E.S., Hotamisligil, G.S., 2010. Defective hepatic autophagy in obesity promotes ER stress and causes insulin resistance. Cell Metab. 11, 467–478.
Yang, H., Rudge, D.G., Koos, J.D., Vaidialingam, B., Yang, H.J., Pavletich, N.P., 2013. mTOR kinase structure, mechanism and regulation. Nature 497, 217.
Yin, Z., Pascual, C., Klionsky, D.J., 2016. Autophagy: machinery and regulation. Microb. Cell 3, 588.
Yorimitsu, T., Klionsky, D.J., 2005. Autophagy: molecular machinery for self-eating. Cell Death Differ. 12 (Suppl 2), 1542.
Yu, J., Lan, L., Lewin, S.J., Rogers, S.A., Roy, A., Wu, X., Gao, P., Karanicolas, J., Aubé, J., Sun, B., 2017. Identification of novel small molecule Beclin 1 mimetics activating autophagy. Oncotarget 8, 51355.
Zatloukal, K., French, S.W., Stumptner, C., Strnad, P., Harada, M., Toivola, D.M., Cadrin, M., Omary, M.B., 2007. From Mallory to Mallory–Denk bodies: what, how and why? Exp. Cell Res. 313, 2033.
Zhang, L., Yu, J., Pan, H., Hu, P., Hao, Y., Cai, W., Zhu, H., Albert, D.Y., Xie, X., Ma, D., 2007. Small molecule regulators of autophagy identified by an image-based high-throughput screen. Proc. Natl. Acad. Sci. 104, 19023–19028.

Zhang, T., Zhang, Y., Cui, M., Jin, L., Wang, Y., Lv, F., Liu, Y., Zheng, W., Shang, H., Zhang, J., 2016a. CaMKII is a RIP3 substrate mediating ischemia-and oxidative stress–induced myocardial necroptosis. Nat. Med. 22, 175.

Zhang, J., Wang, J., Xu, J., Lu, Y., Jiang, J., Wang, L., Shen, H.M., Xia, D., 2016b. Curcumin targets the TFEB-lysosome pathway for induction of autophagy. Oncotarget 7, 75659–75671.

Zhang, S., Stoll, G., San Pedro, J.M.B., Sica, V., Sauvat, A., Obrist, F., Kepp, O., Li, Y., Maiuri, L., Zamzami, N., 2018. Evaluation of autophagy inducers in epithelial cells carrying the ΔF508 mutation of the cystic fibrosis transmembrane conductance regulator CFTR. Cell Death Dis. 9, 191.

Zhou, R., Yazdi, A.S., Menu, P., Tschopp, J., 2011. A role for mitochondria in NLRP3 inflammasome activation. Nature 469, 221.

FSC
www.fsc.org
MIX
Papier | Fördert
gute Waldnutzung
FSC® C083411

Kolibri 360 GmbH
Ferdinand-Jühlke-Straße 7
99095 Erfurt, Deutschland
produktsicherheit@kolibri360.de